CLINICAL REASONING

How Doctors *Should* Think

CLINICAL REASONING

How Doctors *Should* Think

Howard E. Herrell, MD, FACOG
howardisms.com

Sonic Elephant

First Edition.

Medicine is a vibrant and ever-changing discipline. The information provided within this book is for general educational purposes only. While we try to keep the information up-to-date and correct, there are no representations or warranties, expressed or implied, about the completeness, accuracy, reliability, suitability, or availability with respect to the information, products, techniques, or related graphics contained in this book for any purpose. Any use of this information is at your own risk.

The indications and dosages of all drugs in the book have been recommended in the medical literature and conform to the practices of the general medical community. The medications described do not necessarily have specific approval by the Food and Drug Administration for use in the diseases and dosages for which they are recommended. The package insert for each drug should be consulted for use and dosage as approved by the FDA. Because standards of usage change, it is advisable to keep abreast of revised recommendations, particularly those concerning new drugs.

Published by Sonic Elephant
www.sonicelephant.com

ISBN: 978-0-9993696-1-6

10 9 8 7 6 5 4 3 2 1

Acknowledgments

A book that covers the scope of information that this one does is never the result of a single person's individual efforts. The following people deserve credit for anything good you might find here:

Michelle Tanner, MD, *inspired the original concept with a question and a few articles. She contributed much to Chapter 3.*

Antonia Roberts, MD, *spent a month developing the original concepts into something practical. She and* ***Alicia McClintock, MD,*** *have, from the beginning, provided inspiration, encouragement, suggestions, revisions, ideas, and vision. You'll discover by the end of the book that I don't listen well to them. They tried.*

Nancy Elizabeth Wentzel, MS, *is co-author of Chapter 5 and contributed significantly to many other math-heavy portions of the book.*

My partners, ***Jami Goodwin, MD,*** *and* ***Amber Mullins, MD,*** *have provided support and encouragement from the beginning.*

My wife, ***Amy,*** *tried desperately to improve the quality of the writing but I didn't listen, as usual.*

My daughter, ***Sophia,*** *wrote the code for the online calculators that accompany the book.*

My cats, ***Hugo, Leonardo,*** *and* ***Raphael,*** *walked across my keyboard a few times.*

Britney Burleson, FNP-C, *provided threats and a deadline to finish.*

My past and present students have listened to countless iterations of the ideas in this book and helped refine them over the last 12 years.

Most importantly, my patients have provided the reason for the book and the love for the work that went into it.

Small minds are concerned with the extraordinary, great minds with the ordinary.

– Blaise Pascal

CONTENTS

FOREWORD
Bad Thinking, Bad Results

I'm an Obstetrician-Gynecologist. Almost every day, I'm challenged by the same fundamental problem, though it presents itself to me in many different forms:

- The mother who questions the safety of vaccines for herself or her child.
- The patient who had an unnecessary hysterectomy for pelvic pain.
- The woman who follows up from the ED for symptoms that were falsely attributed to a UTI or an ovarian cyst.
- The baby that was delivered too early because of a false diagnosis of ruptured membranes.
- The new mother who had an unnecessary Cesarean delivery due to over-diagnosis of fetal distress or labor arrest.
- The colleague who uses magnesium sulfate or terbutaline to stop preterm labor even though no scientific evidence exists that the practice is effective.
- The expert subspecialist who hasn't changed his recommendation in 30 years (and it wasn't right even then).
- The gynecologist who does a laparoscopic, robotic, or abdominal

hysterectomy on a patient when a vaginal hysterectomy is possible.

- The primary care physician who continues to do unnecessary Pap smears or other unnecessary tests on my patients.
- The researcher who makes a bold, unwarranted claim about some new intervention or treatment.
- The journal that publishes mostly false findings and makes no effort to distinguish the good from the bad.
- The reporter who embellishes or popularizes some snazzy new claim despite a lack of scientific evidence that it works or is meaningful.
- The public health policy expert who wastes money on low-yield or meaningless interventions.
- The lawyer who files a class-action lawsuit against some drug or device that I use every day because of an expected and known side effect.
- The insurance company that won't pay for something my patient needs because so many other doctors over-utilize the intervention for patients who don't need it.

I could go on.

Each of these examples describes the same underlying pathology, and this book is about the fundamental cognitive problems which affect us all – every physician, nurse, research scientist, medical student, and patient. These problems are not something most people want to discuss because they affect us on a very personal level. Indeed, it is painful to admit that we need to improve these flawed thought processes. The core problem is the way we *think*.

There is no easy solution. To ask a person to fundamentally change the way he thinks is doomed for failure, but awareness of the problem can breed change. Real change starts with medical students and residents, who can learn clinical thinking and critical thinking in a different way than previous generations, and slowly replace faulty habits with better habits, faulty thought with better thought, and, eventually, lackluster results with better results.

Don't blame me for dreaming.

INTRODUCTION

The Scope of the Problem

A patient comes to labor and delivery at 30-weeks' gestation after talking to the doctor on the phone and being instructed to do so. She complains of some cramping and leakage of fluid a few hours after sex. She is examined and her vaginal discharge is found to be positive for ferning on microscopy and the fluid is found to be nitrazine positive. She is also diagnosed with a urinary tract infection based on her urinalysis, which shows large leukocyte esterase. She is transferred to a tertiary care center with a diagnosis of ruptured membranes. She is admitted by the high-risk obstetrics service with a plan to administer antenatal steroids for fetal lung maturity. She receives twice-weekly ultrasounds, which reveal normal amniotic fluid levels. Her urinary tract infection is initially treated with IV ceftriaxone and then oral antibiotics. The urine culture comes back with no growth.

She is given magnesium sulfate both for tocolysis and for prevention of cerebral palsy. She is kept on bed rest for the next four weeks and receives ampicillin and erythromycin during the first week to prolong the latent phase of labor. At 34-weeks' gestation, she is induced. She doesn't progress past 4 cm after 18 hours of oxytocin and she is taken for an emergency Cesarean delivery. At the time of Cesarean delivery, the membranes are noted to be intact. She subsequently develops a pulmonary embolism and is stabilized in the ICU. The newborn is taken to the NICU and does well, but does have respiratory complications and is intubated and placed on a ventilator. The newborn, after about two weeks, develops necrotizing enterocolitis, undergoes a colostomy and bowel resection, and is eventually diagnosed with short bowel syndrome.

The mother, family, and referring obstetrician are never told that the patient's membranes were not ruptured. The mother is thankful to the hospital and the doctors who saved her and her baby.

If you're not an obstetrician, the above vignette may be exhausting. Don't worry; I don't expect you to know the subtleties of Ob/Gyn to read this book. The truth is, even if you are a practicing Ob/Gyn, I doubt you noticed the *twelve* critical errors in the above scenario; in fact, many physicians who read this vignette believe that, overall, the bad outcomes described were unavoidable! *And that's why we have bad outcomes...*

Just so we can keep track, let me list these twelve errors. I hope that by the end of our time together, they will all be shamelessly obvious (and you might pick up a few more that I overlooked). I will give the *briefest* of explanations why each error is, in fact, an error, but just enough to pique your interest and perhaps make you a bit defensive.

1. **The phone call.** The probability that the complaint was clinically meaningful was so low that the patient shouldn't have been told to come to the hospital.
2. **The exam.** The pretest probability of ruptured membranes was so low that the exam was more likely to be harmful than helpful.
3. **The ruptured membranes diagnosis.** Given the low pretest probability of ruptured membranes, the positive exam findings gave a very low post-test probability or positive predictive value.
4. **UTI diagnosis.** The patient had no complaints to warrant the urinalysis even being performed, giving a very low pretest probability for UTI.
5. **UTI treatment.** The choice of an IV antibiotic was inappropriate.
6. **Lack of repeat examination.** The receiving physician should have repeated the exam at some point, given the low probability that the diagnosis was correct.
7. **Ultrasounds.** There was no indication for these ultrasounds, and the repeated finding of normal fluid never served to call into

question the original (incorrect) diagnosis of ruptured membranes.

8. **Antibiotic usage.** The UTI antibiotics were continued despite the negative urine culture and were unnecessary in the first place given the concomitant use of the other two antibiotics to prolong the latent phase of labor. This overuse of antibiotics likely contributed to the neonatal necrotizing enterocolitis.
9. **Magnesium sulfate.** There is widespread agreement that magnesium sulfate does not act as a tocolytic. The literature which supports its use to prevent cerebral palsy is statistically insignificant and fundamentally flawed; long-term child follow-up studies show no differences between children exposed to the drug and those not exposed to the drug.
10. **Bed rest.** There is no scientific support whatsoever for bed rest and it is clearly associated with harm, such as thromboembolism. This almost certainly contributed to the maternal pulmonary embolism.
11. **The arrest of labor diagnosis.** Current guidelines for labor-management were not followed in making this diagnosis.
12. **Nondisclosure.** Apart from the ethical malpractice of not telling the family what happened, none of the actors involved likely learned anything from this misadventure due to the subsequent cover-up.

This case is made up, but it is conflated from two or three actual cases. This is how medicine, in all specialties, is unfortunately practiced every day. The result is low-quality care for high cost. Today, these types of cognitive mistakes are the leading cause of harm to patients – not long work hours or medicine interactions or wrong-site surgeries – but the way we *think*.

Cognitive errors and mistakes are the leading cause of medical error. At least 15% of all diagnoses made by physicians are incorrect.[1] Some studies have indicated misdiagnosis rates as high as 26-44%.[2] The diagnoses and subsequent treatments that we currently consider correct are often based on studies and data that are not actually true. The labs we order are often unnecessary and many times lead to more harm than good. Our screening tests are applied to the wrong populations, harming those populations in the process. We are in the midst of an epidemic of over-treatment and over-intervention that threatens the financial security of the healthcare industry.

So, what are some of the problems with the way doctors think? Here is an overview of the main issues that we need to deal with if we are to improve the current state of affairs:

Cognitive biases. Our susceptibility to cognitive biases is enormous and they shape every thought and every clinical encounter that we have. Even when we are supplied with accurate data, the use and interpretation of that data are often compromised and perverted by our cognitive biases. We will talk about some of the most important biases, including *confirmation bias*, *base-rate neglect*, *commission* and *omission bias*, *premature closure*, *anchoring*, *framing*, *availability bias*, *loss aversion* and *extremeness aversion*, *naïve extensional reasoning*, etc.

Under-utilization of physical exam and history. Modern clinicians do not trust their physical exam skills and have relegated the patient history to a laborious chore of documentation necessary for billing. Understanding that the history and physical is essential in determining the pretest probability for subsequent tests that might be ordered can help to reemphasize the importance of these steps. Clinicians need to understand the true value of physical exam, which in some cases is superior to the "objective" tests that they choose to rely upon in its place. Perhaps more important, they need to understand that the tests that they rely on are not interpretable without an assessment of the patient's pretest probability – tests are not simply "positive" or "negative."

Over-utilization of medical tests. Not understanding the actual utility of medical tests leads to their overuse and leads to an over-reliance on them for clinical and diagnostic purposes. Inappropriate diagnosis, in turn, often stems from this inappropriate utilization of medical tests due to false positive results. This over-utilization also promotes quasi- "standards of care" which have deleterious downstream effects, like creating a worse medico-legal environment and driving up the cost of healthcare without a simultaneous and commensurate improvement in patient outcomes.

Misdiagnosis and over-diagnosis. As the history and physical become more and more under-emphasized by clinicians and, in turn, medical tests become more and more inappropriately utilized (without an accurate understanding of post-test probabilities and the true predictive value of the tests), an unacceptable number of patients receive the wrong diagnosis (misdiagnosis) or they receive a correct diagnosis that is clinically inconsequential (over-diagnosis). In a study of 208 patients who had a missed or incorrect diagnosis, 63% of the time this was related to a failure to perform a physical exam.[3] These types of basic mistakes (which may lead to an inappropriate diagnosis) in turn may lead to inappropriate or unnecessary interventions that drive up medical costs while increasing the net risk of harm to patients.

An inappropriate understanding of (and therefore an inappropriate utilization of) statistics. An understanding of statistics is the bedrock of almost every activity of clinical reasoning and the delivery of medical care. You rely on statistics every day – even if you think that you do not. Statistics inform how we think about the impact and importance of various disease processes, the utility of the tests and the therapies that we order, the risk-benefit ratios of our clinical interventions, the magnitude of the importance of the things that we do to and for our patients, how we read and interpret scientific literature, and how we design and perform research studies.

Let me be more technical for a moment. We are currently witnessing a revolution in the field of statistics. In the twentieth century, scientific research was conducted almost exclusively with an orientation towards the Frequentist or Orthodox school of statistics – the traditional form of statistics that deals with things like P-values and confidence intervals. Now, statisticians and researchers across a wide range of disciplines, including medicine, are realizing that P-values have been abused, manipulated, and over-emphasized at the expense of more appropriate, powerful, and relevant forms of inference, such as Bayesian inference. A wide-scale shift to Bayesian statistics, or "probability as logic," in the twenty-first century will transform research methods and the underlying philosophical understanding of research results. The importance of this goes far beyond impacting just the statistical methods we use to analyze data.

Indeed, it changes the entire way we think and even has profound implications for what we accept as true or untrue. It changes the way we conduct experiments and even changes what we call "the scientific method." It should, hopefully, change the way we publish, analyze, and appraise scientific literature. We must, for better or worse, spend some time thinking about this subject in detail.

Discomfort with uncertainty. Clinicians are notoriously uncomfortable with uncertainty. We feel better when we are definitive; we like to tell patients exactly what's wrong with them and precisely what's best for them. We like things to be black or white: one treatment is better than another, one test is better than another, this study *proves* this, that study *proves* that etc. But the reality is, all those types of statements are absurd.

In philosophical terms, there are profoundly critical epistemological issues at play regarding our views of certainty and doubt that are framed in our understanding of probability theory. In practical terms, we are never "certain" about the diagnosis, but we accept some diagnoses as more probable than other diagnoses. Appreciating the uncertainty that exists in every diagnosis and, indeed, in every decision, helps to debias our cognitive processes and de-dogmatize many of the things we hold to be true. It also changes what we think it means to be "wrong" or "right" and helps us think in inferential terms, always assessing new evidence and remaining comfortable with changing our beliefs (diagnoses, etc.) as new data develops. *Plausible inference* or *Bayesian updating* is the key to dealing with uncertainty in a logical manner.[4] We will learn about this, too.

The misunderstanding, underutilization, and perversion of evidence based medicine (EBM). Many believe that EBM is at a crossroads. Its most ardent detractors believe that it is standing in the way of progress. Yet, for its supporters, it's hard to imagine that any alternative to EBM is valid. Should we rely on a single practitioner's common sense and anecdotal experiences when deciding how to diagnose and treat complex medical problems? Or should we rely on our collective experiences? Is the knowledge of one doctor with a few dozen patients more relevant than the knowledge of hundreds of doctors with thousands of patients?

If you agree that the collective knowledge and experience of the many is more valuable than the experience of the one, then you believe in EBM. EBM is simply the organized system for how we collect, analyze, and report those collected experiences and collective knowledge. If we look more closely at the criticisms of EBM, they are not, in fact, criticisms of what I just stated. Rather, they are criticisms of the Frequentist statistical techniques which are at the root of the perversion and manipulation of EBM, and the incorrect use of the results of those techniques.

It is often stated that statistics can be used to prove anything and, in a sense, this is true, depending on what we mean by "prove." However, Bayesian inference cannot be used to *prove* anything. What's missing from our current, often criticized model of EBM is Bayesian updating or probabilistic reasoning. By adding this small but profound component to EBM, all the currently popular criticisms of EBM fall away.

An educational system which does not adequately teach critical thinking skills. When we start to talk about things like *dual process theory, cognitive biases, Bayesian inference, plausibility reasoning, epistemology*, etc., most people just stop reading. It's difficult to convince many medical students, residents, or physicians that a poor understanding of these esoteric topics is the root cause of most medical errors and waste. Or, worse, that a lack of education about these topics stands in the way of scientific progress in medicine and many other fields. It is a struggle to make these topics palatable to folks who would rather be reading about seemingly more "practical" ways to help their patients.

Too often, doctors want to be told to "do this" or to "do that." They want to organize knowledge algorithmically, in the form of "if this, then that." They don't seem to want to think for themselves. Ironically, this type of cookie-cutter patient treatment information often comes to them in the form of an evidence-based practice guideline, which they also reject because it seemingly negates their individual experiences.

Our educational system has selected a certain type of person for medical training by emphasizing certain intellectual attributes that are demonstrable on standardized tests. This unfortunately means that a good memory and efficient recall of "facts" is more important than critical thinking abilities. Philosophy, statistics, and logic are not prerequisites to medical school. Most practicing physicians see little value in the writings of Plato or Socrates, Boole or Cantor, Frege or Russell,

Fischer or Jeffreys, Kant or Nietzsche, Jaines or Cox, Allais or Kahneman, etc.

What's more, the process of medical training does not emphasize critical thinking: there is no thesis to defend, no requirement to do meaningful research, no immersive view of knowledge. Instead, students are taught to memorize "truths" and regurgitate them when asked. This is made more convenient by review-books and exam courses, which teach students lists of "facts," all out of context and without a sense of the level of certainty of those "facts."

All propositions are held to be the same, in simple, Aristotelean terms: either true or untrue. Multiple choice questions define whether we practice medicine or do not. We treat the "truth" of the boiling point of water in the same way as the "truth" of the lipid hypothesis of coronary artery disease; yet, these two "truths" have profoundly different levels of certainty or degrees of plausibility. So, our approach to education needs a fundamental overhaul in order to teach the skills necessary for critical thinking.

Let's dive in and develop some new skills that can help us make a better diagnosis, better utilize tests, read scientific literature more accurately, pick better treatments and interventions, and deliver better healthcare to our patients at lower costs.

1

Problem Solving

I always enjoyed reading and watching Sherlock Holmes as a child. His ability to notice subtle details around him and construct theories and solutions for his cases always seemed to represent a *tour de force* of critical thinking. We solve mysteries every day in medicine. Where Holmes studied murders and burglaries, we study sick people and disease. We examine clues and develop theories and hopefully come to some solution for the case.

Sir Arthur Conan Doyle based the character Sherlock Holmes on a Scottish surgeon called Joseph Bell. Doyle had worked for Dr. Bell earlier in his life and the fictional Holmes shared many characteristics with the real-life doctor, including the uncanny ability to create detailed explanations for a case based on rather minor observations. In more recent years, we've come full circle as the fictional Dr. Greg House on the TV series *House* possessed those same skills. His character was simply a medicalized version of the Sherlock Holmes character.

When I was a kid, I thought that Sherlock Holmes used *deductive reasoning*. He is described as doing so by himself and by observers of the fictional character in the Doyle stories. He also described what he did, however, as *reasoning backwards*. In Chapter 7 of *A Study in Scarlet*,

Holmes states,

> In solving a problem of this sort, the grand thing is to be able to reason backwards. ... In the everyday affairs of life it is more useful to reason forwards, and so the other comes to be neglected.

In *A Scandal in Bohemia*, he says,

> It is a capital mistake to theorize before one has data. Insensibly one begins to twist fact to suit theories, instead of theories to suit facts.

The Sherlock Holmes novels are full of advice on the methods of *inference* and *deduction*. The most famous quote comes from *The Sign of Four*, in which Holmes tells us,

> Eliminate all the factors, and the one which remains must be the truth.

As I said, I grew up thinking that what Sherlock Holmes did was deductive reasoning. But it wasn't. Let me explain.

Types of reasoning

There are three main types of reasoning: *induction, deduction,* and *abduction.*[5]

Induction. We use inductive reasoning or inference every day, even if we don't realize it. Here's an example:

> *Every person I've ever treated with uterine cancer was a woman.*
> *Therefore, the next patient I see with uterine cancer will be a woman.*

In this example, inductive reasoning allows me to form a general rule that is valid. But what about this example?

> *Every person I've ever treated with breast cancer was a woman.*
> *Therefore, the next patient I see with breast cancer will be a woman.*

This second example shows the problem with inductive reasoning. There aren't many absolute rules; most rules have exceptions. Inductive reasoning is dependent upon the breadth of our experience or data. We form biases and stereotypes because we make assumptions through inductive inference. Those assumptions might even be accurate in most cases. Yet, it's the exceptions to the rule that cause our reasoning to fail. Men get breast cancer, too.

As we will explore in this book, we can fix this problem with inductive reasoning by assigning probabilities to our rules. How often is the rule true? Most modern forms of inductive reasoning do this to account for the inherent uncertainty of induction, no matter how good the data that informs the inference might be. Thus, we might state instead,

Ninety-eight percent of patients with breast cancer are women.

Therefore, the next patient I see with breast cancer has a 98% chance of being a woman.

By understanding inductive inference in the context of probability theory, we are not forced to state falsehoods as propositions just because they are *usually* true. Instead, we are forced to consider how likely our proposition is, statistically speaking.

Deduction. Deductive reasoning or inference is just the opposite. An example might be:

All patients with uterine cancer are women.
My patient has uterine cancer.
Therefore, my patient is a woman.

With induction, we surmised a general rule from our observations. With deduction, we applied the rule to draw a conclusion about a specific situation. Deductive reasoning is only as sound as the general rule. Like inductive reasoning, the exceptions to the rule ruin the logic. But also like inductive reasoning, we can state the truthfulness of our propositions with probabilities to avoid drawing incorrect conclusions.

What Holmes did was a little different than either deduction or induction. Conan Doyle wrote the first Sherlock Holmes mystery, *A*

Study in Scarlet, in 1887 and he was unfamiliar with a then relatively new term for a different sort of reasoning called *abduction*.

Abduction. It was an American logician called Charles Sanders Peirce who first described *abductive reasoning* in the 1860s and worked on refining the idea for decades. It didn't become well-known until after 1900, long after Sherlock Holmes had become popular.

In abduction, one makes observations and then attempts to find the simplest and most probable explanation for those observations. A conclusion is reached (an explanation for the observations), but it is not held out to be absolutely certain and can change with new data or observations. In abductive reasoning, one may interrogate or experiment to test a hypothesis and strengthen, weaken, or refine the original hypothesis. This is what we do in medicine, and it's what Holmes did too.

In fact, Holmes treated a crime like a symptom of a disease. He developed theories of the crime (analogous to our differential diagnosis) and then looked for evidence or conducted experiments to test those hypotheses. He had a vast knowledge of previous crimes and believed that the solution for the present case would come from understanding cases of the past. He tended to notice things that others didn't *because he was looking for them*. When we have a good differential diagnosis, we know which clues to look for as well. He was also aware of things that might bias him and trick him into believing a falsehood; he carefully avoided logical errors or other limitations in his thinking that might prevent him from reaching the right diagnosis – *umm*, finding the right criminal.

Probabilistic abduction is mostly what we will talk about in this book, though we usually won't use that term. This is what we get when we apply Bayes' theorem to abductive reasoning. We will talk a lot about Bayes' theorem a little bit later.

After Peirce's work, Karl Popper talked about the hypothetico-deductive method in the twentieth century. This method is essentially the rigorous process of testing the conclusions drawn from abductive reasoning and it is analogous to the scientific method. Essentially, we:

1. Make observations and gather data.
2. Generate a hypothesis.
3. Predict consequences of the hypothesis.

4. Look for evidence that disagrees with our predictions.

In Popper's method, it is a fallacy to look for evidence that confirms our predictions, but this is often how medical reasoning is taught due to the pragmatic demands of clinical medicine. For example, we see a patient with a sore throat and cervical lymphadenopathy. We hypothesize that this is due to Strep throat. We predict that a rapid Strep screen will be positive. Then we do a rapid Strep screen and it is positive. This last step is a breakdown of the hypothetico-deductive method but not of *abductive reasoning* or *probabilistic abduction*. Hypothetico-deductive reasoning is the gold standard and is relevant for rigorous experiments and studies, but it is not what we practically do in clinical medicine on a daily basis.

What's your best guess?

In clinical medicine, we are usually just trying to make the best guess in an efficient and economical way. Let me give an example of abductive reasoning as a method for guessing the best answer.

When my youngest daughter, Adeline, was four years old, she asked me to guess which two numbers she was thinking of and all she would tell me was that the two numbers add up to 10. How do I approach this problem? I guess; but my guess is not entirely uninformed. This is a hallmark of abductive reasoning; the hypotheses are not just pulled out of thin air but, rather, they are based on all available evidence and clues.

First, I would assume that she is talking about integers. I can also assume that she is talking about positive numbers (she probably doesn't know much about the concept of negative numbers). She is likely thinking of two different numbers (since that's how a four-year-old would process the phrase *two numbers*), and I doubt she would consider zero (nothing) as a number. This leaves just four likely options: 1+9, 2+8, 3+7, and 4+6. Knowing that people tend to think of numbers in the middle of a sequence, I picked 4+6 (and I was right).

Am I psychic? No. Lucky? No. I simply bet on the most probable solution. Yet 42.5 and -32.5 as well as $\pi + 6.8584$ could have been the answers as well. Even among rational numbers, there are an infinite number of solutions. But this is how we make any decision. There is a real answer (whatever is in her head) but without her telling me the answer, I need to use all ascertainable data to make my best guess (that is, the most probable solution). When patients present with a problem, there is an actual diagnosis, but they don't know it and neither do you, so we must use all ascertainable data to make a best guess (and change that guess to the *new* most likely solution when we learn new information).

So, what did I do when guessing the two numbers? I divided the infinite number of possible solutions into common solutions with a high probability of being correct, uncommon solutions with a lower probability of being correct, and uncommon solutions that are highly improbable.

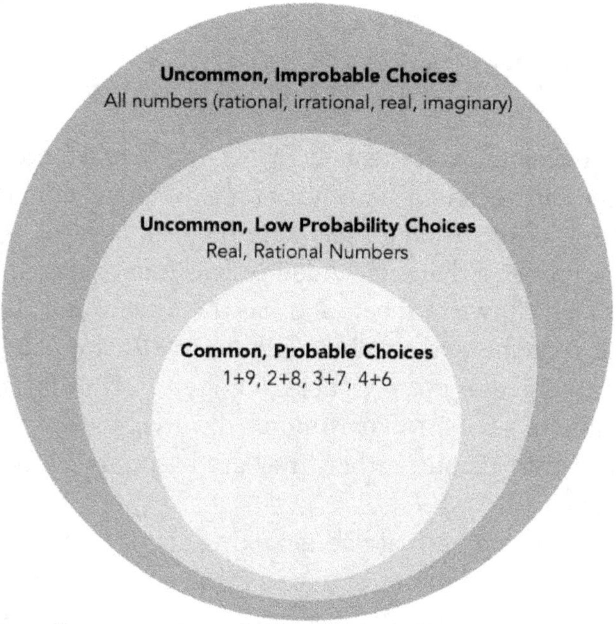

Figure 1. Venn diagram showing probable and improbable guesses to a child's question.

In other words, I made an exhaustive (infinite!) differential diagnosis once I was presented with the chief complaint ("Daddy, guess two numbers..."). In a real clinical encounter, I would then start asking a series of questions to narrow down my differential diagnosis. Note that the questions I ask are based upon the differential diagnosis, so the differential diagnosis is very important and should be considered *before* the questions are asked. For example,

Are they whole numbers?
Are they positive numbers?
Are the two numbers different?
Are the numbers even or odd?

As I gather each additional piece of information, I quickly narrow down that infinite list. If I can get answers to those four questions (don't assume we can always get the answers we need), I can narrow down an infinite number of choices to just two. There is also a possibility that I cannot get *certain* answers to each question; this is more often the case when dealing with clinical medicine. "Does the x-ray show pneumonia?" The true answer might be, "There is an 80% chance that the x-ray shows pneumonia." Nevertheless, I can still use this information to make one diagnosis more likely (pneumonia) and another one less likely (lung cancer).

A test is just another type of question. I can ask my daughter if she is talking about two different numbers; she may not know the answer. I could also perform an experiment. For example, I could have her subtract one number from another and tell me whether the answer is equal to zero. In this same way, I order tests in clinical medicine. A CBC or a CT scan is just another form of interrogation.

In clinical medicine, I like to talk about diagnoses in the same three-tier system. I call them *horses, zebras,* and *Tasmanian tigers*. We all know the classic diagnostic axiom,

When you hear hoofbeats, think horses not zebras.

And it is true: common things are common and rare things are rare. Our knowledge of what is common (the *accessibility bias*; see Chapter 3) saves us time and makes us efficient; plus, we are usually correct because common things explain most things we observe. But sometimes hoofbeats are zebras, and sometimes they are even Tasmanian tigers.

Tasmanian tigers officially went extinct in 1936. They were a marsupial that was hunted to extinction in parts of Australia, New Zealand, and Tasmania by farmers and others. Yet, every now and then, someone spots an animal in the wild they believe is a Tasmanian tiger.

I have seen zebras and I know where to find them (the zoo); but I have never seen a Tasmanian tiger and, as far as I know, I never will. Still, I wouldn't be shocked if someday another one is spotted deep in the wild bushlands of Australia. But, by analogy, 99.7% of hoofbeats in medicine are horses and zebras.

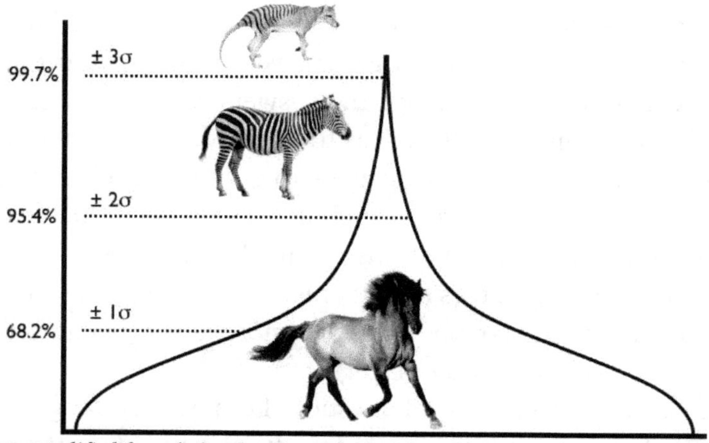

Figure 2. Modified funnel plot showing a normal distribution (σ = standard deviation).

The sum of Tasmanian tigers + zebras + horses is equal to all possible diagnoses. Most of those don't interest me; but as long as something is possible, it is still on the list. We spend most of our time in the green fields of horses, and occasionally we venture off into the goldenrod pastures of zebras. Let's look at a practical example by considering a few choices for postmenopausal bleeding.

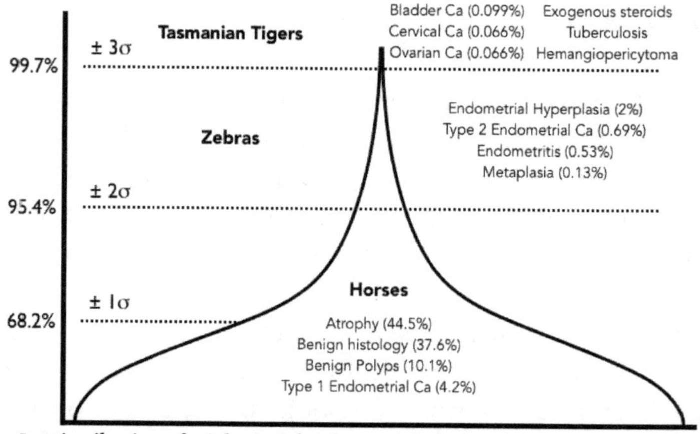

Figure 3. Distribution of etiologies of endometrial causes of postmenopausal bleeding.

I have arbitrarily (and imprecisely) divided this differential diagnosis into three groups based on standard deviations (yes, I know the math isn't perfect). But the general idea is that about 95% of diagnoses are horses, while about 99.7% of all diagnoses are either horses or zebras. That last 0.3% has a ton of stuff in it, not just the ones I have listed in Figure 3, including things like primary lymphoma of the vagina,[6] hypernephroma,[7] ligneous cervicitis,[8] tamoxifen-affected villous papyraceous mass,[9] vaginal neurofibromatosis,[10] uterine angiolipoleiomyoma,[11] etc. It even contains some undiscovered diagnoses.

We spend most of our lives with horses, and occasionally we meet zebras. The next patient you see with postmenopausal bleeding, if you were to tell her not to worry about it and that it's benign, you would be correct about 95% of the time. But some zebras and even Tasmanian tigers are always important to consider (like endometrial, cervical, ovarian, and bladder cancers); so, we do tests to make sure that these exotic animals haven't wandered into our field (like an endometrial biopsy).

How can we put all this together in a cogent approach to diagnosis?

1. Consider the chief complaint and nothing else (e.g., two numbers equaling ten or postmenopausal bleeding).
2. Develop a reasonable differential diagnosis (if you didn't think of villous papyraceous mass, don't feel bad). Developing the differential diagnosis before you become biased and boxed-in by more information allows you to think more broadly and prevents many forms of cognitive bias (like availability heuristics, which we will discuss in detail in Chapter 3).
3. Interrogate the patient (this includes history and physical exam, as well as lab tests and imaging as appropriate).
4. Order the differential diagnosis into a list of working diagnoses (or even a single diagnosis) based on the probability of each diagnosis.
5. Make sure that the "can't be missed diagnoses" are sufficiently unlikely (for example, perform an ultrasound or endometrial biopsy to lower the probability of endometrial hyperplasia and cancer, even though it is an unlikely diagnosis). Note that not every bad diagnosis needs a test performed to "exclude" it. I

don't want to miss ovarian cancer any more than I want to miss endometrial cancer in a patient with postmenopausal bleeding; but the risk of ovarian cancer is already sufficiently low (0.066%) that I don't need to routinely perform an ultrasound to look at the ovaries. However, the risk of endometrial cancer (about 5%) is too high to not do further evaluation (like an ultrasound or biopsy) to reduce this risk. Even after a negative ultrasound or endometrial biopsy, there is still a risk of endometrial cancer (both tests have significant false negative rates), but now the post-test probability (or risk) is similar to the risk of ovarian cancer – so we can stop worrying about it until we have a new reason to worry.

6. When new evidence emerges which calls the working diagnosis into question (like a failed response to treatment or new symptoms or test results), consider less likely diagnoses and even previously unconsidered diagnoses (someone out there has a hemangiopericytoma, but consider more common diagnoses first before spending resources on this Tasmanian tiger). Also, and importantly, reconsider previously "excluded" diagnoses. Let's say that we "ruled out" endometrial cancer with a negative endometrial biopsy, but the patient persists in having bleeding. In that case, a hysteroscopy may be the most appropriate next step because the risk of endometrial cancer in a patient with a negative endometrial biopsy is still higher than the risk of one of those Tasmanian tigers like ovarian cancer. So, we sometimes need to do a better test (or even repeat the same test again). This is also an excellent time to do a literature search and read current literature about the symptom and see if there is anything left to learn.

That's it. We are always giving patients our best guesses; we need to remember that they are all guesses and as such they should be subject to revision as new information becomes available. That's what Bayesian updating is, and the process I have described here is called *Bayesian probabilistic reasoning*. It allows me to give the most accurate answer I can to various questions, ranging from my daughter asking me to guess two numbers to a patient asking me to find out why she has postmenopausal bleeding.

Bayes' theorem

I have mentioned Bayes' theorem a few times already and you will be an expert on how to use it by the end of this book. It is the key to determining the best solutions to many different questions we face as clinicians. We form hypotheses through our abductive reasoning processes. These hypotheses look like these statements:

I believe that...

- You have HIV
- Your child has Down Syndrome
- You have breast cancer
- You have Strep throat
- Elevated systolic blood pressure increases the risk of having a male fetus
- Smoking causes lung cancer
- Listening to Elvis causes brain cancer
- Vitamin C prevents/cures the common cold

We form hypotheses about diagnoses, treatments, pathophysiology, and a lot of other things. When we have any belief, we look for evidence for it and against it, and then we revise our belief according to how strong and good that evidence is. Essentially, that is Bayes' theorem or Bayesian updating. Bayes' theorem is just common sense.

Here is Bayes' theorem where *H* stands for *hypothesis* and *E* for evidence and *P* is *probability* (the vertical bar '|' is read as *given*):

$$P(H|E) = \frac{P(E|H)P(H)}{P(E)}$$

So, the theorem states that the probability of a hypothesis given that some new evidence exists is equal to the probability of the evidence given that the hypothesis is true, multiplied by the prior probability that the hypothesis was true (before we knew about the new evidence), divided by the probability of seeing the evidence under all possible hypotheses.

Let's label the terms of the equation:

Likelihood
The likelihood of observing the evidence (E) if the hypothesis (H) were true

Prior Probability
The probability that the hypothesis (H) is true before the new evidence (E)

$$P(H|E) = \frac{P(E|H)P(H)}{P(E)}$$

Posterior Probability
The probability that the hypothesis (H) is true given the new evidence (E)

Probability of Evidence
The probability that the evidence (E) exists for any hypothesis

Figure 4. Bayes' theorem.

Note the importance in this equation: it tells us the probability of a *hypothesis given some data or evidence*. Traditional statistical methods called **Frequentism** answer the opposite question, that is, the probability of the *data or evidence given that the hypothesis is true*. These are the sorts of methods that result in *P-values*, and these estimations can often inform the *Likelihood* variable of Bayes' theorem in practical terms. But Frequentist methods do not deal with the other terms of the equation. This distinction is important. Frequentism answers this question: What is the probability that the evidence I have would exist if my hypothesis were true, rather than answering whether the hypothesis is true.

Bayes' formula (and this formula alone) answers the fundamental question of every research scientist, physician, philosopher, and person:

What is the probability that my belief is true given this evidence?

Here are different forms of this same question:

- What's the chance that this patient has HIV given a positive screening test?
- What's the chance that this man committed murder given that his DNA was found at the crime scene?
- What's the chance that this drug lowers cardiac mortality given this new study?
- What's the chance that my girlfriend is pregnant given a positive pregnancy test?

- What's the chance that this child has Strep throat given her symptoms?
- What's the chance that this coin is biased given that it rolled heads 8 out of 10 times?
- What's the chance that human activity contributes to global warming given the body of evidence that currently exists?
- What's the chance that my dinner is poisonous given that I am eating at McDonald's?

All questions of this sort ask us to refine our belief in a hypothesis (or postulation) given some piece of evidence. It also demands that we understand what our degree of belief in the hypothesis is before we update our belief, given some new evidence. This prior degree of belief is called the *prior probability*; or, in other settings, the *pretest probability* of a diagnosis or *pre-experimental probability* of a hypothesis. The new probability that is calculated considering the new evidence is called the *posterior probability* (or you might call it the *post-test probability*); it is the chance that the hypothesis you were testing is true, given the new evidence. It is the answer to the original question.

Bayes' theorem is not a radical new idea. Rather, it is accepted science and hundreds of years old. Still, it seems radical to many statisticians who are steeped in Frequentist methods.[12] Some even criticize it. For those who want a deeper dive into Bayes' theorem, please read the Appendix now or later. There I will present a proof of Bayes' theorem and other contextual information that may be useful. But for the general purposes of this book, keep reading.

Bayesian updating

Bayesian updating simply requires that today's posterior probability becomes tomorrow's prior probability. New evidence becomes available for any given hypothesis continuously; we must revise the strength of our belief (our confidence in it, its probability) as new or revised evidence emerges.

There are many other consequences of this equation which we will discuss. The most crucial point to notice is that Frequentist interpretations of inference cannot answer this most basic and fundamental question of the scientist: is my hypothesis likely to be true? As we shall see, **this is one of the chief sources of error in clinical medicine.**

Additionally, Frequentism neither allows for continuous updating based on new evidence nor does it allow for integration of prior information. In fact, it doesn't even accommodate degrees of certainty at all; Frequentist statistical methods demand belief in the absolute acceptance or rejection of a hypothesis (even though it allows no formal method for assessing the truthfulness of a hypothesis). This means that Frequentist methods are inadequate by themselves to help us with abductive reasoning.

Let me give one more example to demonstrate this.

Imagine that you are standing in a room and you hear voices in the next room that sound exactly like Chuck Norris talking to Jean-Claude van Damme. Thrilled, you quickly record the voices. You develop a hypothesis that Chuck and van Damme are in the room next door to you. You use a computer to test whether the voices match known voice samples of the two martial artists. The computer test tells you that the voices are nearly a perfect match, with a probability of error of less than 0.0001.

This sort of analysis is where Frequentist statistical methods excel. This statistical approach tells you the chance that the data you observed would exist if your hypothesis were true. Thus, it tells you the probability of seeing the voice pattern you observed given that JCVD and Chuck are in the room next door; that is, the probability of observing this test result (the evidence) given that the hypothesis is true, or $P(E|H)$.

But it doesn't tell you if your hypothesis is true. What we really want to know is the probability that Chuck and the Muscles from Brussels are in the room next door, given the voices you have heard; that is, the probability of the hypothesis given the test result, or $P(H|E)$. This is the power of Bayesian inference, because it allows us to consider other information that is beyond the scope of the test, like the probability that the two men would be in the room in the first place compared to the probability that the television is just showing a rerun of *The Expendables 2*.

Bayes' theorem tells us the probability of an event based on everything we know that might be related to that event, and it is updated as new knowledge is acquired.

Frequentism tells us the probability of an event based on the limit of its frequency in a test. It does not allow for the introduction of prior knowledge or future knowledge. Too often, researchers assume that their hypotheses are true because they discover a P-value < 0.05, but there are always multiple hypotheses that might explain the observation of the same research data, such as the TV in the room playing the movie versus the men being present in the room. Which of those two hypotheses are more likely?

Our brains naturally work using Bayesian-like logic, so it should come easily.

Only Bayes' theorem allows us to answer this question: What's the probability that my theory or hypothesis is correct?

This question may come in many forms:

- The basic scientist: Is my theory correct?
- The clinical scientist: Does my intervention make a difference?
- The physician: Is my diagnosis correct?
- The patient: Do I have this disease?

These are all the same question; each person has developed a hypothesis, theory, or conjecture, and each person then applies the scientific method, either formally through the hypothetico-deductive model or informally through probabilistic abduction, to answer the original question.

Probabilistic reasoning

Bayes' theorem and abductive reasoning are the basic science principles which support the ideas behind probabilistic reasoning. We are going to use probabilistic reasoning to answer these sorts of questions.

To solve any problem, then, we need to consider a few basic things.

- What's the prior probability of my hypothesis?

- How good is the evidence I'm examining?
- What's the posterior probability of my hypothesis after considering this new evidence?

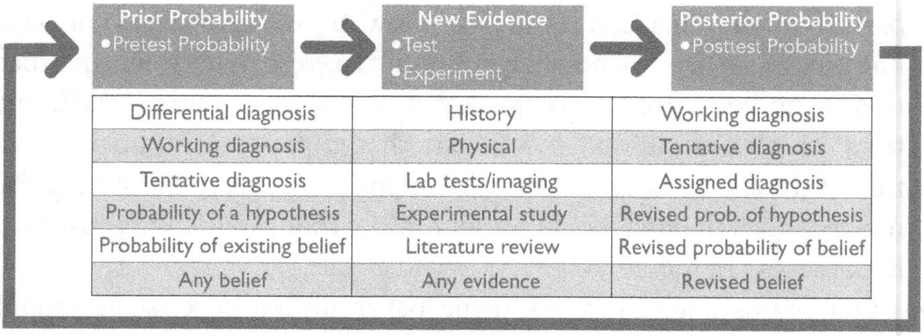

Figure 5. Bayesian updating.

Let's replace the words *prior probability, evidence,* and *posterior probability* with more familiar terms.

Prior probability might be:

- The prevalence of a disease before we apply a screening test (my pregnant patient is 35-years-old and has a 1/270 chance of having an infant with Down syndrome before I do a test).
- The probability that a disease on our differential diagnosis is associated with our patient's chief complaint before we conduct a history and physical (e.g., *hypothyroidism is an unusual finding among women who present with a complaint of fatigue*).
- Our clinical suspicion of a disease before we conduct a test (*I believe based on history and physical examination that this patient is likely to have hypothyroidism*). Note: the purpose of conducting a history and physical is to determine the likelihood of disease.
- The probability that our hypothesis is true before we conduct a scientific experiment to test it (based on available studies and my understanding of the pathophysiology of disease, I believe that it is very unlikely that listening to Elvis Presley music causes a brain tumor).
- Our prior confidence in any belief before looking for evidence (I am very confident that Vitamin C doesn't alter the duration or severity of the symptoms of a cold before I conduct a literature search).

Evidence comes in many forms as well:

- History
- Physical exam
- Risk factors
- Demographic data
- Lab and imaging studies
- Response or lack of a response to treatments

Posterior probability is merely our updated confidence in the belief:

- Positive or negative predictive value
- Probability of having a disease
- Probability that any given hypothesis is true

Let's learn how to use this information clinically.

Exercises

1.1 Identify two recent patients that you have seen (or case scenarios you have studied). What are specific examples of prior and posterior probabilities that you considered? What sort of evidence did you consider?

2

The Diagnostic and Treatment Process

The most important tasks of any physician are making the correct diagnosis and picking the correct treatment. As I have hinted at already, one or both tasks are done poorly or inadequately during almost 80% of all clinical encounters. This book is about how to do those things better. But let's discuss what should happen during the diagnostic and treatment process before we dive into why we fail so often at these tasks. Then we will outline what skills we need to develop to improve our abilities.

Making an accurate diagnosis first requires *gathering information* in an efficient manner (and only gathering the correct and needed information). This process should be as free of bias as possible. The correct and needed information is all the information necessary to adequately consider each important and probable diagnosis in the differential diagnosis.

Gathering information starts with the history and physical examination. Because of this, it is important (as we will discuss in detail later) that the differential diagnosis be constructed as soon as possible, ideally before gathering information begins.

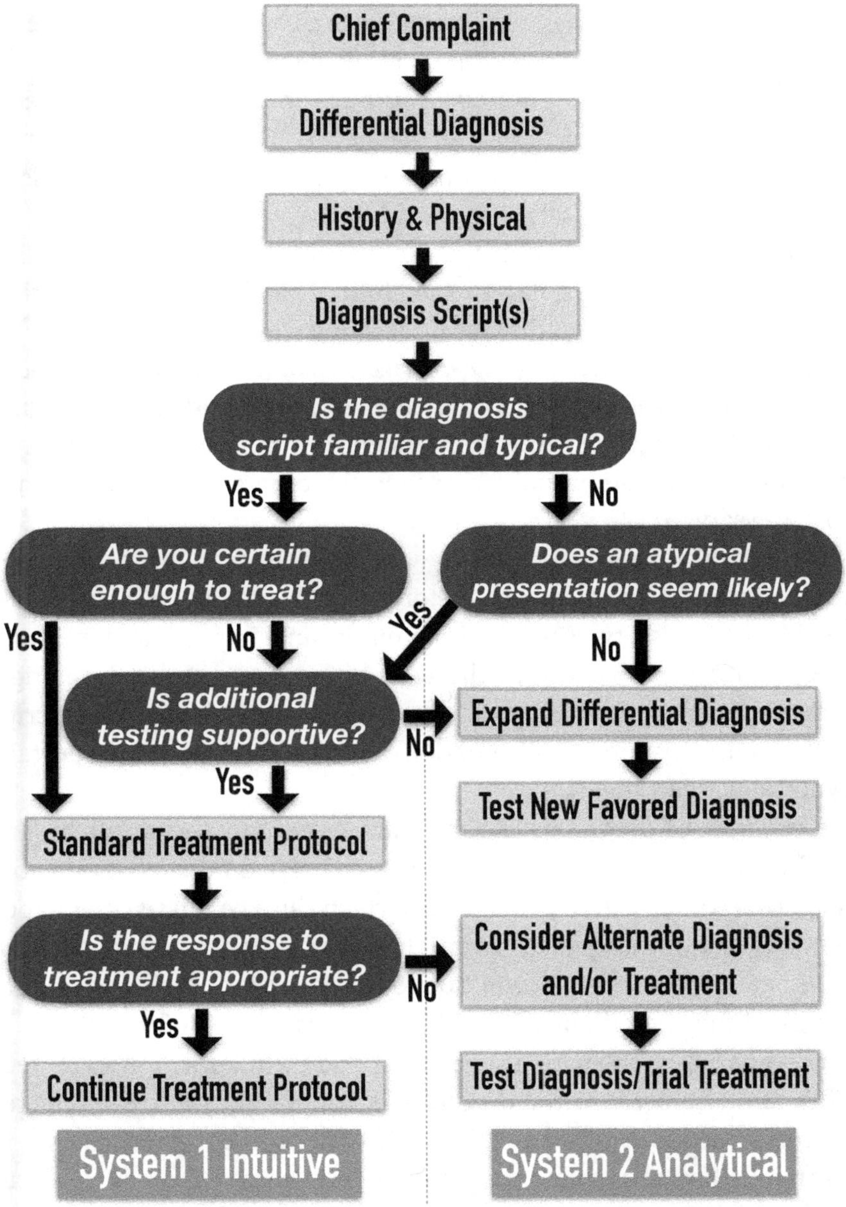

Figure 6. Overview of the diagnostic and treatment process.

After information is gathered that is immediately available to the physician (the history, physical examination, medical records, etc.), the physician will start to organize the patient's presentation and story into what is called a *diagnosis script*. More than one diagnosis script may come to mind, as many diseases mimic one another.

Some of these *diagnosis scripts* may match well with the patient's presentation. The diagnosis script will seem familiar and typical to the clinician. This process is often intuitive; for experienced clinicians, most diagnoses are made in this manner. In *Dual Process Theory*, this intuitive thinking is called System 1 reasoning (we will discuss this in detail in the next chapter).

If the patient's presentation doesn't seem typical or familiar, this should throw up a red flag to the clinician and a slower, analytical diagnostic process should follow. *Could this be an atypical presentation of a diagnosis I have considered? Or have I not even thought of the correct diagnosis yet?* Additional information, often in the form of laboratory or imaging tests, will be necessary to gain confidence in the appropriate diagnosis. Using information tools to expand the differential diagnosis may be necessary.

Additional testing isn't always warranted. In fact, most diagnoses can and should be made with history and physical alone. Sometimes, testing is necessary to exclude common and important mimickers (like endometrial cancer in the case of postmenopausal bleeding). However, in many cases, if a familiar and typical diagnosis script matches the patient's presentation and explains all the important symptoms, then treatment can proceed.

Certainty of diagnosis is rarely achieved and shouldn't be the goal of the clinician. The goal is only to decide on the most probable diagnosis while excluding diagnoses which can't be missed due to potentially serious sequelae. Even then, exceedingly rare diagnoses, whether serious or not, needn't be routinely excluded with additional testing if they remain highly unlikely after gathering information.

In some cases, the patient won't respond to treatment as expected. This could be because the diagnosis is correct but the treatment required is atypical, or, more likely, it may be that the diagnosis is wrong. Failure to respond to treatment as expected should trigger an analytical diagnostic approach and everything (the diagnosis and treatment) should be reconsidered.

Let's consider a common example.

A patient presents to an ambulatory clinic complaining of a sore throat. The physician, *before seeing the patient*, briefly considers the differential diagnosis for a sore throat: viral infections (rhinovirus, coronavirus, adenovirus, herpes virus, influenza, parainfluenza, Epstein-Barr, cytomegalovirus, HIV), bacterial infections (Group A strep, Group C strep, *N. gonorrheae, C. diphtheriae, Mycoplasma pneumoniae, Chlamydophila pneumonia, Fusobacterium necrophorum*), Kawasaki disease, and noninfectious causes (persistent cough, postnasal drip, GERD, thyroiditis, neoplasm, allergies, smoking, foreign bodies).

With this differential diagnosis in mind, the physician gathers information that helps determine the correct diagnosis:

Questions: *How old are you? Do you have a cough? Have you had a fever? How long have you been sick? Do you have other symptoms? Has anything made you better or worse? Have you noticed swelling in your neck? Do you smoke? Do you have acid reflux? Do you have seasonal allergies? Have you been around anyone who is sick? Are you sexually active? Do you have fatigue or malaise? Have you had any rashes? Have you had difficulty swallowing?*

Physical exam: *Is there cervical lymphadenopathy? Does the throat have any distinct lesions? Are there tonsillar exudates? Are the lungs clear? Is there hepatosplenomegaly? Is there conjunctivitis? Are there any rashes?*

After the appropriate information is gathered, the clinician now can synthesize the findings. She may notice immediately that the patient's presentation matches well with a familiar and typical diagnosis script: a child with sick contacts who has had a fever, no cough, sore throat, tender anterior cervical lymphadenopathy, and tonsillar exudates. Recognizing that this fits well with her diagnosis script for group A beta-hemolytic streptococcus (GABHS), she decides to proceed with treatment. She skips the rapid screen, realizing that it won't affect her treatment decision, but does send a culture since the results of the culture might change her management decision. She also pauses to consider what else the child might have that could explain the symptoms, and realizing that no life-threatening or important-to-exclude diagnoses come to mind, she finishes with the case.

Of course, many other diagnosis scripts might have been activated when the physician gathered information. Perhaps instead she found a child with sore throat, fever, bilateral nonpurulent conjunctivitis, anterior cervical node enlargement, erythematous oral mucosa, an inflamed pharynx with a strawberry tongue, and a generalized polymorphous erythematous rash with edema and desquamation of the palms. This information would fit her diagnosis script for Kawasaki disease, though in this case she may realize that there are several overlapping features with GABHS and she may order a rapid strep screen because she realizes that untreated GABHS in a child can have serious sequelae.

Diagnostic failures

Diagnostic failures are common and are the leading cause of preventable medical error. For example, at least 15% of diagnoses made in emergency departments are wrong; the number may be much higher because most disease processes are self-limited (and therefore the patient never knows of or at least is never harmed by a misdiagnosis). Ironically, emergency departments commit more resources to the diagnostic process than nearly any other specialty (labs, imaging, consultation) but are also wrong about the diagnosis more than any other specialty. As we will see in the next chapter, cognitive bias is largely responsible for these errors.

Common reasons for diagnostic errors include:

- The physician didn't consider the diagnosis in the differential
- The physician didn't understand the diagnosis
- The physician didn't obtain necessary history
- The physician didn't conduct sufficient physical exam
- The physician was biased by a red-herring, a similar patient, or the case complexity

Since physicians make so many diagnoses via a quick, intuitive thought process, then faulty activation of familiar diagnosis scripts may often lead to an incorrect diagnosis. Incomplete or incorrect information can also trigger the wrong diagnosis script. Many physicians have inaccurate diagnosis scripts (associating the wrong symptoms or risk factors, for example, with a diagnosis). Additionally, many physicians don't make sure that the diagnosis they have considered satisfies the entire patient presentation (incomplete verification or premature closure).

In the next two chapters, we will explore most of the faulty cognitive processes and biases that lead to these errors. But we also need to learn how to gather information correctly and accurately (Chapter 7), construct a meaningful differential diagnosis (Chapter 6), use physical exam appropriately (Chapter 8), order and interpret tests correctly (Chapters 9 and 10), and finally learn and implement accurate diagnosis scripts to make the right diagnosis (Chapter 11).

Exercises

2.1 Consider a recent failure of diagnosis that you have seen (or suffered). In which parts of the diagnostic process did the failure occur?

2.2 What are some consequences you can think of diagnostic failure? Name at least five consequences.

3

Cognitive Bias

Our human reasoning and decision-making processes are inherently flawed. Faced with so many decisions to be made every day, we take shortcuts (called **heuristics**) that help us make "pretty good" decisions with little effort. These "pretty good" decisions are not always right and often compromise and exchange our *best decision* for one that is just *good enough*.

These heuristics carry with them assumptions which may not be relevant to the individual decision at hand, and if these assumptions are not accurate for the current problem, then a mistake may be made. We call these assumptions *cognitive biases*.

In 2015, the authors of the landmark report *Improving Diagnosis in Healthcare* stated it this way:

> When a heuristic fails, it is referred to as a cognitive bias. Cognitive biases, or predispositions to think in a way that leads to failures in judgment, can also be caused by affect and motivation. Prolonged learning in a regular and predictable environment increases the successfulness of heuristics, whereas uncertain and unpredictable environments are a chief cause of heuristic failure.[13]

More than 40 cognitive biases have been described which specifically affect our reasoning processes in medicine. These biases are more likely to occur with quicker decisions than with slower decisions. The term ***Dual Process Theory*** has been used to describe these two distinct ways we make decisions. Noble Prize winning psychologist Daniel Kahneman, author of *Thinking, Fast and Slow*, has referred to these two processes as System 1 and System 2 thinking.

System 1 thinking is intuitive and largely unintentional; it makes heavy use of heuristics. It is quick and reasoning occurs unconsciously. It is effortless and automatic. It is profoundly influenced by our past experiences, emotions, and memories.

System 2 thinking, on the other hand, is slower and more analytic. System 2 reasoning is conscious and operates with effort and control. It is intentional and rational. It is more influenced by facts, logic, and evidence. System 2 thinking takes work and time and, therefore, is too slow to make most of the decisions we need to make in any given day.

A System 1 decision about lunch might be to get a double bacon cheeseburger and a peanut butter milkshake (with onion rings, of course). That was literally the first meal that popped into my head as I started writing, and each of those items resonates with emotional and reward centers in my brain that recall pleasant experiences and pleasant memories. But not everything that resonates is reasonable.

As the System 2 part of my brain takes over, I realize several things: I am overweight and diabetic (those dietary choices certainly won't help either of those issues); I must work this afternoon (if I eat that meal, I'll probably need a nap); etc. You get the idea. My System 2 lunch might be kale with a side of Brussel sprouts. *Oh, well.*

These two ways of thinking utilize different parts of our brains; they are distinctly different processes. Because System 1 thinking is so intuitive and so affected by our past experiences, we tend to make most cognitive errors with this type of thought. Failures can occur with System 2 thinking to be sure, and not just due to cognitive biases but also due to logical fallacies or just bad data; but, overall, System 2 decisions are invariably more correct than System 1 decisions.

We certainly don't need to overthink every decision. We don't have enough time to make System 2 decisions about everything that comes our way. Yet, the more we make good System 2 decisions initially, the better our System 1 decisions will become over time. In other words, we need good heuristics or algorithms, deeply rooted in System 2 cognition, to make the best of our System 1 thoughts. Thus, I often say:

The mind is like a parachute: it works best when properly packed.

The packing is done slowly and purposefully; the cord is pulled automatically and with very little thought. If we thoroughly think about where to eat lunch using System 2 thinking, it will have a positive effect on all our subsequent decisions about lunch in the future which are made with System 1 thinking.

System 1	**System 2**
Fast	Slow
Unconscious	Conscious
Associative	Rule-based
Heuristic	Analytic
Implicit	Explicit
Experiential	Rational
Undemanding	Tiring
Heavily influenced by bias	Less influenced by bias
Error prone	More reliable
Automatic	Deliberative
Reflex	Reason

***Table 1**. Characteristics of System 1 versus System 2 thought processes.*

I sometimes describe Dual Process Theory in terms of learning to drive a car (with a manual transmission!). When you're first learning, everything requires deliberate and intentional effort. It's slow and buggy. Someone is yelling at you (rule-based) about following too closely, staying in the lines, and not missing a shift. Everything is stated very explicitly. You're very conscious of every part of driving and if you get distracted by anything else, you might crash. But over time, driving becomes effortless. It's unconscious and relatively automatic. It's instinctive and experiential. You respond automatically and reflexively. Because the rules are no longer explicit, you tend to use your turn signal less, drive a little too fast, and follow a little too close. You almost always get away with these errors, too. But, sometimes, those rule violations will get you killed.

Still, it's clear to see that learning to drive is important. You can't skip the System 2 work that goes into developing mastery if you're ever going to master driving a car. Once mastery is achieved, driving is almost exclusively conducted with System 1 thinking. Occasionally, you still need to slow down and use your System 2 skills, like when you're backing up a trailer for example (maybe that's just me).

So, how does Dual Process Theory relate to medicine? We all have cognitive dispositions and biases that may lead us to errors in our clinical thinking. Dual Process Theory is a theoretic model that helps us understand these.

First, we need to be aware of how we make decisions and how our brains may play tricks on us; a thorough understanding of different cognitive biases can help with this. Second, we need to develop processes or tools that help to debias ourselves and prevent us from falling into some of the traps that our cognitive biases have laid for us.

Bias in clinical decision making

Imagine that you are working in a busy ER. A patient comes to the ER and tells the triage nurse that she is having right lower quadrant pain; she says that the pain is just like the pain she had 6 months ago when she had an ovarian cyst rupture. The triage nurse tells you (the doctor) that she has put the patient in an exam room and that she has pain like her previous ruptured cyst.

You laugh, because you have already seen two other women tonight who had ruptured cysts on CT scans. You tell the nurse to go ahead and order a pelvic ultrasound for a suspected ovarian cyst before you even see her. The ultrasound is performed and it reveals a 3.8 cm right ovarian cyst with some evidence of hemorrhage and some free fluid in the pelvis. You quickly examine and talk to the patient, confirm that her suspicions were correct, and send her home with some non-narcotic pain medicine and ask her to follow up with her gynecologist in the office.

Several hours later, the patient returns, now complaining of more severe pain and bloating. Frustrated and feeling that the patient is upset that she didn't get narcotics earlier, you immediately consult the gynecologist on-call for evaluation and management of her ovarian cyst. The gynecologist performs a consult and doesn't believe that there is any evidence of torsion because there is blood flow to the ovary on ultrasound exam. He recommends reassurance and discharge home.

The next day she returns in shock and is thought to have an acute abdomen. She is taken to the OR and discovered to have mesenteric ischemia. She dies post-operatively.

While this example may feel extreme, the mistakes are real and they happen every day. Let's reflect on what cognitive mistakes occurred to lead to this negative patient outcome.

When the patient told the nurse that her ovary hurt, the nurse was influenced by the *framing effect*. The patient suffered from *triage cueing* because of the workflow of the ER. The physician became *anchored* to the idea of an ovarian cyst early on. He suffered from *base-rate neglect* when he overestimated the prevalence of painful ovarian cysts. When he thought about his previous patients that night, he fell prey to the *clustering illusion* and exhibited an *availability bias*. When the ER doctor decided to get an ultrasound, he was *playing the odds* or fell victim to *Sutton's slip*. When the ultrasound was ordered for "suspected ovarian cyst," there was *diagnosis momentum* that transferred to the interpreting radiologist.

When the ultrasound showed an ovarian cyst, the ER physician was affected by *confirmation bias*. The ER doctor's frequent over-diagnosis of ovarian cysts was reinforced by *feedback sanction*. When he stopped looking for other causes of pain because he had discovered an ovarian cyst, he had *premature closure*. When he felt that the patient's return to the ER was due to her desire for narcotics, the ER doctor made a *fundamental attribution error*. When he never considered mesenteric ischemia because she didn't complain of bloody stools, he exhibited *representativeness restraint*. When he consulted a gynecologist to treat her cyst rather than explore other possibilities, he was exploited by the *sunk costs bias*.

Each of these are examples of cognitive biases that affect our reasoning (we'll define these biases in a moment). But what's another way this story might have played out? If this seems far-fetched, you'll have to suspend your disbelief for a few moments and imagine how medicine should be:

The patient presents to the ER. The nurse tells the doctor that the patient is in an exam room complaining of right lower quadrant pain (she orders no tests or imaging before the patient is evaluated and she uses language that does not make inappropriate inferences). The doctor makes (in his head) a differential diagnosis for a woman with right lower quadrant pain (he does this before talking to the patient). While talking to the patient and performing an exam, he gathers information that he can use to rule out certain things on his differential (or at least decide that they are low probability) and determines the pretest probability for the various diagnoses on his list (this doesn't have to be precise – for example, he decides that the chance of acute intermittent porphyria is incredibly low and decides not to pursue the diagnosis, at least at first).

After assessing the patient and refining his differential diagnosis, he decides to order some tests that will help him disprove likely and important diagnoses. He is concerned about her nausea and that her pain seems to be out of proportion to the findings on her abdominal exam. He briefly considered mesenteric ischemia but considered it lower probability because she has no risk factors and she has had no bloody stools (he doesn't exclude it, however, because he also remembers that only 16% of patients with mesenteric ischemia present with bloody stools). Her WBC is elevated. He does decide to order a CT scan because he is concerned about appendicitis.

When the CT is not consistent with appendicitis or mesenteric ischemia, he decides to attribute her pain to the ovarian cyst and discharges her home. When the patient returns later with worsened pain, he first reevaluates her carefully and starts out with the assumption that he has likely misdiagnosed her. This time, he notes an absence of bowel sounds, bloating, and increased abdominal pain on exam. He again considers mesenteric ischemia, even though the previous CT scan found no evidence of it, realizing that the negative predictive value of a CT scan for mesenteric ischemia in the absence of a small bowel obstruction is only 95% – meaning that 1 in 20 cases are missed. This time, he consults a general surgeon, who agrees that a more definitive test needs to be performed and a mesenteric angiogram reveals mesenteric ischemia. She is treated with decompression and heparin and makes a full recovery.

These two examples represent an extreme of very poor care to very excellent care. Note that even when excellent care occurred, the rare diagnosis was still initially missed. But the physician in the second example was not nearly as burdened by cognitive biases as the former physician and the patient is the one who benefited. He used a lot of System 1 thinking, at least initially, but, when it mattered, he slowed down and used System 2 thinking. He also had a thorough understanding of the statistical performance of the tests he ordered and he considered the pretest and post-test probabilities of the diseases on his differential diagnosis. He is comfortable with uncertainty and he doesn't think of tests in a binary (positive or negative) sense, but rather as increasing or decreasing the likelihood of the conditions he's interested in. He used the hypothetico-deductive method of clinical diagnosis, which is rooted in Bayesian inference.

Let's briefly define the most common cognitive biases which tend to affect clinicians and give some examples and potential consequences of these biases.

Aggregate bias: the belief that aggregated data, such as data used to make practice guidelines, doesn't apply to individual patients.

Example: "My patient is special or different than the ones in the study or guideline."

Consequence: Ordering Pap smears or other tests when not indicated in violation of the guideline (which may unintentionally lead to patient harm).

Anchoring: the tendency to lock onto the salient features of a diagnosis too early and not modify the theory as new data arrives.

Example: "Hypoglycemia with liver inflammation is probably acute fatty liver of pregnancy."

Consequence: Ignoring or rationalizing away the subsequent finding of normal fibrinogen levels (which would tend to go against the diagnosis).

Ascertainment bias: this occurs when thinking is shaped by prior expectations, such as stereotyping or gender bias.

Example: "She has pain because she is drug-seeking again."

Consequence: Not conducting appropriate work-up of pain.

Availability: the tendency to believe things are more common or more likely if they come to mind more easily, usually leading to over-diagnosis (it may also lead to under-diagnosis). Things might come to mind more frequently for good reason (maybe it's a common condition among patients you treat); but they may also come to mind for the wrong reasons (you read a case report recently about the condition).

Example: "Ooh, I saw this once in training and it was a twin molar pregnancy!"

Consequence: Not considering statistically more probable diagnoses.

Base-rate neglect: the tendency to ignore the true prevalence of diseases which distorts Bayesian reasoning (by changing the pretest probability of the disease). This may be unintentional or deliberate (for example, when physicians always emphasize the "worst-case scenario").

Example: "It's probably GERD but we need to rule out aortic dissection."

Consequence: Ordering unnecessary tests with high false positive rates and poor positive predictive values.

Clustering Illusion: The tendency to attribute significance to an unremarkable cluster of cases resulting from normal variation.

Example: "That's the fourth case of gastroenteritis we've seen today! Something must be going around." (While it's true that GI bugs spike seasonally, the small sample size doesn't reveal anything of practical significance.)

Consequence: Misdiagnosing the next patient's acute appendicitis as a stomach virus.

Commission bias: the tendency to action rather than inaction, believing action is necessary to prevent harm. This bias is more common in over-confident physicians.

Example: "This trick always works in my patients for that problem" or "It's just a cold, but she made an appointment so she'll be unhappy if I don't give her antibiotics" or "I want you to be on strict bedrest to prevent a miscarriage since you are having bleeding in the first trimester."

Consequence: Overuse of potentially risky or unnecessary therapeutics and perhaps guilt-sanctioning (if, for example, the patient miscarries when she gets up to tend to her crying baby).

Confirmation bias: the tendency to look for supporting evidence to confirm a diagnosis rather than to look for data to disprove a diagnosis. We'll talk about this important bias in more detail later.

Example: "Aha! That's what I suspected."

Consequence: Incorrect diagnosis. We should always look for data to disprove our diagnosis (our hypothesis).

Diagnosis momentum: the effect of attaching diagnoses too early and making them stick throughout interactions with patients, nurses, consultants, etc. and then biasing others' thinking.

Example: "This is probably an ectopic pregnancy" and writing '*suspected ectopic*' on the ultrasound requisition form.

Consequence: Radiologist reads a corpus luteal cyst as an ectopic pregnancy.

Feedback sanction: this occurs when diagnostic errors have no consequence because of a lack of immediate feedback or any feedback at all; it is more likely to happen in acute care settings where there is no patient follow up, which reinforces errors in diagnosis or knowledge.

Example: "I saw this patient with back pain which was due to a UTI."

Consequence: Positive reinforcement of diagnostic errors (such as the belief that UTIs are a common cause of back pain).

Framing effect: the way in which outcomes or contingencies are framed (by family, nurses, residents, or even the patient) greatly influences decision making and diagnostic processes.

Example: "The patients says that her ovary has been hurting for a week."

Consequence: Focusing on ovarian or gynecological sources of pelvic pain rather than other, more likely causes.

Fundamental attribution error: the tendency to blame patients for their illnesses (dispositional causes) rather than circumstances (situational factors).

Example: "Her glucose is messed up because she is noncompliant."

Consequence: Ignoring other causes of the condition (e.g., infection leading to elevated glucose).

Gambler's fallacy: the false belief that an event is less likely to occur in the future if it has occurred several times in the past (such as the belief that a series of coin tosses all showing heads will influence the probability that the next coin toss will be heads).

Example: "My last three MIs have died. I bet we won't lose this one!!"

Consequence: May lead to inappropriate treatment of the current patient, based on facts that are irrelevant (the facts about other patients).

Gender (racial) bias: the belief that gender or race affects the probability of a disease when no such link exists.

Example: "We need to think about osteoporosis since she's white."

Consequence: Under- or over-diagnosing diseases. It has been reported that around two-thirds of published racial predilections in major textbooks, for example, are not supported by evidence.[14]

Hindsight bias: knowledge of the outcome affects perception of past events and may lead to an illusion of failure or an illusion of control.

Example: "Last time I had this, she got better because I gave her ___."

Consequence: Perpetuates error and encourages anecdotal medicine. For example, it is merely an assumption that the intervention affected the outcome, either positively or negatively.

Multiple alternative bias: a multiplicity of diagnostic options can lead to emotional conflict and cognitive uncertainty and then regression to well-known diagnoses to resolve the ambiguity.

Example: "Well let's just focus on what it probably is and not worry about all that other stuff for now."

Consequence: Ignoring other important alternative diagnoses.

Omission bias: the tendency towards inaction, the opposite of a *commission bias,* and more common than commission biases.[15]

Example: "Group B strep infections in neonates are really rare, so I don't see the point in the antibiotic for this mom."

Consequence: May result in rare but serious harms.

Order effects: the tendency to focus on the beginning and the end of a story and fill in the middle part in a way that makes sense (often creating a false narrative or constructing false associations). This bias is worsened by other tendencies like anchoring and is important to consider in patient hand-offs and presentations.

Example: "She had a fever but got better when we treated her for a UTI."

Consequence: May lead to inappropriate causation biases, etc. (The patient got better. This may be due to antibiotics given for a possible UTI or the actual cause of her fever may still be unknown.)

Outcome bias: the tendency to pick a diagnosis that leads to good outcomes rather than a bad outcome, a form of a *value bias.*

Example: "I'm sure it's just a panic attack and not a pulmonary embolism."

Consequence: Missing potentially serious diagnoses.

Overconfidence bias: the belief that we know more than we do, leading to a tendency to act on incomplete information, hunches, or intuitions.

Example: "I see this all the time, just send her home."

Consequence: Grave harm may occur because of missed diagnoses.

Playing the odds: the tendency to opt for more benign diagnoses or simpler courses of action when uncertainty in the diagnosis exists.

Example: "I'm sure that this ovarian cyst is benign; *it's gotta be.*"

Consequence: Potentially devastating outcomes, especially when coupled with an *omission bias* (e.g., not following up with a repeat ultrasound in a few weeks for a questionable cyst).

Posterior probability error: the tendency to believe that what has gone on before for a patient changes the probability for future events for the patient.

Example: "Every time she comes in her bleeding has just been from her vaginal atrophy."

Consequence: Biases current evaluation and work-up (e.g., ignoring the post-menopausal bleeding and not considering malignancy).

Premature closure: the tendency to accept a diagnosis before it has been confirmed when only scant data supports the anchored diagnosis, often leading to treatment failures.

Example: "We know what it is, she just hasn't responded to treatment yet."

Consequence: Ignoring alternative theories; this bias is captured by this famous phrase in medicine: *When the diagnosis is made, the thinking stops.*

Psych-out error: this occurs when serious medical problems (e.g., hypoxia, head injuries, etc.) are misattributed to psychiatric diagnoses.

Example: "She just acts that way because she is bipolar."

Consequence: Ignoring potentially catastrophic physical ailments (e.g., vascular disease or brain tumor).

Representative restraint: the tendency to favor a classical presentation of a rarer condition over an atypical presentation of a common condition.

Example: "The patient doesn't have anorexia so I think porphyria is more likely than appendicitis."

Consequence: Misdiagnosis.

Search satisfying: the tendency to stop looking once something satisfying is found, both on the patient or in the medical literature (a form of premature closure).

Example: "This article says I'm right!" or "That's where the bleeding is coming from!"

Consequence: Ignoring other causes of symptoms or other contradictory evidence or literature.

Sutton's slip: the tendency to *go where the money is*, that is, to diagnosis the most obvious things, ignoring less-likely diagnoses.

Example: "I rob banks because that's where the money is," (This was Willie Sutton's famous response when the judge asked him why he robbed banks).

Consequence: Under-utilizing System 2 thinking and ignoring diseases or presentations that are less common.

Sunk costs fallacy: the more investment that is made in a diagnosis or treatment, the less likely one is to release from it and consider alternatives.[16]

Example: "I've just got to be right; I don't know why this treatment isn't working!"

Consequence: Further delay in pursuing the right treatment or diagnosis. This also results in a lot of false case reports (*We present a case of such-and-such that was refractory to usual treatments but responded to some other crazy treatment* or *We present a case of such-and-such that presented in some weird non-typical way* – in both cases, the diagnosis was likely wrong to begin with).

Triage cueing: the biasing that results from self-triage or systematic triage of patients or presentations, creating tunnel vision.

Example: "I need a Gyn consult because she's a female with pain."

Consequence: Ignoring other organ systems or causes of pain.

Unpacking principle: the failure to elicit all relevant information in establishing a differential, particularly when a prototypical presentation leads to anchoring.

Example: "Anorexia and right lower quadrant pain is classic appendicitis."

Consequence: Not considering all causes of each symptom individually (or collectively). As an aside, Occam's razor and other cognitive processes that favor simplicity over complexity are usually wrong but feel comfortable to human imagination (it's much simpler to blame the MMR vaccine for autism than it is to consider a polygenetic, multifactorial causation theory).

Vertical line failure: this results from routine, repetitive processes that emphasize economy, efficiency, and utility (as opposed to lateral thinking).

Example: "I always do a diabetes screen on all pregnant women" or "When I see *x* I always do *y*."

Consequence: Deemphasizes lateral thinking (e.g., What else might this be?).

Visceral bias: Results from countertransference and other visceral arousal leading to poor decision making.

Example: "That patient is just a troll" or "She is so sweet."

Consequence: Leads to cognitive distortion and augments other biases.

Yin-yang out: the tendency to stop looking or to stop trying once efforts have seemingly been exhausted even though the case hasn't been satisfied.

Example: "We've worked her up to the yin-yang."

Consequence: Leads to errors of omission.

I'm sure you can think of many other examples for these biases, perhaps from your own specialty area, and there are many other biases that have been described apart from those on this list. There is an emerging scientific literature which is examining the effects of bias on diagnostic and therapeutic outcomes and on medical error. The 2015 report *Improving Diagnosis in Healthcare,* which I referred to previously, is a good place to start exploring some of the implications of bias in the diagnostic process.

For now, let's look more specifically at the confirmation bias.

Confirmation Bias

Do you believe what you see or do you see what you believe?

The confirmation bias affects the way we search for, interpret, and recall information. It is the powerful tendency of humans to confirm their preexisting beliefs and give more weight to data that supports their beliefs and less weight to data that tends to refute their beliefs. Francis Bacon wrote,

> The human understanding when it has once adopted an opinion ... draws all things else to support and agree with it. And though there be a greater number and weight of instances to be found on the other side, yet these it either neglects or despises, or else by some distinction sets aside or rejects...

If you believe in ghosts, then strange sounds in your house at night, artifacts in pictures, and other unexplained observations, are taken as "proof" that ghosts exist. If you don't believe in ghosts, then the same phenomena are just noisy pipes, bad lighting, and other things that all have a rational explanation. If you believe that some medical therapy works, say magnesium sulfate for treatment of preterm labor, then you will look for, interpret, and recall data in a way that supports your belief. Your search of the literature in PubMed will be worded in a way to find studies that agree with you (and less likely to find articles that do not); you will less critically analyze any study that has a favorable finding to support the belief, and more likely to criticize papers that do not; and you will recall dozens or hundreds of success stories from years of experience where magnesium "worked" while being less likely to recall times when it did not work.

Some people are more prone to confirmation bias than others. For example, people with poor self-confidence are more likely to seek out only data that supports their beliefs compared to people with higher levels of self-confidence.

When researchers do studies, they tend to design and frame studies in a way that will provide "evidence" that their theories are correct, even though the scientific approach should be to design the most rigorous study possible to disprove their theories. We prefer "positive tests" or confirming tests to "negative tests." For example, if you believe that ghosts are causing the strange noises in your house, you are more likely to look for ghosts with cameras or "paranormal" noise detectors and less likely to examine your noisy pipes. If you perform more "positive tests" than negative tests, then you are more likely to develop false positive data that affirms the belief rather than false positive data that refutes the belief.

Doctors do this not just in the interpretation of literature and in the design of research studies, but also in the diagnosis and treatment of patients. If a woman presents to the emergency department with pelvic pain, and the physician's theory is that an ovarian cyst is the cause of the pain (or at least that the pain is gynecologic in origin), then when an ultrasound shows a 3-cm ovarian cyst, the case is closed (premature closure). Instead, the scientific approach would be to try to disprove that an ovarian cyst is the source of pain, and then, when finding no evidence that goes against the theory, accepting it as a working diagnosis until new data comes along. This is difficult to do practically in the clinical setting, but simply asking and answering "What else could this be?" is an important step to avoid error.

Patients, like all humans, are driven by the confirmation bias. If a patient's preset tendency is to distrust the mainstream medical community, he will look for "evidence" such as websites and blogs and "studies" that tend to undermine or discredit what he perceives as mainstream medical opinions. Unfortunately, search engines are good at pandering to our tendency to confirmation bias. This makes how we search for information on search engines or electronic databases very important.

Let's look at some examples. I performed a Google search in two different ways that explore evidence about a possible link between vaccines and autism.

- First, let's search for "proof that vaccines cause autism." On the first page of results, there are six links that claim to provide evidence that vaccines cause autism, one that is unclear (without reading it), and three which deny that vaccines cause autism. What's more, the webpages are bold with titles like, "Vaccines DO cause autism - Undeniable scientific proof" and "22 medical studies that show vaccines cause autism." Google also could tell by the way the question was asked that the searcher already believed that vaccines caused autism; in fact, Google suggested, in the middle of the search results, that I should search for "Which vaccines cause autism?"
- Next, let's search for "evidence that vaccines don't cause autism." With this query, all 10 results on the first page boldly claim absolutely no link between the two, with titles like: "Massive Study Provides the Best Proof Yet That Vaccines Don't

> Cause Autism" and "75 studies that show no link between vaccines and autism." And Google did not suggest that I search for "Which vaccines cause autism?" because it knows that I don't care.

So, two people with two vastly different beliefs go to Google to do "research" and both people find exactly what they need to confirm their beliefs. Worse, in most instances, Google knows the user's search history and uses previous searches to tailor the results of their next query.

What about the scientific literature? Let's repeat the experiment with PubMed.

- If we first search for "Benefits of Vitamin D," and limit our search to "Review Articles," in the first page of search results, 16 out of 20 results are relevant and all these claim something beneficial about Vitamin D, improving diseases ranging from multiple sclerosis to rheumatoid arthritis.
- If we search instead for "Harms of Vitamin D," we find that 18 of the 20 results are relevant and about half of these are frankly critical of Vitamin D usage and screening in contexts outside of the prevention of osteoporosis.

Pick almost any medical topic and similar results on Google and PubMed can be had. Do you search for "Fetal benefits of magnesium sulfate" or "Fetal harms of magnesium sulfate"? The truth is, neither search method is appropriate. A better query might be to search for specific outcomes, like "magnesium sulfate AND cerebral palsy" or "magnesium sulfate AND neonatal mortality." A generalized version of the search might be, "magnesium sulfate AND neonatal outcomes."

What happens if we repeat the Google search with this strategy? Searching for "autism AND vaccines" (no bias implied) actually produces the best results yet, with 12 results, 10 of which are excellent rebuttals of the purported link between the two and also suggestions from Google Scholar for a number of excellent scientific studies. But we also see two results that make the case for a link between vaccines and autism, giving balance to both sides.

Notice that one of the benefits of understanding biases is that it allows us to design specific strategies to mitigate them. In this case, it's hopefully obvious that we need to be deliberate to avoid the confirmation bias in the way we search for evidence. But, let's explore some other specific strategies to mitigate our biases.

Debiasing strategies

> All men make mistakes, but a good man yields when he knows his course is wrong, and repairs the evil. The only crime is pride.
>
> – Sophocles, *Antigone*

The long list of different biases is presented to raise awareness and provoke thought. Learning the individual names of each is cumbersome and unnecessary; indeed, many of them masquerade by other names as well. Rather than even use the word *bias*, many authors, including one of the thought-leaders in this field, Pat Croskerry,[17] prefer the phrase **cognitive dispositions to respond** (CDR) to describe the many situations where clinicians' cognitive processes might be distorted, including the use of inappropriate heuristics, cognitive biases, logical fallacies, and other mental errors. The term CDR is thought to carry less of a negative connotation, and indeed, physicians have been resistant to interventions aimed at increasing awareness of and reducing errors due to cognitive biases because they tend to take the errors too personally and negatively.

After the publication of the Institute of Medicine Report's *To Err is Human* in 2000,[18] which attributed up to 98,000 deaths per year to medical errors, many efforts were made to reduce errors in our practices and systems. Development of multidisciplinary teams, computerized order entry, clinical guidelines, and quality improvement task forces have attempted to lessen medical errors and their impact on patients. Since the publication of the report, we have seen an increased emphasis on things like medication safety cross-checking, reduction in resident work hours, using checklists in hospital order sets, and 'time-outs' in the operating room. But most serious medical errors stem from misdiagnosis.

Yes, occasionally, a patient might have surgery on the wrong side or receive a medication that interacts with another medication, but at any given time, up to 15% of patients admitted to the hospital are being treated for the wrong diagnosis – with interventions that carry risk – while the actual cause of their symptoms remains unknown and often untreated. *To Err Is Human* noted that causes of death diagnosed in the postmortem were different from antemortem diagnoses 40% of the time! How many of those deaths might have been prevented if physicians had been treating the correct diagnosis?

Most of these failures of diagnosis (probably two-thirds) are related to CDRs[19] and a lot of work has been done since 2000 to elucidate various causes and interventions.[20] But, as I said before, physicians have been resistant to being told that there might be a problem with how they think. Physicians love to blame medical errors on someone or something else – thus, the focus has been on residents' lack of sleep or deleterious medication interactions. Seeking to reduce physicians' resistance due to a feeling of being criticized is a prime reason why Croskerry and others prefer to use the term *cognitive disposition to respond* rather than negative words like *bias* or *fallacy*.

I'm happy with either terminology because I'm not sure that relabeling the problem will change the main issue: physicians tend to be a bit narcissistic and therefore resistant to the idea that all of us are biased and all of us must actively work to monitor those biases and avoid making decisions that are overly-influenced by them.

We make poor decisions for one of two reasons: either we lack information or we don't apply what we know correctly. Riegleman, in his 1991 book *Minimizing Medical Mistakes: The Art of Medical Decision Making*, called these two issues 'errors of ignorance' and 'errors of implementation.' One of the goals of *To Err is Human* was to create an environment where medical errors were attributed to systematic rather than personal failures, hoping to make progress in reducing error by deemphasizing individual blame.

The focus here, of course, is to evaluate the errors of implementation. Graber and colleagues, in 2002, further categorized diagnostic errors into three types: No-fault errors, system errors, and cognitive errors.[21] No-fault errors will always happen (like when our hypothetical physician failed to diagnose mesenteric ischemia despite doing the correct work-up). System errors have been explored heavily since the publication of *To Err is Human* (things like medication interactions and problems with hand-offs). But the cognitive errors remain largely unexplored and gaining a better understanding of our CDRs (our biases, etc.) is the first step to reducing this type of error.

Croskerry has divided the CDRs[22] into the following useful categories which can help us focus on the key problems central to each bias:

Errors of over-attachment to a particular diagnosis
Anchoring, confirmation bias, premature closure, sunk costs

Errors due to failure to consider alternative diagnoses
Multiple alternatives bias, representativeness restraint, search satisfying, Sutton's slip, unpacking principle, vertical line failure

Errors due to inheriting someone else's thinking
Diagnosis momentum, framing effect, ascertainment effect, bandwagon effect

Errors in prevalence perception or estimation
Availability bias, ambiguity effect, base-rate neglect, gambler's fallacy, hindsight bias, playing the odds, posterior probability error, order effects

Errors involving patient characteristics or presentation context
Fundamental attribution error, gender bias, psych-out error, triage cueing, contrast effect, yin-yang out

Errors associated with physician affect, personality, or decision style
Commission bias, omission bias, outcome bias, visceral bias, overconfidence, vertical line failure, belief bias, ego bias, sunk costs, zebra retreat

This last category also includes some important things we might not classically think of as cognitive bias. For example, a physician who is under great personal stress (problems with family life or finances) or a physician who has a physical or psychological impairment (substance abuse, health problems, etc.) may also respond differently given a challenging clinical scenario. Distraction, burn-out, cynicism, substance abuse, fatigue, an abusive work environment, depression, anxiety, stress, too heavy a workload, interpersonal conflict, and a variety of other issues can all affect a physician's cognitive function in a very negative way, leading to sometimes profound mistakes.

Some additional biases mentioned above include the *bandwagon effect* (doing something just because everyone else does, like giving magnesium to women in premature labor despite a lack of scientific evidence of efficacy), *ambiguity effect* (picking a diagnosis or treatment because more is known about it, like the outcome), *contrast effect* (minimizing the treatment of one patient because, in contrast, her problems pale in comparison to the last patient), *belief bias* (accepting or rejecting data based on its conclusion or whether it fits with what one already believes rather than on the strength of the data itself), *ego bias* (overestimating the prognosis of your patients compared to that of others' patients), and *zebra retreat* (not pursuing a suspected rare diagnosis out of fear of being viewed negatively by colleagues or others for wasting time, resources, etc.).

We are all vulnerable to cognitive dispositions that can lead to error. Just being aware of this is meaningful and might make us less likely to make these mistakes, but we need to do more. We need to actively work to debias our processes. Let's look at some strategies for this (adapted from Croskerry):[23]

Develop insight/awareness: Education about CDRs is a crucial first step to reducing their impact on our clinical thinking, but it cannot stop with reading an article or a book. We must look for examples of them in our own practices and integrate our understanding of CDRs into our quality improvement processes. We need to identify our biases and how they affect our decision-making and diagnosis formulation. An analysis of cognitive errors (and their root causes) should be a part of every peer review process, quality improvement meeting, and morbidity and mortality conference. Most cases that are reviewed in these formats are selected because a less-than-optimal outcome occurred; the root cause (or at least a major contributing cause) in most cases was a cognitive error. Even when things go well, mistakes often happen and systematic analysis of good outcomes is also important.

Consider alternatives: We need to establish forced consideration of alternative possibilities, both in our own practices and in how and what we teach. Considering alternatives should be a part of how we teach medicine routinely. Always ask the question, "What else could this be?" Ask yourself, ask your learner, ask your consultant. The ensuing conversation is perhaps the most educational thing we can ever do. Even when the diagnosis is obvious, always ask the question. This needs to become part of the culture of medicine.

Metacognition: We all should continually examine and reflect on our thinking processes actively, and not just when things go wrong. Even when things go right, it is a meaningful and important step to consider *why* things went right. We focus too much on negative outcomes (this is a form of bias); consequently, we develop a skewed sense of what contributed to the negative outcome. So, try thinking about what went right as well, reinforcing the good things in our clinical processes.

Decrease reliance on memory: In the pre-computer days, a highly valued quality in a physician was a good memory. Unfortunately, medical schools today still emphasize this skill, selecting students who might excel in rote memorization but fall behind in critical thinking skills. In the 1950s, memory was everything: there was no quick way of searching the literature, of comprehensively checking drug interactions, of finding the latest treatment guidelines, etc. But today, memory is our greatest weakness. Our memories are poor and biased, and there is more data that we must master than ever before to be a good doctor. So, stop relying on your memory. We need to encourage the habitual use of cognitive aids, whether that's mnemonics, practice guidelines, algorithms, or computers. If you don't treat a particular disease every week, then look it up each time you encounter it. If you don't use a particular drug all the time, then cross-check the dose and interactions every time you prescribe it. Even if you do treat a particular disease every day, you should still do a comprehensive literature search every 6 months or so (yearly at the very least).

Many physicians are sorely dated in their medical knowledge. What little new information they learn often comes from the worst sources: drug and product reps, industry-sponsored throw-away journals, popular media, and even TV commercials.[24] Education is a life-long process. Young physicians need to develop the habits of life-long learning early. Today, this means relying on electronic databases, practice guidelines, etc. as part of daily practice. I, for one, use PubMed at least five times a day on average (and I feel like I need to do this to be up-to-date in my areas of expertise).

Our memory, as a multitude of psychological experiments have shown, is among our worst assets. Stop trusting it.

Specific training: We need to identify specific flaws in our thinking and specific biases and direct specific efforts to overcome them. For example, the area that seems to contribute most to misdiagnosis relates to a poor understanding of Bayesian probabilities and inference, so specific training in Bayesian probabilities might be in order, or learning from examples of popular biases, like distinguishing correlation from causation, etc. We will spend time doing this later.

Simulation: We should use mental rehearsal and visualization as well as practical simulation or videos exhibiting right and wrong approaches. Though mental rehearsal may sound like a waste of your time, it is a powerful tool. If we appropriately employ metacognition, mental rehearsal of scenarios is a natural extension. Remember, one of our goals is to make our System 1 thinking better by employing System 2 thinking when we have time to do so (packing the parachute correctly). So, a practical simulation in shoulder dystocia, done in a System 2 manner, will make our "instinctual" responses (the System 1 responses) better in the heat of the moment when the real shoulder dystocia happens. A real shoulder dystocia is no time to learn; you either have an absolute and definitive pathway in your mind for how you will deliver the baby before it suffers permanent injury or you don't. But this is true even for things like making differential diagnoses (we will talk about this in detail later).

Another axiom of mine: practice does not make perfect, but good practice certainly helps get us closer. A corollary of this axiom is that bad practice makes a bad doctor; unfortunately, a lot of people have been packing the parachute incorrectly for many years and they have gotten lucky with the way the wind was blowing when they jumped out of the plane.

Cognitive forcing strategies: We need to develop specific and general strategies to avoid bias in clinical situations. We can use our clinical processes and approaches to force us to think and avoid certain biases, even when we otherwise would not. Always asking the question, "What else could this be?" is an example of a cognitive forcing strategy. Our heuristics and clinical algorithms should incorporate cognitive forcing strategies. For example, an order sheet might ask you to provide a reason why you have elected not to use a preoperative antibiotic or thromboembolism prophylaxis. It may seem annoying to have to fill that out every time, but it makes you think.

Make tasks easier: We must strive to reduce task difficulty and ambiguity. We need to train physicians in the proper use of relevant information databases and make these resources available. We need to remove as many barriers as possible to good decision-making. This may come in the form of evidence-based order sets, clinical decision tools and nomograms, or efficient utilization of evidence-based resources. Bates and colleagues[25] have recommended 10 principles for effective clinical decision support in electronic health records and other information management tools. They emphasize that any decision support system must be fast, fit into the normal workflow, anticipate the physician's needs and occur in real-time while being clear, simple, and highly useable. They also point out that it's not enough for the system to tell a physician not to do something (like to not prescribe a medication given a patient's allergy), but the system should also provide an alternative. As they point out, physicians will strongly resist stopping; changing directions is much easier than stopping. Many currently available clinical decision support systems and electronic health records fail miserably.

Minimize time pressures: Our workflows must provide sufficient quality time to make decisions. We fall back to System 1 thinking when we are pressed for time, stressed, depressed, under pressure, etc. Hospitals and clinics should promote an atmosphere where appropriate time is given, so that System 2 critical thinking can occur when it is necessary, without further adding to the stress of a physician who already feels over-worked, under-appreciated, and behind. I wouldn't hold my breath for that to occur. Perverse economic incentives and payment systems are major obstacles. But clinicians can make a difference. Don't be afraid to tell a patient "I don't know" or "I'm not sure" and then get back to her after finding the data you need to make a good decision. We should emphasize this idea even on simple decisions. Our snap, instinctive answers are usually correct (especially if we have been packing the parachute well) but we need to always take the time to slow down, research, and consider if it is necessary.

Accountability: Establish clear accountability and follow-up for decisions. Far too often, physicians don't learn from cases that go wrong. They circle the wagons around themselves and go into an ego-defense mode, blaming the patient, nurses, the resident, or anyone but themselves. While others may have some part in contributing to what went wrong, you can ultimately only change yourself. We must keep ourselves honest (and when we don't, we need honest and not-always-punitive peer review processes to provide feedback). Physicians, unfortunately, often learn little from "bad cases," or the "crashes," but they also learn very little from the "near-misses." Usually, for every time a physician has a "crash," there have been several near-misses (or, as comedian George Carlin called them, "near-hits"). Ideally, we would learn as much from a near-miss as we might from a crash, and, in doing so, hopefully reduce the number of both. We cannot wait for things to go wrong to learn how to improve our processes.

Using personal or institutional databases for self-reflection is one way to be honest about outcomes. I maintain a database of every case or delivery I do; I can use this to compare any number of metrics to national, regional, and institutional averages (like primary cesarean delivery rates, for example). We also need to utilize quality improvement conferences, even in nonacademic settings. Remember, even when things go right, we can still learn and improve.

Feedback: We should provide rapid and reliable feedback so that errors are appreciated, understood, and corrected, allowing calibration of cognition. Calibration is the key to refining and improving our System 1 decisions. We need to do this for ourselves, our peers, and our institutions. Peer review processes should use modern tools like root-cause analysis, and utilize evidence-based data to inform the analysis. Information about potential cognitive biases should be returned to physicians with opportunities for improvement. Also, adverse situations and affective disorders that might lead to increased reliance on CDRs should be assessed, including things like substance abuse, sleep deprivation, mood and personality disorders, levels of stress, emotional intelligence, communication skills, etc.

The Ten Commandments

Leo Leonidas has suggested the following "ten commandments" to reduce cognitive errors[26] (I have removed the *Thou shalts* and modified the wording slightly):

1. Reflect on how you think and decide.
2. Do not rely on your memory when making decisions.
3. Have an information-friendly work environment.
4. Consider other possibilities even when you are sure you are right.
5. Know Bayesian probability and the epidemiology of the diseases (and tests) in your differential.
6. Rehearse both the common and the serious conditions you expect to see in your specialty.
7. Ask yourself if you are the right person to make this decision.
8. Take time when deciding; resist pressures to work faster than accuracy allows.
9. Create accountability procedures and follow-up for decisions you have made.
10. Use a database for patient problems and decisions to provide a basis for self-improvement.

Let's implement these commandments with a few clinical examples:

1. Reflect on how you think and decide.

Case: A patient is in labor with a history of diet-controlled gestational diabetes. She has been complete and pushing for the last 45 minutes. The experienced nurse taking care of the patient informs you that she is worried about her progress because she believes the baby is large. You and the nurse recall your diabetic patient last week who had a bad shoulder dystocia. You decide to proceed with a Cesarean delivery for arrest of descent. You deliver a healthy baby weighing 7 lbs. and 14 ounces.

What went wrong?

- The decision was made with System 1 instead of System 2 thinking.
- Ascertainment bias, framing effect, hindsight bias, base-rate neglect, availability, and probably a visceral bias all influenced the decision to perform a Cesarean delivery.
- This patient did not meet criteria for an arrest of descent diagnosis. Available methods of assessing fetal weight (like an ultrasound or even palpation) were not used and did not inform the

decision. Negative feelings regarding the last case influenced the current case.

2. Do not rely on your memory when making decisions.

Case: A patient is admitted with preeclampsia with severe features at 36 weeks' gestation. She also has Type IIB von Willebrand's disease (vWD). Her condition has deteriorated and the consultant cardiologist has diagnosed cardiomyopathy and recommends, among other things, diuresis. You elect to deliver the patient. Worried about hemorrhage, you recall a patient with von Willebrand's disease from residency, and you order DDAVP. She undergoes a Cesarean delivery and develops severe thrombocytopenia and flash pulmonary edema and is transferred to the intensive care unit where she develops ARDS (and dies).

What went wrong?

- Overconfidence bias, commission bias. The physician treated an unusual condition without looking it up first, relying on faulty memories.
- DDAVP is contraindicated in patients with cardiomyopathy/pulmonary edema. DDAVP may exacerbate severe thrombocytopenia in Type IIB vWD. It also may increase blood pressure in patients with preeclampsia.

3. Have an information-friendly work environment.

Case: You're attending the delivery of a 41 weeks' gestation fetus with meconium stained amniotic fluid (MSAF). The experienced nurse offers you a DeLee trap suction. You inform her that based on recent randomized trials, which show no benefit and potential for harm from deep-suctioning for MSAF, you have stopped using the trap suction, and that current neonatal resuscitation guidelines have done away with this step. She becomes angered and questions your competence in front of the patient and tells you that you should ask the Neonatal Nurse Practitioner what she would like for you to do.

What went wrong?

- Hindsight bias, overconfidence bias on the part of the nurse.
- The work environment is not receptive to quality improvement

based on utilizing data, and instead values opinion and anecdotal experience. This type of culture likely stems from leadership which does not value evidence based medicine, and institutions that promote ageism, hierarchy, and paternalistic approaches to patient care. An information-friendly environment also means having easy access to the appropriate electronic information databases; but all the databases in the world are useless if the culture doesn't promote their routine utilization.

4. Consider other possibilities even though you are sure you are right.

Case: A previously healthy 29 weeks' gestation pregnant woman presents with a headache and she is found to have severe hypertension and massive proteinuria. You start magnesium sulfate. Her blood pressure is not controlled after administering the maximum dose of two different antihypertensive drugs. After administration of betamethasone, you proceed with Cesarean delivery. After delivery, the newborn develops severe thrombocytopenia and the mother is admitted to the intensive care unit with renal failure. Later, the consultant nephrologist diagnoses the mother with new onset lupus nephritis.

What went wrong?

- Anchoring, availability, confirmation bias, premature closure, overconfidence bias, Sutton's slip or perhaps search satisfying. In popular culture, these biases are summed up with the phrase, *if your only tool is a hammer, then every problem looks like a nail.*
- The physician failed to consider the differential diagnosis.

5. Know Bayesian probability and the epidemiology of the diseases (and tests) in your differential.

Case: A 20-year-old woman presents at 32 weeks' gestation with a complaint of leakage of fluid. After taking her history, which sounds likes the fluid was urine, you estimate that she has about a 5% chance of having ruptured membranes. You perform a ferning test for ruptured membranes which is 51.4% sensitive and 70.8% specific for ruptured membranes. The test is positive and you admit the patient and treat her with antibiotics and steroids. Two weeks later she has a failed induction leading to a Cesarean delivery. At that time, you discover that her membranes were not ruptured.

What went wrong?

- Premature closure, base-rate neglect, commission bias.
- The physician has a poor understanding of the positive predictive value of the test that was used. The PPV of the fern test in the case is very low, but when the test came back positive, the patient was treated as if the PPV were 100%, not considering what the post-test probability of the hypothesis was.

6. Rehearse both the common and the serious conditions you expect to see in your specialty.

Case: You are attending the delivery of a patient who has a severe shoulder dystocia. Your labor and delivery unit has recently conducted a simulated drill for managing shoulder dystocia and though the dystocia is difficult, all goes well with an appropriate team response from the entire staff, delivering a healthy newborn. You discover a fourth-degree laceration, which you quickly repair, using chromic suture to repair the sphincter. Two months later, she presents with fecal incontinence.

What went wrong?

- Under-emphasis of the seemingly less important problem. This is a form of contrast bias. We are biased towards emphasizing "life and death" scenarios sometimes at the expense of other unusual but less important problems. Simulation was a benefit to the shoulder dystocia but rehearsal could have been a benefit too for the fourth-degree laceration.

7. Ask yourself if you are the right person to the make this decision.

Case: Your cousin comes to you for her prenatal care. She was considering a home birth because she believes that the local hospital has too high a Cesarean delivery rate. She says she trusts your judgment. While in labor, she has repetitive late decelerations with minimal to absent variability starting at 8 cm dilation. You are conflicted because you know how important a vaginal delivery is to her. You allow her to continue laboring and two hours later she gives birth to a newborn with Apgar scores of 1 and 4 and a pH of 6.91. The neonate seizes later that night.

What went wrong?

- Visceral bias.
- In this case, due to inherent and perhaps unavoidable bias, the physician made a poor decision. This is the reason we shouldn't treat family members, for example. But this commandment also applies to the use of consultants. Physicians need to be careful not to venture outside their scope of expertise (overconfidence bias).

8. Take time when deciding; resist pressures to work faster than accuracy allows.

Case: A young nurse calls you regarding your postoperative patient's potassium level. It is 2.7. You don't routinely deal with potassium replacement. You tell her that you would like to look it up and call her back. She says, "Geez, it's just potassium. I'm trying to go on my break." Feeling rushed, you order 2 g of potassium chloride IV over 10 minutes (this is listed in some pocket drug guides!). The patient receives the dose as ordered and suffers cardiac arrest and dies.

What went wrong?

- Overconfidence bias.
- Arguably, the physician's main problem is a lack of knowledge, but feeling pressured, he deviated from what should have been his normal habit and did not look it up (if this scenario seems far-fetched, it was taken from a case report from Australia).

9. Create accountability procedures and follow-up for decisions you have made.

Case: Your hospital quality review committee notes that you have a higher than average Cesarean delivery wound infection rate. It is also noted that you are the only member of the department who gives prophylactic antibiotics after delivery of the fetus. You change to administering antibiotics before the case, and see a subsequent decline in wound infection rates.

What went wrong?

- Nothing went wrong in this case. Peer review worked well, but it required the physician to be receptive to it and to be a willing participant in continuous quality improvement processes. It also required the non-malignant utilization of peer review. The situation might have been avoided if the physician had better habits of continuous education.

10. Use a database for patient problems and decisions to provide a basis for self-improvement.

Case: You track all your individual surgical and obstetric procedures in a database which records complications and provides statistical feedback. You note that your primary Cesarean delivery rate is higher than the community and national averages. Reviewing indications, you note that you have a higher than expected number of *arrest of dilation* indications. You review current literature on the subject and decide to reassess how you decide if a patient is in active labor (now defining active labor as starting at 6 cm) and you decide to give patients 4 hours rather than 2 hours of no cervical change to define arrest. In the following 6 months, your primary Cesarean delivery rate is halved.

What went wrong?

- Again, nothing went wrong. This type of continuous quality improvement is the hallmark of a good physician. But it must be driven by data (provided from the database) rather than a subjective recall of outcomes. We must promote a culture of using objective data rather than memory and perception to judge the quality of care that we provide. Additionally, we must be open to the idea that the way we have always done things might not the be the best way and look continuously for ways to improve. This is another skill that is strengthened with metacognition.

We must actively change the way we teach students and residents if we hope to reduce the rates of serious diagnostic errors in medicine. Robert Trowbridge offers these twelve tips for teaching avoidance of diagnostic errors:[27]

1. Explicitly describe heuristics and how they affect clinical reasoning.
2. Promote the use of 'diagnostic timeouts.'
3. Promote the practice of 'worst-case scenario medicine.'
4. Promote the use of a systematic approach to common problems.
5. Ask why.
6. Teach and emphasize the value of the clinical exam.
7. Teach Bayesian theory as a way to direct the clinical evaluation and avoid premature closure.
8. Acknowledge how the patient makes the clinician feel.
9. Encourage learners to find clinical data that doesn't fit with a provisional diagnosis. Ask, "What can't we explain?"
10. Embrace Zebras.
11. Encourage learners to slow down.
12. Admit one's own mistakes.

The differential diagnosis as a cognitive forcing tool

I believe that the *differential diagnosis* can be one of our most powerful tools in overcoming bias in the diagnostic process. But to do this, the differential diagnosis must be made at the very beginning of a patient encounter in order to provide mental checks and raise awareness of looming cognitive errors before we are flooded with sources of bias. The more information that is learned about the patient, the more biased we potentially become. Even looking at the patient is a source of bias.

The traditional method of making a *differential diagnosis* is to construct it as the patient's story unfolds, usually after the history and physical; yet this may lead to multiple cognitive errors. *Triage cueing* from the patient's first words may lay the groundwork for *availability, anchoring, confirmation bias,* and *premature closure*. The most recent and common disease processes will easily be retrieved from our memories, limiting the scope of our thinking merely by their availability.

With bias occurring during the patient interview, by default – through System 1 thinking – we may begin to anchor on the first and most likely diagnosis without full consideration of other possibilities.

This causes us to use the interviewing process to seek confirmation of our initial thoughts and it becomes harder to consider alternative possibilities. Scientific inquiry should not seek confirmation of our hypothesis (or our favored diagnosis), but rather evidence for rejection of any possibilities being considered (the differential diagnosis). The question should be, *Why isn't it what I think it is?* Most physicians settle upon a diagnosis in the first few seconds after a patient begins to speak; this is both useful and dangerous.

Once we've gathered enough data to confirm our initial heuristic thinking, we close in quickly, becoming anchored to our diagnosis. A simple strategy to prevent this course of events is to pause before every patient interview and contemplate the full scope of possibilities; that is, to make the differential diagnosis after learning the chief complaint but *before* interviewing the patient.

By using the chief complaint given on the chart, a full scope of diagnostic possibilities can be considered including the most likely ones, the most common ones, as well as the rare and the life-threatening diagnoses. This differential diagnosis can then help shape the interview with a larger availability of possibilities and encourage history-taking that works to exclude other diagnoses.

Another of my sayings:

You can't diagnose what you don't think of first.

Having taught hundreds of medical students how to make differential diagnoses, I have always been impressed how easy it is to bias them to exclude even common and likely diagnoses. For example, a patient presents with right lower quadrant pain. The student is biased (because I am a gynecologist), so the differential diagnosis focuses only on gynecologic issues. When taking the history, the student then fails to ask about anorexia, migration of the pain, etc., and fails to consider appendicitis as a likely or even a possible diagnosis. The history and physical becomes limited because the differential diagnosis was not broad enough. In these cases, triage cueing becomes devastating.

If bias based merely on my specialty is that profound, imagine what happens when the student opens the door and sees the patient (making assumptions about class, socioeconomic status, drug-dependency, etc.), then hears the patient speak (who narrows the complaint down to her ovary or some other source of self-identified pain), then takes a history too narrowly focused (not asking a broad review of system questions, etc.). I have considered lead poisoning as a cause of pelvic/abdominal pain every time I have ever seen a patient with pain, but, alas, I have never diagnosed it nor have I ever tested a patient for it. But I did exclude it as very improbable based on history.

This is not the traditional teaching regarding the construction of differential diagnoses. Most physicians have taught their students and residents to construct the differential diagnosis after interviewing the patient and performing relevant physical exam. But I believe this is a fatal flaw in our teaching.

Remember Croskerry's list of cognitive dispositions to respond:

- Errors of over-attachment to a particular diagnosis
- Errors due to failure to consider alternative diagnoses
- Errors due to inheriting someone else's thinking
- Errors in prevalence perception or estimation
- Errors involving patient characteristics or presentation context
- Errors associated with physician affect, personality, or decision style

I believe that three of these are strongly counteracted by forming the differential diagnosis before seeing the patient; namely, errors due to failure to consider alternative diagnoses, errors due to inheriting someone else's thinking, and errors involving patient characteristics or presentation context.

I must add that to do this correctly means generating the differential diagnosis before *anything* is known about the patient except the chief complaint. Obviously, we can't always do that. But we can control this to a large degree. In addition to doing this before interviewing and examining the patient, it should also be done before reading anything about the patient. Residents and medical students have become quite good at reading all about a patient on the electronic health record before ever meeting the patient. I have known residents and fellows who have essentially written their entire history and physical or consult note before ever even seeing the patient, by copying other peoples' work; talk about inheriting other people's thinking!

It seems that we have lost sight of some very basic clinical skills and perhaps lost sight of the purpose of things like the history, physical, differential diagnosis, consultation, etc.

We will explore each of these skills in the following chapters.

Exercises

3.1 Reflect on recent examples you've seen in clinical practice or in life where you have observed bias, in others or in yourself. List at least five examples and identify the type of bias specifically.

3.2 In your next ten clinical encounters, after making a diagnosis, ask yourself or your learners, "What else could this be?"

3.3 Identify structural elements in your personal or institutional workflow that promote cognitive bias in the diagnostic process. List at least two.

4

Cognitive Fallacies

Biases are things that push us in one direction or another in our reasoning. As we have observed in the last chapter, these directions are often the wrong directions. Some biases are good. In fact, bias isn't a flaw – it's a feature. It promotes efficiency. We would die without bias. But fallacies are actual mistakes in reasoning. We need to identify them and remove them from our thinking to avoid cognitive errors.

When people think of cognitive fallacies, they usually think of what are called logical fallacies. We will briefly discuss some of these, but we will also consider some other fallacies that affect the way we care for patients and interpret the medical literature.

Logical fallacies

Here are few of the most important logical fallacies that affect medical decision making

Ad hominem. This is simply an attack against a person, organization, or group rather than an attack against a proposition. A vaccine-denier might do this by attacking the CDC or doctors as a class rather than present evidence of vaccine injury. Physicians might do this when they attack Andrew Wakefield for being unethical and losing his medical license rather than discuss the problems with his paper which suggested a link between the MMR vaccine and autism. It is true that an individual or organization's biases and agendas may cause them to misrepresent evidence, but the important thing is to analyze the evidence, not the personalities.

Ad ignorantiam. The *argument from ignorance* asserts that something is true simply because we don't know that it's not true. For example, a vaccine-denier might claim that the current vaccine schedule increases the risk of autism since there is no study that shows that it does not. Obviously, some things are true for which no affirmative evidence has yet been found. As Nassim Nicholas Taleb says in *The Black Swan*, "Doctors most commonly get mixed up between absence of evidence and evidence of absence." Just because we might not have evidence that our patient has a disease doesn't mean that the patient doesn't have the disease. However, it is a fallacy to claim that the patient has the disease without evidence. **If you suspect something, prove it.**

***Post hoc ergo propter hoc* fallacy.** This common mistake assumes that because one event follows another, there must be an association between the two or even causation. If the patient got sick after eating a taco, those two events might be related but they might not be. You need more evidence to make this connection. In its most popular form, the *post hoc* fallacy takes the form of inappropriate claims that correlation equals causation. The vaccine denier might claim that because her child developed autism after the MMR vaccine, then the MMR vaccine is both correlated with *and* is the cause of autism (despite overwhelming evidence to the contrary). This is a logical fallacy and more evidence than the temporal association is needed. But researchers and clinicians make the same mistakes, too. Researchers publish papers every day which show a statistical correlation between two things and then they assume or imply that there is a causal relationship, such as, perhaps, low Vitamin D levels and preterm labor. Clinicians correlate an intervention with an outcome and assume causation in the same way, like noting that a woman's contractions stopped after ordering her to bed rest and assuming causation. This fallacy is the chief source of anecdotal medicine. We will talk more about this in a minute when we discuss *regression to the mean*.

***Tu quoque* fallacy.** This phrase means *you too* and it describes an attempt to justify one wrong action by pointing out that alternative actions are also flawed. So, in medicine, one might justify one questionable intervention by pointing out that other interventions used for the disease are also questionable. This is fallacy, of course, and luckily evidence-based medicine tells us not to use either of those interventions (both treatments can be wrong).

Special pleading. This fallacy is sometimes called *moving the goalpost*. If the evidence suggests that a theory is incorrect, then a person may choose to make-up exceptions or change the parameters of the argument. In medical studies, this happens when researchers fail to find the outcome they expected from the studied intervention but then search for any outcome which might be statistically significant even though the study wasn't necessarily powered to find that different outcome. They desire to "win" by any means; that is, they want to publish a positive and novel finding, so they move the goalpost until they find one.

Bandwagon fallacy. It is never an acceptable argument that it's appropriate to do something because most people do it. In medicine, non-evidence-based interventions are frequently supported by claiming that the intervention is the "standard of care." But standards of care change with new evidence. If they did not, we would still be using leeches to release the dark humors or using frontal lobotomy to treat mental illnesses.

Appeal to nature. This fallacy is common among proponents of alternative medicine and countless medical interventions. If something is natural, it must be harmless and safe. Many patients talk about wanting to try "natural" remedies before accepting medical intervention. But there is no agreed upon definition of what is "natural;" even if there were, it is a fallacy to assume that what is natural is better or safer than whatever is "not natural." Asbestos, anthrax, and arsenic are all "natural."

The Texas sharpshooter fallacy. This occurs when people cherry-pick some piece of data that supports their beliefs while ignoring the larger body of evidence. It may occur when a person cites the one article that agrees with them while ignoring the 99 that do not or it may occur when the authors of a study create parameters that select for certain patterns in the data; this is called *data mining* or *data fitting*.

Authors do it by playing with inclusion and exclusion criteria or by playing with outcomes or combinations of outcomes until a statistically significant finding appears (and this happens very, very frequently in published literature). It is very important when conducting scientific studies that inclusion criteria, statistical methods, and outcomes all be selected before any data is collected and that the analysis be blinded and performed by a third party (triple blinding), yet all those things only rarely happen in any given study.

Illusion of causality

Nearly 1 in 3 Europeans believe that homeopathic medicines are effective, despite there being no scientific evidence for this belief. Cancer patients who use homeopathic and alternative remedies are twice as likely to die from the disease.[28] Nearly 1 in 3 Americans believe that there is a causal relationship between vaccines and autism, despite hundreds of large, exhaustive studies suggesting there is no such link.

The illusion of causality (*cum hoc ergo propter hoc*) is the belief or perception that two events which might be temporally or statistically related have a causal relationship. Examples might be the statistical association between rates of autism and sales of organic food, or the rates of ice cream sales in the 1940s and rates of polio, or the per capita consumption of cheese and the rates of people dying by becoming strangled in their bed linens. All these events are highly correlated but that doesn't mean they have a causal relationship.

Matute and colleagues have argued that the illusion of causality is responsible for most pseudoscientific beliefs and most medical quackery.[29] These illusions are strong and occur irrespective of education or intellect. Once they occur, they tend to prevent the believer from seeing evidence to the contrary. People tend to believe what they see with their own eyes, and they prefer anecdotal and experiential knowledge compared to scientific and systematic knowledge. Thus, there is a growing rejection of science and a growing embrace of pseudoscientific beliefs. The illusion of causality spurs strong beliefs in things ranging from vaccine denial, essential oils, and homeopathy to astrology, crystal healing, and Reiki massage. Beliefs form because a person witnesses something and then looks for antecedent or parallel events to which they might attribute the outcome. *"I was sick, then used the essential oil and now I feel better. Therefore, essential oils cure tummy aches."*

Matute also reports that we are more likely to assume causation from correlation when a supposedly causative factor is present in more than 75% of cases. For example, it would be easy to assume that McDonald's might be a cause of obesity because more than 75% of obese people have eaten at a McDonald's before. Similarly, most children who develop autism have been vaccinated. With rare events, any possibly correlated or antecedent factor may appear to be a cause. Level of personal involvement also pushes people to be more likely to assume false causation. If you are making money selling snake oil, you are more likely to see causal relationships between snake oil and good health outcomes.

Of course, correlation rarely equals causation. To determine a causal relationship between two events usually requires rigorous testing. Other variables, even unknown variables, need to be eliminated. Even in high-quality scientific studies, causation is difficult to tease out.

But the illusion of causality has a much greater impact than just your patients believing nonsense. It has just as much impact on doctors and researchers. A great deal of the published biomedical literature merely reflects observed correlations. For example, hundreds of studies have been published in the last 15 years which associate low levels of Vitamin D with a variety of diseases – indeed, almost *any* disease you might think of. But are low levels of Vitamin D a cause of, say, depression, impotence, hypertension, preterm labor, and dozens of other disease processes? The strength of the correlation doesn't matter either. Something could be associated with another variable 100% of the time and still not be causal. Many times, two variables have common causes; or perhaps there is a causal relationship, but in the direction opposite the intuitive one. For example, low Vitamin D levels might be a consequence of not going outside into the sunlight very much (they are), and not going outside into the sunlight might be dramatically more common among depressed, obese, hypertensive, or impotent patients (it is).

One way of evaluating causation is to do an interventional study. So, for example, give patients Vitamin D supplements versus a placebo and then study the presence of the disease and its severity prospectively, in a blinded manner. Even this, though, is not necessarily enough.

The Bradford Hill criteria are one commonly used approach to determining causation from correlation. Sir Austin Bradford Hill suggested these 9 criteria to consider:[30]

1. **Strength**: The stronger the correlation, the more likely that a causal relationship exists. However, a small or even absent correlation doesn't exclude a causal relationship just as a large correlation doesn't guarantee one either.
2. **Consistency:** This is another way of saying that the results of a correlative study have been replicated, in various settings and by various researchers. The more consistent the findings during replication, the more likely there is a causal relationship.
3. **Specificity:** A theorized causal effect should be very specific. In other words, a specific and predetermined effect should occur from a cause, if there is a relationship.
4. **Temporality:** One would expect a cause to precede an effect.
5. **Biological gradient:** In medicine, most of the time there should be a specific and measurable dose effect. The more the cause (or dose), the more the effect.
6. **Plausibility:** A plausible explanation based on other known science that would explain the cause-effect relationship makes it more likely that a causal relationship exists.
7. **Coherence:** If an effect can be demonstrated in one type of experiment and also observed in another type of experiment, this increases the likelihood of a relationship (e.g., a lab experiment and an epidemiologic study).
8. **Experiment:** An interventional trial, in medicine, that shows the expected effect, goes a long way to showing causation.
9. **Analogy:** If other, similar cause-effect relationships are known to exist, this increases our confidence in a causal relationship.

What if we were to apply the Bradford Hill criteria to the observed relationship that insulin lowers blood glucose when administered subcutaneously? 1. Lower glucose is observed almost every time insulin is administered compared to placebo, so the **strength** of the correlation is extremely high. 2. This effect has been **consistently** observed thousands of times in different settings, populations, and species.

3. The effect of lowered glucose is very **specific**, not part of a larger generalized outcome, like *the patient improved* or *felt better*. 4. The lowered glucose follows **temporally** in a predictable time after administration of insulin. 5. There is a **biological gradient**, because the more insulin given, the greater the decrease in glucose. 6. The causal relationship is **plausible** based on other experiments and observations related to glucagon, insulin receptors, etc. 7. The effect is **coherent** across different types of studies, including large population studies and *in vitro* tissue experiments. 8. Carefully conducted, highly-controlled **experiments** have demonstrated that insulin is the causal agent rather than some unknown variable. 9. The effect of insulin in lowering glucose is **analogous** to other know endocrine hormone-receptor interactions and intracellular transport models.

Ergo, insulin lowers blood glucose when administered subcutaneously.

Compare this to the "relationship" between vaccines and autism. Since autism occurs with the same frequency among unvaccinated children as vaccinated children, then there is no correlation, let alone strength. I'll stop there, but if you went through each of the Bradford Hill criteria, you would find that none are satisfied.

These criteria are important to think about both when reading literature about "novel" associations and when making sense of your own observations.

In 2018, a new study in Obstetrics received a lot of attention. The study, called the ARRIVE Trial, received considerable publicity for its novel finding: that induction of labor at 39-weeks' gestation (compared to expectant management until 41-weeks') resulted in a *lower* risk of Cesarean delivery. Many patients and doctors were bewildered by this finding, while those obstetricians who love to induce every patient at 39-weeks' latched onto the study without any skepticism whatsoever. Without going through boring details, let's quickly apply some of the criteria to the "novel" finding and see what happens.

The strength of the association (lower Cesarean rate with induction) was statistically small. This correlation has not been consistently replicated – in fact, dozens of other studies have shown the exact opposite (making it a "novel" finding). The observed effect seems specific (reduction in Cesarean rate), but it isn't. A specific effect would be something like *a reduction in Cesarean rate due to fetal distress.* Cesarean rate is a complex outcome with many different causes and a specific effect would focus on a particular cause. Of course, there is a temporal relationship (there usually is when people see causation from correlation). The study cannot be used to determine a biologic gradient (*is there an even lower Cesarean rate with induction at 38 weeks?*). The observed relationship isn't plausible based on our current understanding of the factors influencing parturition. The finding also lacks coherence or analogy with other concepts. The experiment itself (the study) uses a poor comparator (a population with an excessively high Cesarean rate) and therefore holds little value.

So, what's the conclusion? The finding of a lower Cesarean rate in the ARRIVE Trial is, at best, a false positive finding. We will look at other, more formal methods of determining the validity of a study like this later.

Now, let's apply these criteria to an observation that you might make in the clinical setting. Let's say you observe that a patient's symptoms of dysuria improve after treatment with an antibiotic for a UTI. Go through each criterion and you will see most of the Bradford Hill criteria are fulfilled. What if you applied these criteria to the observation that eating turmeric improves cancer survival rates? None of these criteria are fulfilled. We will talk more about the fallacies of anecdotal medicine in a few minutes.

What are some other things that can be done to diminish the illusion of causality both for yourself and for patients?

- Emphasize times when the outcome doesn't follow the suggested cause. Snake oil salesmen and practitioners of anecdotal medicine always emphasize the times when the intervention produced the desired result, but rarely talk about when it didn't. The one person who took turmeric and had an unexplained improvement in his terminal cancer should be contrasted to the tens of thousands of patients who died of cancer after taking turmeric.
- Point out potentially negative consequences of the intervention to make the cost of believing in the false association higher. An example of this is pointing out the horrors of vaccine-preventable illnesses when a patient is considering not vaccinating a child.

Illusion of control

There are a couple of cognitive biases that conspire together to create this fallacy, **motivated reasoning** and the **therapeutic imperative**.

Motivated reasoning occurs when people go to great lengths to rationalize and justify their beliefs in things that have been clearly shown by evidence to be false. They are motivated to construct elaborate explanations and workarounds about why all the evidence must be wrong because they *need* to justify the false belief in order to avoid cognitive dissonance and deconstruction of their identities which have become closely linked to the false belief. The longer a person has held a false belief and the more central that belief is to her identity, the more likely it is that she will use such motivated reasoning. It's more than just confirmation bias, which involves accepting inadequate evidence that tends to confirm one's belief; motivated reasoning causes a person to invent evidence and create rationalizations.

An example of this, of course, is the belief that vaccines cause autism. A large industry exists among passionate vaccine-deniers which publishes elaborate explanations and rationalizations about the science-based evidence which disagrees with their premise and many work tirelessly to create "evidence" illustrating complex theories of causation or other data that can be used to justify the belief.

But these same characteristics exist among physicians. I have written extensively elsewhere about the lack of scientific evidence supporting the use of tocolytic drugs (like magnesium sulfate or nifedipine) to treat preterm labor. The body of evidence is overwhelming. Yet, physicians continue to do so. They twist and rationalize evidence and ignore contrary evidence using elaborate mechanisms. Many are older physicians with some degree of power in terms of making policies and guidelines; these older physicians have spent their entire careers using tocolytic drugs to treat patients who had preterm labor. They invested a lot of themselves into these patients and they genuinely cared for them; it is difficult to accept evidence that all those efforts were misguided and, worse, that the interventions might actually have harmed the patients. So, they search desperately for some modicum of evidence that they did right. Today, that has resulted in the equally non-evidenced belief that magnesium sulfate may prevent cerebral palsy in children born prematurely (without increasing total mortality).

This sort of motivated reasoning is present in all medical specialties. **Medical reversal** happens frequently, where we determine that a once-believed treatment is not effective based on better evidence. Examples include cardiac stents, HgA1c goals in diabetes management, vertebroplasty, or arthroscopic knee surgery, etc.[31] Physicians who have dedicated large portions of their careers to these sort of efforts (and in many cases made a lot of money doing so) quickly adopt motivated reasoning to justify what they do. Worse, they are often considered the "experts" about the interventions. Based upon their "expertise" alone, practices without any scientific support become pervasive throughout medical practice and it is difficult to stop them.

Therapeutic imperative. Kathryn Montgomery describes the therapeutic imperative as "Always do everything for every patient forever,"[32] but she juxtaposes this to our ethical standard of *primum non nocere* (first, do no harm), which she cleverly translates as "Don't just do something, stand there." It feels wrong to just stand there and do nothing, and this feeling is the origin of the therapeutic imperative. In logic, we might call this an ***appeal to emotion***. Physicians feel like they must do something for patients, even when there is no reasonable intervention. This leads to over-treatment, over-diagnosis, and ultimately harm as every intervention will be associated with some negative risk for the patient.

Physicians feel like they must do *something* (even if the evidence for the intervention is weak or nonexistent) when, for example, a patient presents in preterm labor; so, they order a tocolytic like magnesium sulfate. Ironically, simply by using the intervention repeatedly they will soon adopt motivated reasoning to justify its use. Any harms that result because of the intervention will be ignored (like ARDS from tocolytic-induced pulmonary edema or pulmonary embolus from bed rest).

Together, the feeling that *something must be done* combined with our ability to rationalize evidence supporting that intervention lead to an **illusion of control.** The physician begins to believe that he actually has control over preterm labor (or any other disease process) which he does not have. Ellen Langer first noted this illusion and defined it as the tendency for people to overestimate their ability to control events. This illusion further perpetuates nonscientific medical practices and anecdotal medicine. It also creates significant medicolegal risk for physicians who attempt to practice in an evidence-based manner since in the courts they are often compared to the community practice standards, whether they are legitimate or not. This perceived medicolegal risk then often becomes the reason why physicians persist in incorrect practices.

Anecdotal medicine fallacy

The *appeal to experience fallacy* and the concept of *regression to the mean* are both important concepts to understand when considering why so many physicians practice anecdotal medicine and why so many patients are vulnerable to anecdotal medicine and other anecdotal practices (like various alternative medicine products which are sold on testimonial or anecdotal evidence). Anecdotal medicine (practicing medicine based on one's own experiences and observations) is the opposite of evidence-based medicine (practicing medicine based on our collective and systematically analyzed experiences). Let's spend a little more time thinking about these important fallacies.

Appeal to experience

A common fallacy that we all are sometimes influenced by is the *appeal to accomplishment*. This occurs when we claim that an opinion or idea is invalid because the person making the claim doesn't have the accomplishments, experiences, or perhaps moral standing to make such a claim. Example: "How can you stand there and criticize the chef's cooking? Have you ever been a chef?" I prefer, however, to call this fallacy an *appeal to experience.*

The appeal to experience comes in many forms. For example,

- Your ideas about education are invalid, because you've never been a teacher.
- Your ideas about some area of science are wrong, because you're not that type of scientist/you're not a scientist.

- Your ideas about a certain disease are wrong, because you don't specialize in that disease.
- Your ideas about a certain disease are wrong, because you don't personally have that disease.
- Your ideas about some aspect of sociology are wrong, because you don't belong to that social group.

I'm sure you can think of more. Each of these arguments are logically absurd. You may be wrong about your ideas, and your lack of experience may even contribute to why you are wrong, but you are not wrong simply because you lack the experience. Arguments should stand on their own merits, not the merits of the ones making the arguments. An argument should only be discounted because it lacks merit.

But this logical fallacy is important for a different reason, and it relates in particular to the practice of anecdotal medicine. Here's one of my axioms:

Our experiences lead us to awareness, not insight.

People often confuse awareness and insight. Let's say you were a victim of a violent crime. This experience might make you a passionate advocate for reducing such crimes and helping victims of such crimes. But your experience alone will not tell you how best to do those things, and, worse, your experience may bias you in a way that you might never be the best person to solve the problem. You might want to focus all our resources on ineffective measures related only to this one problem that has so viscerally affected you, and by doing so you might unintentionally cause increased crime elsewhere.

Extend this metaphor to healthcare. Someone may be passionate about a certain disease (say, ovarian cancer) because she has been personally affected by it or has taken care of many patients who have died from it. This visceral bias may lead to selective interpretation of data and a desire to do whatever possible, no matter the unintended consequences or costs, to prevent ovarian cancer. Such passion may lead for the advocacy of screening CA-125 tests or screening ultrasounds, both of which have been found to be ineffective and potentially harmful. But often it takes an objective observer to see the real risk-benefit ratio of a given intervention.

Our experiences make us aware of problems, they allow us to have empathy for others, and they focus our attention, but they also make us biased in a deep, visceral way; sometimes our experiences may intellectually cripple us, making us unaware of larger, more global issues. The person with experience is often not the right person to make decisions about the things he has personally experienced. When our personal identities are so closely tied to a specific thing, say a way of practicing medicine or teaching or anything else, we may have a problem seeing the forest as we focus too much on our one blade of grass. This is called **inattentional blindness**. Inattentional blindness leads us to not see something very obvious because our perception is directed elsewhere and it's one of the negatives of bias and triggering.

So, how do we gain insight? Through deliberate analysis and study. This needs to be dispassionate and objective, considering all factors fairly. A person with "experience" can do this and her experience can be valuable, but she must carefully guard against her own biases. We don't need to experience something to gain insight into it. I don't need to damage a ureter during a hysterectomy to gain expertise at preventing ureteric injury. I don't need to perform a series of Zavanelli maneuvers in order to know how to do one when called upon to do so (a surgical rescue maneuver for stuck shoulders during a delivery).

In medicine, many discussions (a form of argument) with our colleagues or patients are begun or ended with *"I once had a patient who…"* or *"In my experience..."* or *"When I do this procedure..."* etc. These sorts of statements may serve the purpose of providing context for the argument and that's fine, but too often these anecdotes *are* the argument. It is a claim that the experience with that patient or procedure is sufficient to end the debate, and this is the hallmark of anecdotal medicine.

Patients often find this appealing, and many will seek out the most *experienced* physician, assuming that this is the same as the most *competent* physician. This idea assumes that experience is a predicate of insight. Patients often ask, *"Have you ever seen a patient with this diagnosis before?"* when they mean to ask, *"Do you have competency in treating this diagnosis?"* These are often not the same question.

One of my favorites (as a male gynecologist) is, *"I know you're a man so you don't understand what it's like to have a period."* That's true, but I am an expert at treating problems associated with the menstrual cycle and I have discussed the menstrual cycles with thousands of women; I've read about the menstrual cycle in dozens of books and papers, and I possess the shared and accumulated knowledge of hundreds of years of scientific research. Even better, I'm not personally biased about the menstrual cycle and I don't believe it causes my toes to hurt, as someone might anecdotally believe who notices that her toes hurt once a month. So, while I lack the joy of personally having a period, I am qualified to treat problems related to one.

This *appeal to experience* fallacy is inappropriately extended to almost every issue in modern society: Unless you have a child with autism, you can't decide whether vaccines are related to autism; unless you are a victim of gun violence, you aren't entitled to have an opinion about gun laws; unless you've walked a mile in my shoes, don't tell me how to walk.

But worse than this, the argument in medicine is, *"I don't care what the evidence says, I know from my own personal experience that it works."* I have heard this more times than I care to mention. Or statements like these: *"I don't need to read about it, I treat it every day!" "I know the literature says that this doesn't work, but in my hands, I have excellent outcomes."*

For all its faults, evidence based or science based medicine is lightyears ahead of anecdotal medicine. It is odd, to me at least, that most physicians still practice anecdotal medicine. Even those physicians who claim to believe in evidence based medicine tend to pick and choose the literature that confirms to their biases and ignore the literature that disagrees with them. This is disingenuous at best and dishonest at worst. But humans, especially doctors, like to believe that their experiences are more valuable than those of other people. It makes us feel special to think that we can intuit how to do something and use our genius, rather than just blindly follow best practices.

When you practice anecdotal medicine, you are basing your patient's health on a few personal experiences. Whether your past experience with the disease is one patient or 1,000, you are still operating on assumptions based on far fewer patients than our collective experiences of thousands or tens of thousands of patients. What's more, your patients were "collected" in a biased way, without controls, subject to faulty memories, and the prior patient or patients you recall may not match the current patient in front of you.

Even worse, you are likely committing the *post hoc* logical fallacy, which states that since some second thing happened after some first thing, then the first thing must have caused the second. *"I gave her an antibiotic for a UTI and her pain went away; therefore, her pain was due to a UTI."* This is an illogical statement. Her pain went away, yes, but you don't know that it wouldn't have gone away anyway if you hadn't diagnosed her with a UTI and treated her. In fact, for all you know, she might have gotten better even sooner if you hadn't misdiagnosed her and treated her with an antibiotic. An incredible amount of medical practice is based on this fallacy alone.

People do strongly believe in anecdotes, though. Michael Shermer has said,

> Anecdotal thinking comes naturally; science requires training.

I have read numerous defenses of anecdotal medicine over the years. Each of these must, ultimately, take the position that our experience with one or two patients is better than our experience with one or two thousand patients. This quote by Italian surgeon Nicola Basso I think sums up the argument:

> The fact that a stone, thrown out of the window, rises instead of falls is not statistically significant, but it is a powerful observation.

I've always liked that quote and Basso was a wonderful surgeon. But it's total bunk. There is so much evidence that gravity is a true thing that if you throw a stone out the window and it goes up instead of down, you better start coming up with alternative explanations:

- You are high on psychotropic drugs
- Someone has placed a huge fan under your window
- It's a magnetic stone and there is a huge magnet above your window
- Someone put a wire on the stone and is playing a cruel trick on you

Observation of the extraordinary or outlandish doesn't overturn good evidence. The observation of the stone is not statistically significant for a reason, and this is not a powerful observation but a dangerous one. This is the same type of thinking that leads to the conclusion that vaccines cause autism. If the stone really went up, then set up some controls and replicate the finding sufficiently and it will become significant data and we will revise our theories appropriately. Until you disprove gravity, I won't jump out of any windows just because you had a strange experience with a stone. As a side note, Bayesian updating saves us from such fallacies because we would account for the existing evidence and revise our confidence accordingly considering any new evidence. Extraordinary claims require extraordinary evidence.

It is true that collected anecdotes may eventually amount to useful data, but we shouldn't act on that data until it becomes sufficiently collected. Marc Bekoff has said, "The plural of anecdote is not data." We must collect the data systematically and analyze it blindly and without bias, until a reasonable probability of factuality is reached. A lot of good science starts with a keen observation of something interesting, but a lot of bad patient care also starts with such observations that go untested. We don't apply anecdotes to patients until the information has been controlled, tested, replicated, and validated. Bertrand Russell commented:

> It is not what the man of science believes that distinguishes him, but how and why he believes it. His beliefs are tentative, not dogmatic; they are based on evidence, not on authority or intuition.

Of course, we don't have evidence based answers to many, many questions. This leads us often to decide what to do based on weak evidence or even "expert opinion." But we must be careful to regard such opinions as low level evidence and not become overly dogmatic.

Just because we don't have evidence that something works doesn't mean that it doesn't work. Many things have not been appropriately studied (or cannot be studied ethically given the nature of the problem). In this case, we do resort back to our experience, but not just our experience. We also have analogy, inference, and other powerful tools that can lead us to the best approach. But please don't do something just because it seemed to work one time before.

Regression to the mean

In life, we don't tend to notice the mediocre or the average; rather, we notice extremes. We never think about the weather on an average day, but we always take note when it's very hot or very cold, very dry or very wet, etc. In medicine, we don't care too much about patients who have average or predictable outcomes; rather, we focus on those patients who seem to do miraculously well or who have poor outcomes that we would like to avoid. Our intuition teaches us that there is something to be learned from those patients who have outcomes that are at one extreme or another. Being intuitive beings, we ask questions. Why did the patient with the great outcome have such a good outcome? Why did the patient with the bad outcome have a bad outcome? Our intellects always seek to supply these answers so that we can learn from these experiences. We want to distinguish the extremes from the average.

These anecdotal observations are the foundation of anecdotal medicine. For many physicians, these types of experiences are powerful evidence that dictate future practice. If a bad outcome was observed, and it was accounted to a specific cause or intervention, that physician may decide to carefully avoid that intervention for the rest of his career. For example, a surgeon might experience a postoperative bleed after administration of ketorolac; he then avoids using the drug in his patients in the future and dogmatically evangelizes against the medication to his students and residents. He also notices that his problem with postoperative hemorrhage does not return and this serves as evidence to him that ketorolac increases the risk of postoperative hemorrhage and convinces him even more of the correctness of his actions in blaming the ketorolac for any prior problems he had with bleeding. This type of anecdotal evidence often leads to the publication of case reports which further and often incorrectly perpetuate anecdotal medicine. Ketorolac, by the way, does not increase the risk of postoperative bleeding,[33] though avoiding it likely does contribute to mortality associated with opioid misuse.

When a series of good or bad outcomes occur, our ideas of cause and effect become stimulated and reinforced. The surgeon in the previous example might have noticed that he had a series of hemorrhages early in his career and that these hemorrhages went away when he stopped using ketorolac; it is easier to blame the medication than it is to introspectively acknowledge that other factors might have reduced his personal risk of hemorrhage, including better technique or patient selection. A series of negative outcomes might encourage a change or a new intervention in clinical practice; when that change is made, if the negative outcome is reduced or improved, this is often accepted as hard evidence of success and the continuing need for the intervention. This accumulation of anecdotal evidence often leads to the publication of a case series.

In March 2017, a study published from Eastern Virginia Medical School stole headlines around the world for claiming that Vitamin C infusions were the miracle cure for patients with sepsis. Many notable physicians immediately called for Vitamin C to be used routinely in the treatment of patients with sepsis due to the amazing outcomes reported in *Chest* by Paul Marik.[34] Marik reported that of 47 successive patients with sepsis to whom he administered Vitamin C, only four died. He compared this to the previous 47 patients treated at the same hospital for sepsis and 19 of those patients had died prior to the new intervention. With such a dramatic change, many began calling for the immediate implementation of Vitamin C into the treatment algorithm for sepsis. Some physicians pointed out that he had not just added Vitamin C but had also added thiamine and a corticosteroid, meaning that there was uncertainty about which of the three actually made a difference. Many called for adding all three interventions due to this ambiguity.

But there is a bigger, more fundamental problem with Marik's observation: **regression to the mean.** In fact, regression to the mean is one of the fundamental problems with anecdotal medicine in general.

The concept of regression to the mean is simple: over time, extremes will average out and return to the expected mean. If a group of students answers questions on a true or false test randomly, the average student will make a 50; but, out of a group of students, there will be some high scores and some low scores. In other words, we expect the average to be 50 but we also expect some variation around this number if the choices are truly random. If the students are sorted from the highest grade to the lowest grade and then the top and bottom 10% of students are given the same test again, then we would expect the test scores of those individual students to be closer to the mean. In other words, the top-performing students will usually perform more poorly (closer to a score of 50), and the bottom-performing students will usually perform better (again, closer to a score of 50). This is true because their original scores were based on luck or probability, not on talent or intelligence.

Look at Figure 7. The line represents any random distribution of data, whether it be students randomly answering questions on a test or a collection of labs or blood pressures from a population. Taken together as a data set, we understand that the data points in *A* and *B* represent the extremes of random data. It's only when we focus on *A* and *B* separately that we begin to believe that they are meaningfully different from one another, even though they are from the same random data set. If *A* represents the group of students who tended to score well on the exam, the next time they take the test their scores will move towards the middle. Same for *B*. Again, this assumes that the answers are given randomly.

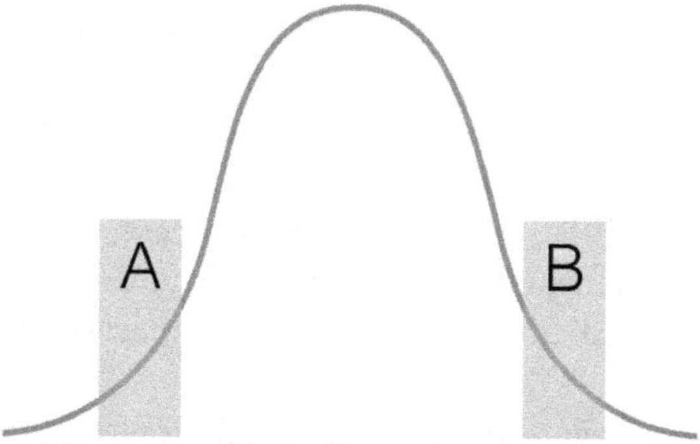

Figure 7. Normal distribution of data with samples sets A and B taken near the tails.

Even with random answers, the next time the test is administered, there will still be some student in Group *A* who scores the highest score and some student in Group *B* who scores the lowest. Does this mean that they are the smartest and dumbest kids? No. It was random. We need more information than this to decide who the bright and dull kids are.

This concept explains a lot of poorly understood and non-intuitive phenomena such as the sophomore slump in sports. If a rookie has an outstanding performance in his first year, you can expect that his performance will go down the next year – and most of the time it does. Because the performance was so outstanding, there was probably a bit of good luck in the outcomes. Don't be surprised when he does more poorly in the next season, meaning that he begins to return to his true performance mean. His average performance may still be better than that of most players (his great season was likely due to some combination of skill and luck), but it almost certainly will not be as good as the outstanding performance in his rookie year which was aided by some balls bouncing the right way and other improbable elements.

The point is, in any large collection of random (and therefore not meaningfully different) data, there will naturally be extremes. Not all people have the exact same course when they catch a cold or have a surgery, for example. We must be careful not to see causes for randomness when none exists. Look at the graph below.

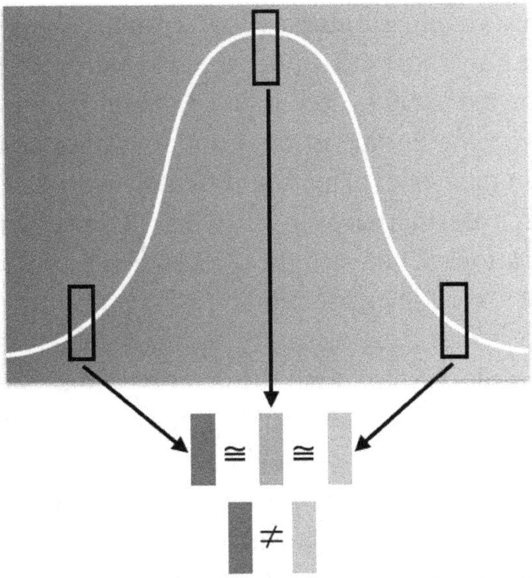

Figure 8. Comparisons of extremes in the same random distribution.

Figure 8 shows different shades of the gray. When a sample from the left tail is compared to the median, there isn't a significant difference. Similarly, when a sample from the right tail is compared to the median, we are more likely to admit that the grays belong to the same set. But when we compare the two samples from the tails to *each other* (rather than the median), then the difference becomes more dramatic and we are more likely to see significance that may not exist (or at least that might be due to random chance).

If you have a bad year in the ICU, and 19 of 47 patients with sepsis die (left side of the graph), your next year is likely going to be better – maybe even significantly better, as your sepsis related mortality rate returns to the mean. It may even be the case that your next year is improbably good, with only four of 47 patients dying of sepsis (right side of the graph). While this change seems like it must be significant, it may be related to chance alone (or to other factors like less morbid patients or maybe that corticosteroid that you gave). The two years vary not from each other but from the mean in different directions. This means that something that looks statistically significant may look that way because the comparator is incorrect. In other words, one year might be statistically different compared to another year, while neither year may statistically differ from the mean.

This simple principle explains why most anecdotal observations (and even many systematic studies) are wrong.

Think about how most anecdotes start: someone observes an unusual finding (an extremely good or bad outcome), then he tries to explain it. The extreme finding is likely due to chance alone, but even if it is not, the thought-of explanation is usually wrong and its apparent cause-and-effect association is misleading (association usually does not equal causation). There are thousands of variables to account for in any given observed phenomenon.

An anecdote is one data point. In the literature, it is called a case report. In the clinic, it starts with the doctor saying, *"This one time I had a patient who ..."* If you have several data points, this is a case series. A case series is a collection of unusual cases, but there are still way too many variables to assume that the one variable you are interested in is the one (or the only one) that matters. The Vitamin C report is a case series. It is uncontrolled. In addition to the interventions that the authors focus on (Vitamin C, steroids, and thiamine) there may be an unlimited number of other differences (different nurses, different strains of flu, less legionella in the air conditioning system, fewer smokers, more flu shots, less resistant strains of bacteria, more attentive nurses and doctors after suffering through a year where almost everyone died, more hand washing, etc.). In October, 2019, a randomized, prospective trial evaluating Vitamin C for the treatment of sepsis in the ICU was published and, little surprise, it makes no difference.[35] The VITAMINS RCT in 2020 also found no effect from Vitamin C.[36]

A lot of data points may suffer from the same problems. A retrospective study is just a big and fancy case series. Just because there are two big swaths of data doesn't mean that they don't come from two extremes of the same larger, random data set.

This is where the randomized, blinded, placebo-controlled trial enters in. It is, unfortunately, our best way to control for known and unknown variables and to attempt to eliminate regression to the mean as a source of perceived difference. It's not perfect, and we will talk about that in detail later.

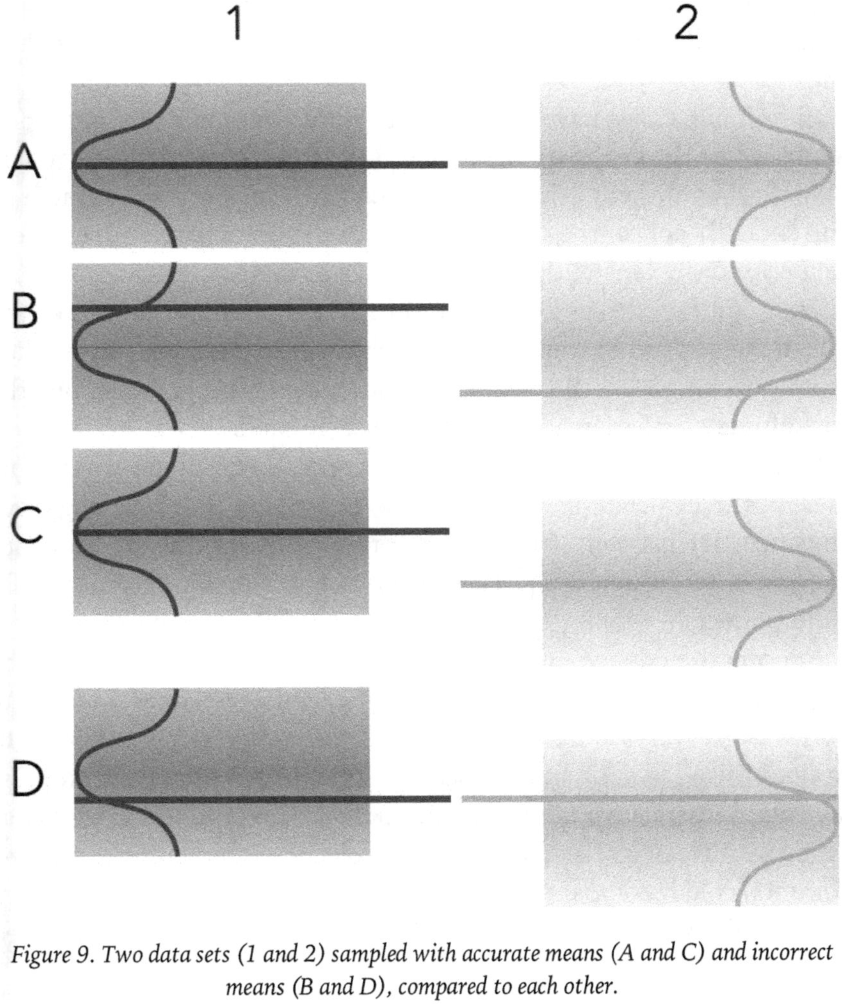

Figure 9. Two data sets (1 and 2) sampled with accurate means (A and C) and incorrect means (B and D), compared to each other.

Look at the four examples in Figure 9. Imagine that we are comparing two data sets to one another to look for statistical differences (1 and 2). The lines in 1 and 2 represent the means of our observed, experimental data sets. We don't know how our data sets comport with the real data sets (represented by the bell-shaped curves in the boxes) because our data collection is almost always incomplete or flawed.

In example A, our experimental data set means (the lines) match the mean of the actual data sets (the curves); what's more, the data sets match each other, meaning that there is no statistical difference between the two and this observation is authentic. This is a true negative (if the assumption was that the two data sets would be different). In example B, the two actual data sets still match each other, but our experimental data sets were taken from different extremes of those data sets and appear different statistically. This is a false positive finding.

In example C, we have once again captured data sets that are representative of the real data distribution. In this case, the data sets are statistically different. This is a true positive. However, in example D, the data sets appear the same because our experimental data sets were taken from the tails of the curves. This is a false negative.

These four examples don't represent all the possibilities. For example, imagine if in example C the experimental data sets were taken from opposing tails. We would still identify a true positive, but the magnitude of effect would be greatly exaggerated.

In a perfect experiment, with perfect controls and perfect matching, false positives (example B) happen 5% of the time and false negatives (example D) happen 10-20% of the time. We will talk more about experimental design later. The world of experimental science is far from perfect.

Anecdotal evidence, of course, is even worse and is dominated by regression to the mean. The average state for most people is *absence of disease*, so regression from disease to no disease is common with self-limited illnesses. Not all diseases are self-limited. Many are progressive without effective intervention. There are no type 1 diabetics alive today because of alternative medical or placebo interventions. But the common cold has apparently been cured by just about everything from Vitamin C to zinc. This is due to the natural course of the untreated diseases. If type 1 diabetes goes untreated (by insulin), death is inevitable; therefore, only insulin therapy appears to work (either anecdotally or by systematic analysis). Conversely, most disease states like the common cold resolve (or at least get better) on their own; therefore, any number of interventions (Vitamin C, homeopathy, Z-packs, essential oils, etc.) will cause it to improve since the condition will regress back to the mean anyway.

There's a lot of money to be made in treating diseases that get better on their own. Do you feel bad? Take my magic elixir and most of you who have self-limited illnesses will feel better. The key is learning which diseases are self-limited.

In palliative care, there is a concept called an **illness trajectory**. Illness trajectories can be used to illustrate how chronic illnesses will eventually turn out and then understand how interventions might change that trajectory.

In Figure 10, the gray line at the bottom shows a healthy person who dies suddenly. To change the course of this death would require prevention of the proximate cause, like seat-belts for car accidents or depression screening and treatment for prevention of suicide. The dark line at the top represents a self-limited disease, like the common cold.

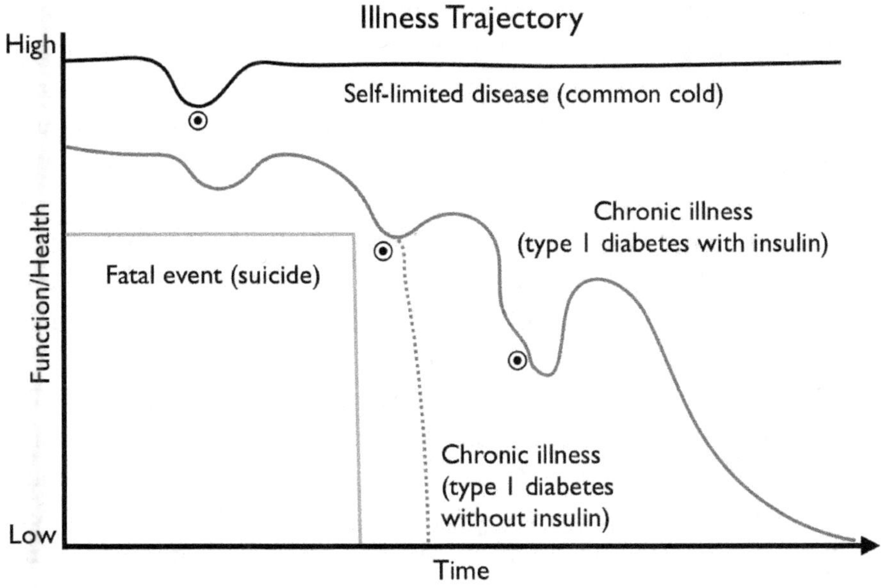

Figure 10. Illness trajectory comparing sudden, chronic, and self-limited illnesses.

The common cold lasts just a few days. By the time patients seek treatment (at the bullseyes), when the symptoms are at the worst, there are usually just a few days of symptoms left. Any intervention started would seem to work unless the illness trajectory is carefully compared to a placebo-controlled disease trajectory. Dozens of products exist to treat the common cold. None have solid evidence that they change the illness trajectory but patients swear by them because of their own anecdotal experiences (think Vitamin C, which has been tested in nearly 30 randomized trials and found to be ineffective for treatment of the common cold). Physicians will prescribe an antibiotic at the same point and patients will pay to come and be mistreated in this manner (most commonly a Z-Pack, which is dosed and indicated for treatment of community acquired pneumonia or perhaps cervicitis due to *Mycoplasma genitalium*, but not a viral cold).

The line in the middle represents a typical chronic disease. There are ups and downs of chronic diseases that may be random, may be related to an intervention, or may be related to less obvious things. For example, a newly diagnosed diabetic may be motivated to eat well and exercise; after some time, she may become depressed and her diabetes suffers because less attention is paid to diet and exercise. Eventually, the depression subsides and the condition improves. Lots of extrinsic and intrinsic factors may affect the course of the disease. If it is insulin-dependent diabetes, then without insulin she will die. The dotted line, starting at the second bullseye, shows a disease trajectory for type 1 diabetes without insulin. When we compare this disease trajectory to the one of diabetes with insulin, clearly insulin is an effective intervention.

Another intervention for diabetes is given at the third bullseye; the intervention did precede an improvement in the course of the disease but did not change the disease trajectory overall. Anecdotally, this new intervention seems to work well, but hopefully a large systematic study would show that it is no better than placebo. Imagine that a diabetic is doing poorly and buys some chromium off a website. The hope that this miracle will make a difference gives the diabetic new optimism and the disease does improve – either by chance or because of a change in hopefulness. Either way, this is the placebo effect: simply a regression to the mean. We will talk about the placebo and nocebo effects in detail later.

The whole field of alternative medicine and non-evidence based medicine is largely based upon people's inability to recognize regression to the mean. Desperate people seek any treatment when they are at their worst. If they have a self-limited disease, any treatment will probably work. Even if they have a chronic disease, they will probably notice some period of improvement.

This fallacy is related to the **Clustering Illusion,** which was introduced in the previous chapter. This illusion occurs when people assume that random clusters or streaks of data in small samples of data are not random. People see associations that aren't real all the time. Inadequately designed studies will declare these random clusters of data statistically significant when they are not.

Unfortunately, most published scientific studies that focus on the association between two variables are also flawed by this concept. Remember, around 80% of medical studies published this year are wrong and will eventually be reversed. We will discuss these issues in more detail in later chapters when we discuss analysis of the scientific literature.

Exercises

4.1 Name three disease processes that you treat which are largely self-limited. How does regression to the mean impact their clinical courses? What are some treatments used for them which appear to work only due to regression to the mean?

4.2 Select a recent study in a journal that you read which found a statistically significant association between a variable or risk factor and some outcome. Now apply the Bradford-Hill criteria to this association.

4.3 Pick three alternative medical therapies used commonly in your specialty. What logical fallacies are used to market them or justify their use?

5

Necessary Statistics

Before we go much further, we need a brief refresher on some important statistical concepts. Unfortunately, statistics gets a bad reputation as a dull and tedious subject to be endured in medical school before we can get on to the "real" business of treating patients. However, **the misuse of statistics is one of the largest sources of error and wrong thinking in medicine**, and these basic concepts apply to everything from making differential diagnoses to ordering and interpreting lab tests to making sense of the medical literature.

So, commit yourself to the discipline and time required to master basic statistical concepts. Recognize the power of statistics to inform your clinical decision-making and efficacy as a physician. Your patients will be the ones who benefit.

Statistics which describe the performance of clinical tests

One of the most important tasks a physician has is to order appropriate clinical tests and to properly interpret them. Unfortunately, this is not commonly done well. We must learn a few things about statistics to understand how tests perform. Don't skip this section. I'll remind you of some terms you've learned before, but hopefully make them more meaningful and show how they are used and misused practically.

Imagine we design a test to detect a given disease. Suppose we administer the test to 500 patients, whose test results are depicted in Figure 11:

- Some patients have a positive test result and have the disease; this is a **True Positive (TP)**.
- Some patients don't have the disease but nevertheless have a positive test result; this is a **False Positive (FP)**.
- Most patients do not have the disease and, accordingly, have a negative test result; this is a **True Negative (TN)**.
- A few patients do have the disease but, unfortunately, had a negative test result; this is a **False Negative (FN)**.

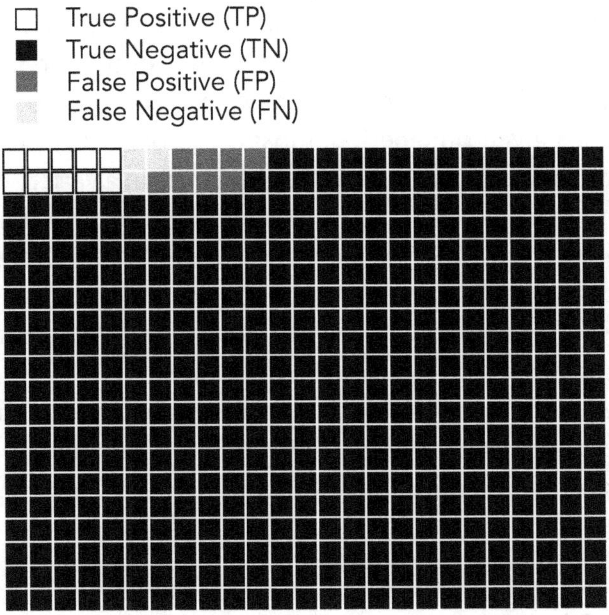

Figure 11. Visualization of experimental results.

We can make a simple table summarizing these results for our test group of 500 patients. Table 2 demonstrates the results of our experiment in a traditional 2x2 table:

	Has Disease	No Disease	Totals
Positive Test	True Positive (TP) 10	False Positive (FP) 8	18
Negative Test	False Negative (FN) 3	True Negative (TN) 479	482
Totals	13	487	500

Table 2. Two by two table of experimental results.

Let's look at other terms that are derived from these basic test results:

Prevalence. Prevalence is the fraction of people who have the disease at any given time.

$$Prevalence = \frac{Total\ number\ of\ patients\ with\ disease}{Total\ number\ of\ patients}$$

Notice that the total number of patients with disease is equal to the True Positives plus the False Negatives. In our sample data, the prevalence of the disease = 13/500 or 2.6%. Don't confuse prevalence with **incidence**, which is the number of new cases of the disease that are diagnosed in a given time period, most commonly *one year*.

We can talk about how well our test performs in a variety of different ways. Some of these terms are mistakenly interchanged and can be grossly misinterpreted if you don't understand them correctly. Let's look at each briefly.

Sensitivity. Sensitivity is the same as the **true positive rate** and predicts how well the test can detect the disease if it's there. It's sometimes called the *probability of detection.* The more sensitive a test is, the fewer false negatives. Sensitivity is equal to the number of people who had true positive results divided by the number of people who have the disease:

$$Sensitivity = \frac{True\ Positives}{Total\ number\ of\ patients\ with\ disease}$$

In our sample data, Sensitivity = 10/13 = 76.9%. It means that 76.9% of patients with the disease were correctly identified as having it (but this doesn't tell us how many were *incorrectly* identified as having the disease).

Specificity. Specificity then is the **true negative rate** and predicts how well a test can predict which patients *do not* have a disease; that is, How well does the test accurately predict healthy patients?

$$Specificity = \frac{True\ Negatives}{Total\ number\ of\ healthy\ patients}$$

In our sample data, Specificity= 479/487 = 98.4%. It means that 98.4% of people who do not have the disease were correctly identified as not having it (the true negative rate).

As we shall see, ***sensitivity and specificity do not tell you intuitively how a test should be interpreted.*** This is a common pitfall. Try this: tell a group of doctors that a test is 99% sensitive and 99% specific for detecting a cancer. Then tell them that a patient has a positive test result and that the patient is sitting in front of them. Ask them to tell the patient what her chance is that she truly has cancer. I've done this sort of experiment hundreds of times and most doctors (and medical students and residents) will tell you that it is 99%. But this is absolutely incorrect. Instead of sensitivity or specificity, we need to consider a value called the positive predictive value.

Precision or **Positive Predictive Value (PPV).** The PPV is the proportion of patients with a positive test who truly have the disease.

$$PPV = \frac{True\ Positives}{Total\ number\ of\ patients\ who\ test\ positive}$$

In our sample data, PPV = 10/18 = 55.6%. Thus, if an individual in our test group of 500 has a positive test result, he or she has a 55.6% chance of having the disease. Notice that there is **no numerical relationship** between the PPV and either sensitivity (which was 76.9%) or specificity (which was 98.4%).

Negative Predictive Value (NPV). The NPV is the proportion of patients with a negative test who actually do not have the disease. In other words, if the patient's test is negative, what is the probability that she is truly disease-free?

$$NPV = \frac{True\ Negatives}{Total\ number\ of\ patients\ who\ test\ negative}$$

In our sample data, NPV = 479/482 = 99.4%. So, if a patient in our test group has a negative test result, there is a 99.4% chance that he or she truly does not have the disease. Again, there is no numerical relationship between this number and either the sensitivity or specificity

So, as we have seen, the sensitivity and specificity of a test don't help you understand the test in the clinical setting, because those numbers don't tell you or your patients what you both actually want to know: do they really have the disease if the test is positive, or are they really disease-free if the test is negative? PPV and NPV are much more important clinically, as they indicate what a positive or negative test result means for a patient whose risk profile is similar to that of the patients in the test group. (We will see later that if our individual patient has a significantly higher (or lower) underlying risk of disease than the members of the test group, PPV and NPV will need to be "adjusted" to account for this difference.)

		Disease		**Predictive Value**	
		Positive	*Negative*		
Test	*Pos*	**A** True Positive (TP)	**B** False Positive (FP)	***PPV*** $\frac{TP}{TP+FP}=\frac{A}{A+B}$	*Total Positives* (A+B)
	Neg	**C** False Negative (FN)	**D** True Negative (TN)	***NPV*** $\frac{TN}{FN+TN}=\frac{D}{C+D}$	*Total Negatives* (C+D)
		Sensitivity $\frac{TP}{TP+FN}=\frac{A}{A+C}$	***Specificity*** $\frac{TN}{FP+TN}=\frac{D}{B+D}$		
		All diseased patients (A+C)	All non-diseased patients (B+D)		

Table 3. Summary of equations.

Here are some other terms sometimes used to attempt to communicate the clinical performance of tests:

False Positive Rate. The false positive rate or **fall-out rate** tells us what proportion of healthy patients had a false alarm.

$$False\ Positive\ Rate = 1 - Specificity$$

In our sample data, 98.4% of healthy patients had a negative test result, so 1.6% of healthy patients had a positive test result. They had a false alarm.

False Negative Rate. It follows then that the false negative rate is the **miss rate** and tells us what proportion of patients with the disease were incorrectly identified as not having the disease:

$$False\ Negative\ Rate = 1 - Sensitivity$$

In our sample data, 76.9% of diseased patients had a positive test result, so 23.1% of diseased patients had a negative test result, in which case the diagnosis was *missed.*

Since both the false positive and the false negative rates are based on sensitivity and specificity, they don't help us understand what a test result means for a patient whose pretest probability of having the disease is different than the prevalence of the disease in the study.

False Discovery Rate. Recall that *precision* was equal to the positive predictive value (PPV) and that value told us what portion of patients who tested positive actually had the disease. The false discovery rate tells us what percentage of patients who tested positive didn't have the disease. Thus,

$$False\ Discovery\ Rate = 1 - Precision$$

In our sample data, 55.6% of patients testing positive actually had the disease, so 44.4% of people with a positive test result did not have the disease.

Accuracy. This term is commonly used to market tests and simply describes what proportion of test results were accurate; that is, what percentage of the time did the test result match the true condition of the patient.

$$Accuracy = \frac{True\ Positives + True\ Negatives}{Total\ number\ of\ patients}$$

In our sample data, the accuracy was (10+479)/500 = 97.8%.

But note how incredibly misleading this term is and why it is largely irrelevant in clinical practice. Before we administer the test, we know there is a 97.8% chance that its results will be accurate. However, *once we administer the test*, we have a patient sitting in front of us with either a positive or negative test result, and now the relevant facts are these:

- If the test result were positive, there is a 55.6% chance the patient actually has the disease (PPV/ precision).
- If the test result were negative, there is a 99.4% chance the patient does not actually have the disease (NPV).

Of course, this is assuming the patient's pretest probability of disease is approximately equal to the prevalence of disease in the sample that was tested – we will see more about this later.

In other words, while it's technically true that the test will be accurate 97.8% of the time, in practice this value means little. The PPV (55.6%) and the NPV (99.4%) are the most clinically useful results. But since most results are negative, then the *accuracy* of the test (which lies between the PPV and the NPV) is very close to and approaching the NPV. Companies exploit this fact in marketing a test, claiming a high overall accuracy rate that is of little benefit to a patient with positive test results.

Imagine we have a test that looks for a rare disease with a prevalence of 1/1000. Our test correctly identifies 998 of 999 patients who don't have the disease. It also identifies the one person who has the disease incorrectly, with a false negative result. In other words, it was wrong just two times out of 1,000 results and therefore would have an accuracy of 99.8%. Still, it failed to identify the one person with disease. With rare diseases (or diseases with a low *pretest probability*) most tests will be "highly accurate." But this is usually clinically meaningless.

Note: The popular mnemonics SPIN and SNOUT that medical students use to remember specificity and sensitivity are misleading. SPIN states that a highly SPecific test (when positive) rules IN a disease, while SNOUT states that a highly SeNsitive test, when negative, rules OUT a disease. This is misleading because both sensitivity and specificity influence the utility of a test to predict or exclude a disease, as we will see next. Unfortunately, it is rare to see a review book or medical school biostatistics class which doesn't perpetuate this falsity.

There are two additional terms used in the literature to communicate information about test results:

Positive Likelihood Ratio (or LR+). This is the probability of a person who has the disease testing positive (the sensitivity or true positive rate), divided by the probability of a person who doesn't have the disease testing positive (the false positive rate):

$$LR+ = \frac{Sensitivity}{1 - Specificity} = \frac{True\ Positive\ Rate}{False\ Positive\ Rate}$$

In our sample data, LR+ = (.769)/(.016) = 48.

Negative Likelihood Ratio (or LR-). Similarly, this is the probability of a person who has the disease testing negative (the false negative rate), divided by the probability of a person who does not have the disease testing negative (the specificity or true negative rate).

$$LR- = \frac{1 - Sensitivity}{Specificity} = \frac{False\ Negative\ Rate}{True\ Negative\ Rate}$$

In our sample data, LR- = (0.231)/(0.984) = 0.23.

Notice that all the characteristics we have considered so far, other than PPV and NPV, are **independent of the prevalence.** We know what the prevalence of the disease in the 500 patients we studied is, but this might not be the actual prevalence in the population at large or in the population of patients that we see, or it may not be the probability that the individual patient sitting in front of us has the disease.

The most useful descriptors clinically of the test that we have learned about so far are the PPV and the NPV. These tell us what we need to know when the probability of the disease is equal to the prevalence of the disease in our study. But, unfortunately, they do not tell us how the test will perform in the real world. Clinically, when we order a test, the questions we should ask ourselves are, first: *How likely is it that my patient has the disease before I order this test?* Secondly, after the test result comes back: *How likely is it that my patient has the disease now that I know the result of this test?*

These questions relate to the **pretest probability** and the **post-test probability.** This process is the fundamental process that should be utilized when making clinical and diagnostic decisions, and is part of what we have called **Bayesian inference**.

Pretest and post-test probability

The chief purpose of clinical medicine is to determine a patient's pretest probabilities for the diseases on the differential diagnosis. All the rest of the diagnostic process follows from this assessment.

So, how do we determine the **pretest probability** of a given disease?

- It may be the known *prevalence* of the disease in a population; for example, the rate of breast cancer in white women in their 40s, or the rate of HIV in patients who have tuberculosis, or the rate of fetal Down Syndrome in 40-year-old pregnant women.
- It may be based on the results of a previous test, that is, the post-test probability of a test already performed.
- It may be a rough estimation based on clinical experience and the history and physical. This is often very imprecise, and may be divided into *low, medium,* or *high probability*. It can be more precise with the aid of clinical decision tools, such as the Centor Criteria for Strep throat. The reliability of these clinical estimates improves the more we learn about the relative risks of various risk factors for the disease.

The pretest probability is essential to consider routinely in the clinical setting. It informs a variety of decisions:

- Selecting the correct test to support a clinical diagnosis
- Determining whether a test is worth ordering at all (the **test threshold**)
- Deciding whether to start therapy without further testing (the **treatment threshold**)
- Interpreting a test result, that is, determining the post-test probability that the patient has the disease of interest and whether this is sufficient to make a diagnosis (the **diagnostic threshold**)

Once we know the pretest probability and we have decided that a test is likely to be useful, we must then determine the post-test probability after we receive the test result.

How do we determine the **post-test probability**?

In some cases, we can determine the post-test probability by using the positive predictive value of the test in published literature. If the pretest probability of our patient is the same as the prevalence of the disease noted in the published study (for our sample data, the prevalence was 2.6%), then the post-test probability that a positive result means the patient has the disease is equal to the published PPV (which was 55.6% for our sample data).

Such a situation might occur where the inclusion criteria of a study closely match your real-life patient. For example, if a study determined the performance of a given test for myocardial infarction in a population of white, male smokers with a history of coronary artery disease who present with typical chest pain to the emergency department, and you have a patient just like that in the emergency department, then the PPV of that test is likely very close to the post-test probability of that test for myocardial infarction for your patient.

However, it is rare in the real world that the patient is front of you has the same probability of having the disease as the prevalence used in the study. This is even more true when one realizes that most studies are carefully designed to make the test look good by selecting a patient population which is at particularly high risk for the disease in question. You cannot apply, for example, the positive predictive value of a fetal nonstress test determined in a study of pregnant women with type 1 diabetes and hypertension to your patient with diet-controlled gestational diabetes.

Therefore, the more useful method of calculating the post-test probability is to use following formula:

$$\begin{aligned}&\textit{Positive post-test probability}\\&= \frac{(\textit{Pretest Prob})(\textit{Sensitivity})}{(\textit{Pretest Prob})(\textit{Sensitivity}) + (1 - \textit{Pretest Prob})(1 - \textit{Specificity})}\end{aligned}$$

In our sample data set, we can use the prevalence of the disease as our pretest probability, which was 2.6%. Using sensitivity = 76.9% and specificity = 98.4%, we find the positive post-test probability is

56%, identical to our calculated positive predictive value of 55.6% after rounding. So, apart from rounding errors, the PPV in the study is equal to the positive post-test probability when the pretest probability of our patient having the disease is roughly the same as the prevalence of the disease in the study.

What if we did not assume that the pretest probability of the disease was equal to the prevalence in the study? **What if, instead, we assumed that the patient had a 60% pretest probability** (compared to a prevalence of 2.6% in the study) based on risk factors, history, physical exam, and other preliminary evaluations? Then, if you work through the math, you'll find that **the positive post-test probability that our patient has the disease is 98.6% when the test comes back positive.** Thus, the same test has much more clinical predictive value if the pretest probability is high compared to when it is lower.

To understand intuitively where this formula comes from, consider making a new hypothetical sample chart based on the idea of a disease prevalence equal to 60%. To make the calculations easy, suppose we give the test to 10,000 patients:

	Has Disease	*No Disease*	*Totals*
Positive Test	True Positive (TP)	False Positive (FP)	
Negative Test	False Negative (FN)	True Negative (TN)	
Totals	6000	4000	10000

The sensitivity and specificity do not depend on prevalence, so we can still use 76.9% and 98.4% for these values. This tells us the number of True Positives would be (0.769) x (6000) = 4614 and the number of True Negatives would be (0.984) x (4000) = 3936. We can then subtract to find the number of False Negatives and False Positives:

	Has Disease	*No Disease*	*Totals*
Positive Test	True Positive (TP) 4614	False Positive (FP) 64	4678
Negative Test	False Negative (FN) 1386	True Negative (TN) 3936	5322
Totals	6000	4000	10000

So now, we can see the Positive Post-test Probability is (number of true positives) / (total number of positive tests) which is 4614/4678, or 98.6%, exactly the value we found using the formula above.

But, what if the pretest probability instead were lower, say 1 chance in 1000? In this case, if you do the math, the positive post-test probability is only 4.6%! So, the same test, depending on how likely you believe it is that your patient has the disease, performs dramatically differently. A positive result for one patient is a near guarantee that she has a disease while a positive result for another means almost nothing. We can show this concept graphically with probability trees:

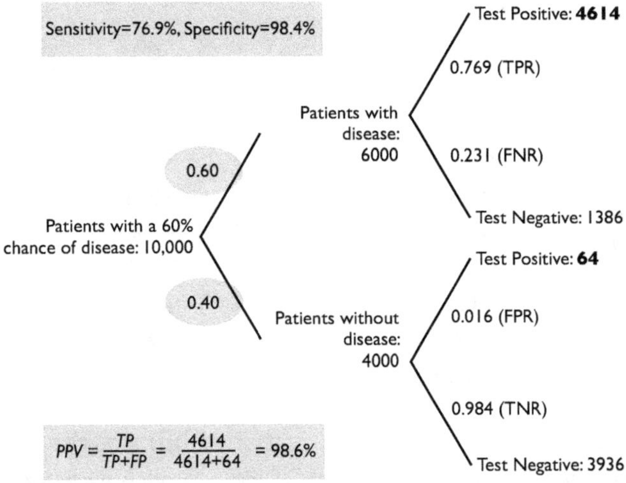

Figure 12. Probability tree with 60% pretest probability using example test performance.

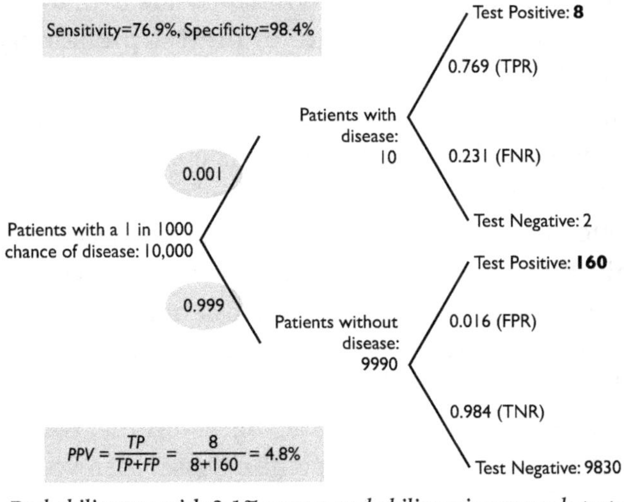

Figure 13. Probability tree with 0.1% pretest probability using example test performance.

Here is the incredible importance of Bayesian inference in clinical reasoning: The same test, with the same performance characteristics, has two entirely different meanings (post-test probabilities) when applied to patients with different pretest probabilities.

To emphasize this point even further, here's a chart summarizing various values of pretest probability and post-test probability, using the example above (the values we've already calculated are **bolded**):

Pretest Probability	*Post-test Probability*
0.1%	**4.6%**
1%	33%
2.6%	56%
5%	72%
10%	84%
20%	92%
30%	95%
40%	97%
50%	98%
60%	**99%**
70%	99%

Table 4. Pretest probabilities compared to post-test probabilities for sample data.

Notice the wide variation in post-test probability as the pretest probability increases from 0.1% to just 10%: a positive test result could mean anything from a 4.6% chance of disease to an 84% of disease. **This example should make it abundantly clear that interpreting test results without any estimate of pretest probability is virtually meaningless!**

Probability trees as shown in Figure 12 and Figure 13 can be very intuitive ways of understanding how tests apply to real patient populations and how important the pretest probability of having the disease is. I often find myself drawing these out on pieces of scrap paper in the clinic. All you need to fill one out is the sensitivity and specificity of the test you are using plus some idea of the pretest probability of the disease. If you are using a screening test, then the pretest probability will be the incidence of the disease under consideration in a population like your patient. If it is a clinical test that you are ordering for a patient suspected of having some disease, then you will have to use an estimate of your clinical suspicion that the patient has the disease, which we will discuss in detail in later chapters. Don't get caught up in an exact number for these clinical estimations. I often just think in terms of low, intermediate, and high with numbers like 10%, 50%, or 90%.

Here is a how a probability tree works:

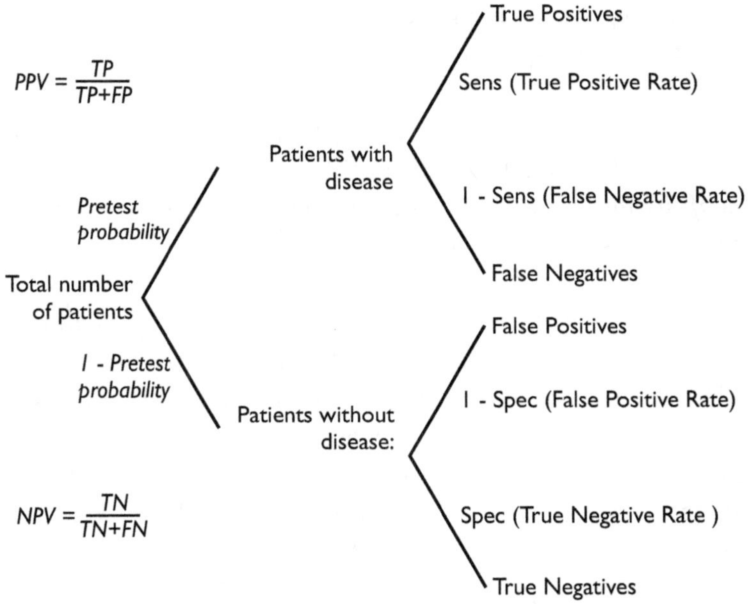

Figure 14. Blank probability tree.

It is simple multiplication then to determine the values at each branch point of the tree. You could do this with a 2 x 2 table as well, but for some, the tree is more intuitive. More importantly, the probability tree graphically shows the importance of pretest probability.

What should we take away from what we've learned about the statistics describing the performance of clinical tests?

When tests are arbitrarily applied to patients without knowledge of the pretest probabilities of the condition being tested for, then serious diagnostic mistakes can and do happen. For example, let's say that you order a "screening" CBC on an asymptomatic patient; if the test reveals some abnormality, it will be very unlikely that the result has any clinical meaning (there was a low pretest probability that the patient had a disease). Tests perform differently when applied to different populations or to patients with different presentations.

I have found that clinicians struggle with this idea almost more than any other concept in medicine. It seems so non-intuitive. We will see this mental struggle again when we use the same statistics to talk about interpretation of scientific studies, where the hypothesis being tested is analogous to some disease and the clinical study is analogous to some clinical test; we will see that the pretest (or prestudy) probability of the hypothesis is incredibly important.

Let me try an example to show why all this is true:

Imagine that we develop a test to determine if a man is a hipster or not. Our test is, "Is he a white male with a full, thick beard?" We determine the sensitivity and specificity for this by sampling 10,000 men randomly across the country. We find that our test is 85% sensitive and 90% specific for being a hipster.

From this, we can determine the PPV and the NPV and all the other characteristics that we have looked at so far. But we need to use this test clinically and understand how to interpret a positive result when we get one. Will the post-test probability for this test be the same if we apply it at a Starbucks in Seattle as compared to a bar at a lumberjack camp outside of Juneau, Alaska? Clearly not. This reflects the fact that even though we can determine the true prevalence of being a hipster, nevertheless hipsters are not distributed equally throughout America. We do not use tests in a vacuum. Data must be interpreted in context. We use tests with some prior knowledge and that prior knowledge (pretest probability) affects the test's meaning.

We will discuss later the process for deciding which test to order and how to interpret it; but these statistics are the bedrock of that decision-making process.

Relative risk versus odds ratio

Every medical student at some point has been taught the difference between relative risk (RR) and an odds ratio (OR). Yet, these statistical terms are confused and misused every day while creating and interpreting scientific literature (which we'll talk more about later if you get bored by the math). They are unfortunately assumed to be the same thing and this can lead to significant misunderstanding.

Odds are derived from the world of gambling and are simply an expression of how many times an event is likely to occur compared to how many times the event is not likely to occur. If you roll a dice, there are 6 possible outcomes. The odds of rolling a 3 is 1:5 (or 1 to 5) because there is only one 3 on the dice and there are five other options. This 1:5 could be expressed as 0.2.

The **probability** of rolling a 3 is 1/6 since there is one 3 among the 6 faces of the dice. This 1/6 can also be expressed as 0.167.

From this example, it should be apparent that odds and probability are not the same thing. Odds are not intuitive. When someone asks, *What are the odds?* they usually mean *What is the chance?* or the probability. We use probability to talk about chance or risk. We use odds to gamble.

This gets more confusing when we compare odds or risks. We can compare two sets of odds with an **odds ratio (OR)** and we can compare two sets of risks or probabilities with the **relative risk (RR)**. Since odds are not intuitive, the odds ratio will also not be intuitive. Risk or probability is more intuitive and in the same way relative risk seems to be more intuitive. The OR and RR are not the same thing, yet we tend to substitute them for one another, as we will see.

An odds ratio then is simply the comparison of two odds,

$$Odds\ ratio = \frac{Odds\ (A)}{Odds\ (B)}$$

The relative risk (RR) is simply the comparison of two risks or probabilities,

$$Relative\ risk = \frac{Probability\ (A)}{Probability\ (B)}$$

Relative risk is sometimes referred to as the **risk ratio,** analogous to the odds ratio.

Let's look at this graphically. In Figure 15 is an experiment where we are attempting to pick a black card. In group A, there are five cards, one of which is black. In group B, there are also five cards, but now three cards are black. The difference in the groups could be anything you like. Think of an intervention like a surgery or a medication that changes the number of black cards; or think of a risk factor that is different in the two groups. It doesn't matter. But imagine two groups, A and B, and we are interested in picking the black card or outcome:

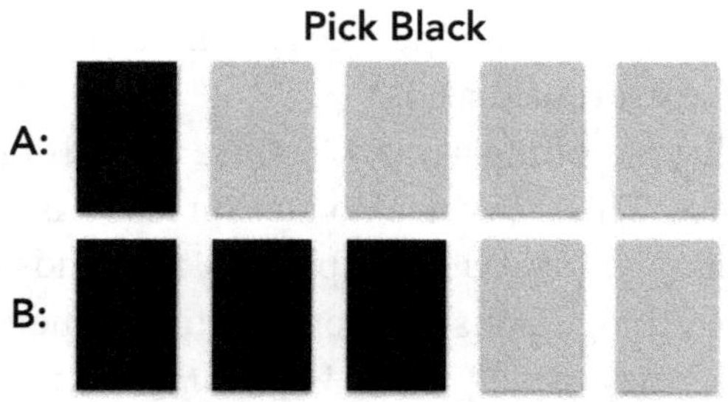

Odds of Black(A) = 1:4 (1 Black:4 Gray) or 0.25
Odds of Black(B) = 3:2 (3 Black:2 Gray) or 1.5
Probability of Black(A)= 1/5 or 0.2
Probability of Black(B) = 3/5 or 0.6
Relative Risk (A/B) = 0.2/0.6 or 1/3 or 0.33
Odds Ratio (A/B) = 0.25/1.5 or 1/6 or 0.16

Figure 15. Odds and probabilities in lower probability events.

So, we see that odds makes sense when we say there is a 1 to 4 chance of picking a black card in A versus a 3 to 2 chance of picking a black card in B. Yet odds don't make sense when we convert those odds to 0.25 odds and 1.5 odds.

Probability makes wonderful sense, however. There is a 20% chance of picking black in A and a 60% chance of picking black in B. We can all understand this intuitively. Also, we can relate odds back to probability if we think about it. Odds of 1:4 means there are five possible outcomes. If you are interested in the black card, it happens once among those five outcomes, or 20% of the time.

Likewise, relative risk makes a lot of sense. The relative risk of picking a black card in group A compared to group B is 1/3 or 0.33. We can understand that by looking at the picture. The relative risk of picking a black card in group B compared to group A is 3. Stated differently, you are 3 times as likely to pick a black card in B as you are in A. This makes perfect sense.

But the odds ratio makes no sense (to non-statisticians). The odds ratio of picking black in A relative to B is 0.16. *What!?* Exactly. The odds ratio of picking black in B relative to A is 6. People who misunderstand odds ratio might then state that you are 6 times as likely to pick black in B relative to A. Yet we understand intuitively (and from the RR) that you are only 3 times as likely to pick black in this situation, not 6 times as likely.

It should be obvious that you cannot substitute ORs for RRs in your thinking. You cannot say with an OR of 6 that you are 6 times as likely to pick a black card in situation B. Let's look at picking gray to see how the numbers change for more likely events:

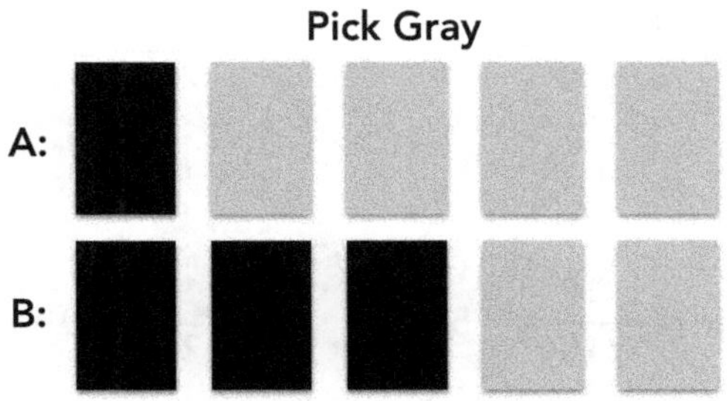

Odds of Gray(A) = 4:1 (4 Gray:1 Black) or 4
Odds of Gray(B) = 2:3 (2 Gray:3 Black) or 0.66
Probability of Gray(A)= 4/5 or 0.8
Probability of Gray(B) = 2/5 or 0.4
Relative Risk (A/B) = 0.8/0.4 or 2
Odds Ratio (A/B) = 4/0.66 or 6

Figure 16. Odds and probabilities in higher probability events.

In this case, the numbers seem even more exaggerated. We understand intuitively that you are twice as likely to pick gray in A as in B; and this is conveyed by the RR of 2. But the OR is even more misleading since it is 6. In B, you are as half as likely to pick gray as in A, or a RR of 0.5, but the OR for this event would be 0.166!

Thus, you should never use an OR to imply the magnitude of risk.

It is true, however, that, in many cases, the OR is approximately (though never exactly) equal to the RR. This is true for rare events. But what does rare mean? Let's look at another example.

If an event happens in a patient 1/1000 times with an intervention but 2/1000 times without an intervention, then the RR for the intervention group would be:

$$RR = \frac{1/1000}{2/1000} = \frac{1}{2} = 0.5$$

While the OR would be:

$$OR = \frac{1/999}{2/998} = 0.4995$$

In practical terms, these are the same number. How rare is rare? Let's do the same math for events happening on the scale of hundreds and tens:

For the n per 100s:

$$RR = \frac{1/100}{2/100} = 0.5 \qquad OR = \frac{1/99}{2/98} = 0.495$$

For the n per 10s:

$$RR = \frac{1/10}{2/10} = 0.5 \qquad OR = \frac{1/9}{2/8} = 0.444$$

Still not much of a difference, but also imprecise to use the OR result in the sentence, "You are x times as likely to experience the event if you have the intervention."

If the OR < 1, it will always be smaller than the RR. Conversely, if the OR > 1, it will always be larger than the RR by some amount. To demonstrate a case where the OR > 1, let's look at the inverse of that last problem:

$$RR = \frac{2/10}{1/10} = 2 \qquad OR = \frac{2/8}{1/9} = 2.25$$

Again, 2 and 2.25 are very close, but it is clearly not fair to say that something is 2.25 times as likely to happen in this experiment. The more common the problem, or the greater the effect, the more misleading an OR might be. For example, let's imagine that our study relates to obesity (a common problem). Pretend that if you eat below a certain number of calories per day, you have a 1/15 chance of being obese, but if you eat above a certain number of calories per day, you have a 1/3 chance of being obese. Here are the calculations:

$$RR = \frac{1/3}{1/15} = 5 \qquad OR = \frac{1/2}{1/14} = 7$$

So, these numbers are different enough to be misleading (you are not 7 times as likely to be obese, only 5 times as likely) but close enough that the average non-statistician (that is, doctors, reporters, politicians, patients, other researchers) will assume that 7 times as likely makes sense.

These are very basic concepts, but unfortunately misuse of these terms has become common in research articles, textbooks, journalism, and clinical practice. In 2001, Holcomb and colleagues studied the misuse of the OR in OB/GYN literature published in *Obstetrics and Gynecology* and the *American Journal of Obstetrics and Gynecology* in the years 1998-1999.[37] They studied 107 articles and found the following results:

- 44% of the articles published ORs that were more than 20% different than the appropriate RRs, almost all magnifying the reported effect.
- In 26% of the articles, the ORs were explicitly interpreted by the authors as a RR without justification, meaning that the authors stated, "There is an X-fold increased risk..." Only one study clarified that this was done and they explained to the readers the inherent assumptions they were making by using the OR as a RR.
- In one study on the familial occurrence of dystocia, the authors stated, "The risk is increased by more than 20-fold (odds ratio 24.0, 95% confidence interval 1.5 to 794.5) if one sister had dystocia..."[38] The baseline risk was 11% and the actual RR was 6.75.

Articles have been published in almost every scientific field examining the inappropriate use of ORs in the literature, and the conclusions are striking.[39] Grimes and Schulz (in an excellent article) have argued that the use of ORs should be limited to case-control studies and logistic regression, where they are necessary mathematically.[40] Unfortunately, the incidence of publication of ORs is increasing and so too are the false conclusions sometimes based on them. Often, error begets error and other derivative articles and guidelines make incorrect assumptions based upon distorted risks.

Risk distortion pervades our daily life. Our decision to do anything in life is affected by our perception of the risks associated with the activity. Both patients and providers tend to estimate risk incorrectly, either because of a lack of knowledge about the risk or often because absolute and relative risk are confused. We also tend to place more emphasis on risks that are perceived as more drastic, which leads to over-emphasizing catastrophic risks and under-emphasizing more common and statistically meaningful risks and benefits.

Absolute risk and relative risk

Absolute risk is the chance that you will experience an outcome over some period of time. **Relative risk** is the chance that you will experience an outcome compared to someone else, based on individual risk factors or circumstances.

The ***absolute risk*** that you might develop, say, melanoma, is 1 in 50 in your lifetime, or 2%. If you have had 5 or more sunburns in your lifetime, your risk of melanoma is 80% higher than someone who has not had 5 or more sunburns. So, your ***relative risk***, if you have had 5 or more sunburns, is 80% higher *relative* to someone who has not had the sunburns.

If a person who hasn't had 5 or more sunburns has a 2% risk of melanoma (it's actually somewhat lower, as the 2% figure cited above includes everyone), then the *absolute* risk of someone who has had 5 or more sunburns would be about 3.6%. We could also compute the risk ratio as described above: risk ratio = 3.6% / 2% = 1.8, so we could say you are 1.8 times as likely to develop melanoma if you have had 5 or more sunburns.

It is very useful to think about relative risk when assessing the impact of various health determinants, but it is essential to think clearly about and communicate clearly to our patients the difference between absolute risk and relative risk. Most patients are alarmed to hear a figure like 80% in the same sentence as "cancer," and while most people understand the statement doesn't imply an 80% risk of cancer for those with at least 5 sunburns, there's still a strong psychological effect. Of course, in context it is appropriate for our patients to understand that a fairly simple and inexpensive measure (e.g., preventing sunburns with sunscreen) can significantly reduce the risk of melanoma, but it is still important to communicate that the absolute risk of melanoma with or without sunburn remains fairly low (especially compared to other major preventable health threats, like cardiovascular disease.)

Relative risk is often used in alarmist headlines reporting the latest research finding about some protective or dangerous health determinant, and the disproportion of the sense of alarm is only magnified when the condition discussed has an extremely small absolute underlying risk to begin with. For example, a headline may enthusiastically tout a new study showing that eating kale every week reduces risk of a certain condition by an astounding 200%; but, if the underlying absolute risk of that condition were very low (say, 0.001%), then the reduction in absolute risk might not really be worth the intervention (especially for someone who can't stand the taste of kale.)

Every risk reduction is attached to certain costs – albeit sometimes very small costs, like putting on sunscreen or wearing a seatbelt or helmet. Many risks undertaken have certain benefits – like the joy of riding a motorcycle or paragliding. These activities have increased relative risk of head injuries, but the benefits may outweigh the increased risk. It is our job as physicians and patient educators to help our patients understand absolute and relative risks in a way that is both accurate and meaningful in the context of their personal values. We will discuss this more in Chapter 17.

Another source of confusion in understanding how risk relates to the use of conditional probability and total probability. Recall from Chapter 1 that conditional probability means the probability that one event occurs assuming that another event occurs. The probability of event A, given that event B occurs, is written $P(A|B)$ or P(A given B).

Someone who is more afraid of flying than driving may be influenced by the perception that death is more likely in plane crashes than in car crashes. Well, it is in a sense: you are much less likely to survive a plane crash than a car crash – although, surprisingly, 95.7% of people do survive plane crashes.[41] Stated in terms of conditional probability, we could say that the risk of dying in a plane crash, assuming a crash occurs, is much greater than the probability of dying in a car crash, assuming a crash occurs. However, the more relevant consideration is the *absolute* risk of dying in a car crash vs. a plane crash, not the risk of dying assuming that a crash occurs. Because plane crashes occur much less often than car crashes, the *absolute* risk of dying in a plane crash is much lower, leading to the conclusion that flying is safer than driving.

So, what about a clinical example that affects patients and physicians? There are countless numbers. Here is one example.

What is the risk of having an ectopic pregnancy with various forms of birth control? I have seen many women who have declined or at least been leery of an intrauterine device (IUD) because of the increased risk of ectopic pregnancy that they have been warned about online or even by other doctors. The following shows the risk of having an ectopic pregnancy **IF** you get pregnant while using each of the following methods (in other words, the conditional probability of ectopic pregnancy, given that you conceive):

- No birth control – 2% risk of ectopic
- Birth control pills/patch/ring – 3.5% risk
- Levonorgestrel IUD – 10% risk
- Copper IUD – 5% risk
- Tubal occlusion/ligation – 25% risk

(Note that these are rough estimates; surprisingly there are no good data to inform these statistics.)

Notice what happens when we compute the relative risk of a Levonorgestrel IUD compared to no birth control, assuming in each case that pregnancy has occurred:

$$Relative\ risk = \frac{P(Ectopic \mid Pregnancy\ with\ IUD)}{P(Ectopic \mid Pregnancy\ with\ no\ IUD)} = \frac{10\%}{2\%} = 500\%$$

Similarly, we see that the Copper IUD has a 250% relative risk of ectopic compared to no birth control, assuming a woman conceives. Given these very significant relative risks, should women be counseled that they are at an increased risk of ectopic pregnancy if they choose to get an IUD? Should OBs encourage the use of other forms of birth control, to avoid this potentially deadly complication?

On the contrary; what matters is the total (absolute) risk of ectopic pregnancy for each method, which must also take into account the probability of getting pregnant in the first place. Here are the one-year risks of pregnancy for each method:

- No birth control – 85%
- Birth control pills/patch/ring – 8.4%
- Levonorgestrel IUD – 0.1%
- Copper IUD – 0.8%
- Tubal occlusion/ligation – 0.3%

Obviously using no birth control results in a very high chance of getting pregnant, but patients may be unaware that their chance of conceiving while on the pill is much higher than with an IUD. To compute the absolute risk of ectopic pregnancy with each method, we must multiply the probability of getting pregnant by the probability of ectopic given a pregnancy, for each birth control choice:

- No birth control 0.85 x 0.02 = 17/1,000 women
- Birth control pills/patch/ring 0.084 x 0.035 = 2.94/1000 women
- Levonorgestrel IUD 0.0015 x 0.10 = 0.15/1000 women
- Copper IUD 0.008 x 0.05 = 0.4/1000 women
- Tubal occlusion/ligation 0.003 x 0.25 = 0.75/1000 women

This example illustrates the importance of understanding absolute risk! Earlier we stated that, assuming conception has occurred, the Levonorgestrel IUD has a five-fold relative risk of ectopic pregnancy compared to no birth control. However, when we assess absolute risk of ectopic – the statistic that actually matters – we see the relative risk of ectopic with the Levonorgestrel IUD is 0.15/17 or 0.0088 compared to no birth control.

Stated another way, *if you do not use birth control your overall risk of ectopic pregnancy is approximately 113 times higher than if you have a Levonorgestrel IUD*. This fact should offer enormous reassurance to women who are considering an IUD but who might be concerned about ectopic pregnancy. (Again, most of the increased risk associated with 'no birth control' is simply because the chance of pregnancy is astronomically smaller with an IUD; if you DO get pregnant with the IUD, the chance of ectopic pregnancy is increased.) It is also worth noting that your overall chance of ectopic pregnancy using birth control pills is roughly

20 times as high as your chance of ectopic while using the Levonorgestrel IUD. IUDs are the safest and most effective birth control on the market today, and physicians must understand and be able to communicate these risks appropriately to help patients make informed choices.

To interpret data in ways that are meaningful for our patients, we need to focus on the statistics that matter, like number needed to treat (NNT), absolute risk reduction (ARR), and absolute risk. We also cannot be confused by odds ratios and other clever things authors do to distort data. We will talk about communicating risks and benefits to patients in Chapter 17.

Exercises

5.1 Using the data set available at the QR code, calculate the following values:

a. Sensitivity, specificity, false positive rate, false negative rate, and accuracy.
b. Prevalence (in sample data), positive predictive value, negative predictive value, false discovery rate, positive likelihood ratio, and negative likelihood ratio.
c. Suppose a patient in your office has just received a negative test result. Which of the values in part (b) would be the most relevant to share in discussing the test result with your patient? What consideration should you have made regarding your patient in comparison to the sample data before administering the test?
d. Suppose a patient has no known risk factors for the disease, but this test has been administered as part of a larger screening panel. The patient receives a positive test result and is very concerned, commenting "I read that this test is 95% accurate." How would you advise the patient regarding the test result? (Think about the specific wording you would use.)
e. Suppose your patient has a first-degree relative with this condition, which leads you to estimate his pretest risk of disease as 10%. What would this individual's positive post-test probability be? Negative post-test probability? (Hint: you may use an app, use the formula stated in the chapter, or fill in the following

chart using the specificity and sensitivity you calculated above.) Would a positive test result be more or less valuable for this individual than for a member of the general population?

	Disease	No Disease	Total
Positive			
Negative			
Total	1000	9000	10000

5.2 Consider examples of risk in your area of interest. What are three examples in your area of interest where relative risk and absolute risk are potentially confused?

5.3 Select a recent study from a journal you read. Did the authors use odds ratios or relative risk? Was the choice correct? Did the conclusions appropriately utilize the term or make a false inference?

6

The Differential Diagnosis

The aim of medical care is to improve the patient's health, prolong life, increase the quality of life, treat disease, and relieve symptoms. Proper diagnosis isn't essential, but it sure helps. Without the correct diagnosis, we cannot accurately select the best interventions which provide the most benefit weighed against the least risk.

As discussed already, diagnoses are frequently wrong. Wrong diagnosis happens because of a lack of relevant clues, failure to consider the appropriate diagnosis, or cognitive biases that steer us towards inappropriate conclusions. There are other reasons, too, but these are the most important ones and the one over which we have the most control.

A good differential diagnosis helps with all three. First, if conducted before the patient interview and the information gathering process, it can guide us to look for the correct clues and information and to ask the right questions. Second, since we can't make a diagnosis that we don't think of first, it's important to have an appropriately broad differential diagnosis. Third, the differential diagnosis can help us combat cognitive bias.

Indeed, the differential diagnosis, used appropriately, can serve as one of the most powerful debiasing tools in our armamentarium. When we interview a patient, review records, perform physical exam, etc., we begin to focus on certain areas and eliminate other things from our consideration. Much of this is good but some of it is due to inappropriate bias. For example, consider these three important cognitive biases in a patient who presents with a complaint of pelvic pain:

- **Framing effect:** The patient tells us initially that her ovary is hurting, and that the pain is similar to when she had an ovarian cyst before. The patient (or family member, referring physician, etc.) has framed the condition for us and potentially made us blind to other possibilities.
- **Triage cueing:** Since the patient was referred by her primary care physician to a gynecologist for evaluation of her pelvic pain, we tend to focus only on gynecologic sources of pain (or a GI doctor might only focus on GI causes of pain, etc.).
- **Anchoring:** The patient tells us in the first few sentences of the interview that she has right lower quadrant pain, a low-grade fever, and nausea; we immediately become worried about appendicitis even after we learn that she does not have anorexia and also does have a three-day history of diarrhea. Anchoring is the tendency to lock onto the salient features of a diagnosis too early and not modify our theory as new data arrives.

There are many other types of bias important in this scenario as well, which we discussed in Chapter 3. One tool that can be used to avoid these and other biases is the differential diagnosis. Traditionally, the differential diagnosis is developed after the history has been obtained, records reviewed, and perhaps even after the physical exam has been performed; then the differential is narrowed down with further testing. But delaying the creation of the differential diagnosis until after information gathering opens the door to many of these biases which may lead to an incorrect diagnosis (or no diagnosis).

However, if the differential diagnosis is made at the earliest opportunity, there is less chance that we will be biased by what we learn. Practically speaking, this means making the differential diagnosis after learning of the chief complaint, then using the history, physical, etc., to arrange the differential diagnosis in order of most probable to least probable. If you have never done this consciously, you'll find that when you do you will ask different (and more meaningful) questions when taking the history than when you form your differential diagnosis later in the process. You will ask questions and seek information relevant to the diagnoses on your list.

As an aside, sometimes it is beneficial to organize a differential diagnosis in order of most dangerous to least dangerous rather than most probable to least probable; this sort of approach is often used in emergency departments. However, since diagnosis is a Bayesian process, arranging the list according to the probability of each diagnosis is more helpful. We can still note on that list which diagnoses can't be missed due to their importance. Thinking of diagnoses in terms of probability (at least categorizing them as likely, indeterminate, and unlikely, or perhaps high probability, medium probability, and low probability) helps us with selecting the right tests and the appropriate next steps in management since the probability of the diagnoses on our list will serve as the pretest probability for any test we order.

Our biased thought processes (System 1 thinking), of course, serve us well for most patients. *"Common things are common and rare things are rare." "When you hear hoofbeats, think horses, not zebras."* But as nicely sounding as those axioms are, it doesn't mean that rare things don't exist or that there is no such thing as a zebra. Our System 1 thinking fails the patients who are outliers because they tend to get misdiagnosed and potentially harmed in the process. We won't diagnose something that we never think of, so we need to think broadly and then hone in on the most probable diagnoses as new information becomes available. Therefore, we should think of a broad differential before we learn more about the patient.

In fact, most physicians tend to think of only about 3-5 things on their differential diagnoses. For pelvic pain, a gynecologist might consider only endometriosis, ovarian cysts, pelvic inflammatory disease, urinary tract infection, and uterine fibroids. None of those, by the way, are common causes of female pelvic pain. Because of this, the rates of misdiagnosis (or at least false attribution of symptoms to the wrong diagnosis) are high.

Good doctors make good differential diagnoses. Period.

Constructing the differential diagnosis

There are many ways to make a differential diagnosis, and there is no one way that is always appropriate. Here are some tips. We will consider female pelvic pain as an example.

Anatomy and physiology. Think about the anatomy or physiology related to the problem. Indeed, a good differential diagnosis starts with what you already know about the normal function of the body. Think about how to disrupt this normalcy to cause the symptom. You don't even need to know the name of the disease to do this. For pelvic pain, for example, ask yourself some basic questions:

- **What are the mechanisms that cause us to feel pain?** For example, think about how a noxious stimulus in the pelvis is transmitted to the brain. Thinking this way will lead you to consider problems in other places, like the spinal cord or brain.
- **What anatomy is in the place where the pain is?** For the pelvis, you need to consider each organ (or system) in the vicinity: uterus, ovary, fallopian tube, large and small bowel, bladder, ureter, pelvic floor muscles, abdominal wall muscles, ligaments, etc. And since pain must be transmitted through nerves to the brain, consider the pathways of those nerves all the way to the brain, as well as issues that affect the perception of pain, such as psychiatric conditions.

Pathophysiology. Once we have decided what structures are relevant to the complaint, next we must decide how those structures might be affected. In our example, we might ask, **What types of processes might affect those organs to cause pain?** Pick an organ and apply a process: inflammation, infection, neoplasm, vascular, mass effects, etc. For example, you consider the fallopian tube and then you go through these pathological processes and consider perhaps salpingitis, hydrosalpinx, pyosalpinx, tuboovarian abscess, tuboovarian complex, tubal cancer, tubal torsion, etc.

One useful mnemonic for remembering the relevant processes is **VINDICATE'M:**

- **V**ascular
- **I**nflammatory/**I**nfectious
- **N**eoplastic
- **D**egenerative/**D**eficiency/**D**rugs
- **I**diopathic/**I**ntoxicant/**I**atrogenic
- **C**ongenital
- **A**utoimmune/**A**llergic/**A**natomic
- **T**raumatic
- **E**ndocrine/**E**nvironmental
- **M**etabolic/**M**ass effect

When a learner first considers a new complaint, like pelvic pain, it is useful for her to spend some time thinking about an exhaustive differential diagnosis. This is part of the process by which System 2 thinking will make for better System 1 thinking in the future (packing the parachute correctly). Table 5 shows an example of this exercise using just two organs, *uterus* and *bowel*. Consider each organ or tissue that may be associated with the symptom (in this case pelvic pain) and then think of at least one example of a disease of that organ or tissue for each VINDICATE'M process.

	Uterus	*Bowel*
Vascular	Pelvic congestion syndrome	Mesenteric ischemia
Inflammatory *Infectious*	Endometritis	Diverticulitis, appendicitis, infectious colitis
Neoplastic	Uterine sarcoma	Colon cancer
Degenerative *Deficiency/Drugs*	Degenerating fibroid	Use of anticholinergic medications
Idiopathic *Intoxicant* *Iatrogenic*	Embedded IUD	Radiation colitis
Congenital	Non-communicating horn	Hirschsprung disease
Autoimmune *Allergic* *Anatomic*	Uterine prolapse	Intussusception
Traumatic	Uterine rupture	Colon rupture
Endocrine *Environmental*	Dysmenorrhea	Gastroparesis
Metabolic *Mass effect*	Large fibroids	Constipation

Table 5. Anatomy versus process differential diagnosis table for female pelvic pain.

A learner should also be able to have at least a basic understanding of each of the diagnoses listed. Again, you can't diagnose something you've never heard of or thought about. A physician needs a *diagnosis script* for each diagnosis listed. Eventually, this process will become quick and less formal, but work needs to be done while learning and any time that an unusual presentation or unexplained complaint is discovered.

Multiple symptoms. Patients obviously present with more than one symptom most of the time. Usually, multiple symptoms are explained by one diagnosis, but not always. Consider each symptom independently at first. If a patient presents with right lower quadrant pain, nausea, fever, and anorexia, consider the differential for each of those things. Then investigate the diagnoses that are the same on each list. Obviously, the most probable diagnoses are those which appear on the differential for each independent symptom.

When you start to ask more questions of the patient, you will learn a lot very quickly that will make your list narrow down into a small list of probable diagnoses. We will discuss in the next chapter this process for narrowing down the list through information gathering.

Non-anatomic or process complaints

So far, we have been considering a complaint that is focused around anatomic structures. This technique is useful for complaints like pain or weakness but less helpful for complaints like infertility or anemia. I call these **process diseases**: something has gone wrong with a process, like getting pregnant or maintaining a normal level of hemoglobin.

To make a differential diagnosis for issues like this, consider the normal process in as much detail as possible, and think about ways to disrupt that process. Over the years, I have asked students to give me a differential diagnosis for infertility numerous times, and an individual student will usually list 5 or 6 things (purely from recall) and a small group of students might think of 10. Then I change the question and ask them to tell me the process of how a pregnancy occurs. We learned these normal processes in the first two years of medical school. The description might look like this:

Sperm is made. During sex, it travels out the male urethra eventually. It travels up the female cervix, through the uterus, down the tube and finds an awaiting egg. Fertilization occurs and the fertilized conceptus travels back into the uterus and implants into the endometrium.

I'll then encourage them to think about each step. For example, what conditions must be present for sperm to be made? Now we can apply VINDICATE'M to each small step. Soon, they are considering the vascular, endocrine, traumatic, congenital and other abnormalities that might affect spermatogenesis. I encourage them to ask follow-up questions. If FSH isn't present, what might be wrong with the pituitary gland? Apply VINDICATE'M to the anterior pituitary gland. When a small group of students goes through this process, it's not unusual for them to make a list of over 100 causes of infertility without consulting any references. Eventually, this can be turned into a useful approach to diagnosis that is quick and efficient.

Try this exercise with anemia. Every student has been prompted at some time to consider the differential diagnosis for anemia. Think about the process. Either hemoglobin isn't being made adequately, it's being lost somewhere, or it's being destroyed too fast (or there is a lab error). Now break those processes down in as much detail as you remember based upon your understanding of physiology.

Gaining efficiency

When we first learn about a complaint (e.g., pelvic pain), it is part of the learning process to be as exhaustive as possible in making a differential diagnosis. As I said, if you have never thought about a condition, you will never diagnosis it. This is a System 2 cognitive process. In a busy clinic, you will simplify this differential down to include the most common issues for speed and efficiency (System 1 thinking). It would be impossible to make an exhaustive differential diagnosis for every complaint you see each day. But, of course, most doctors see lots of the same complaints day in and day out. We just need to pack the parachute well and the process will become quick and automatic. We will think automatically of the most probable diagnoses, but in the back of our minds we will still think about zebras when the data that we gather do not fit the diagnoses we are pursuing (i.e., when our diagnosis scripts aren't appropriately triggered).

Our speedy System 1 heuristics will work better if we take the time in the beginning, when learning, to be comprehensive and analytical. We also need to understand how common or rare the diseases on our list are, and how to diagnosis them. Don't worry about treatments; you can and should look up treatments when you make the diagnosis.

Below I will give two differential diagnoses. First, the practical differential diagnosis that I use every day when I see patients with pelvic pain. This list reflects the common and likely diagnoses and it helps me keep perspective of the most likely diagnoses. I also go through this shorter differential diagnosis with the patient to educate her about my thinking. This is a heuristic or a System 1 differential diagnosis, but it is the product of a lot of System 2 work. Then, I will show a more exhaustive differential of things that are still in the back of my mind (my System 2 differential diagnosis).

System 1 differential diagnosis for chronic pain in a non-pregnant woman

- **Gastrointestinal** (about 1/3 of all causes of female "pelvic pain")
    - Constipation
    - Irritable Bowel Syndrome
    - Diverticular Disease
    - Celiac Disease
- **Urologic** (about 1/4 of causes)
    - Interstitial cystitis/Painful Bladder Syndrome
    - Chronic UTI
- **Gynecologic** (about 1/5 of causes)
    - Endometriosis
    - Ovarian cysts
    - Adhesions
    - Dysmenorrhea
    - Adenomyosis
    - Fibroids
- **Peripheral nervous system/musculoskeletal**
    - Abdominal wall hernia
    - Myofascial pain syndrome
    - Pudendal Neuralgia
    - Piriformis syndrome
- **Central nervous system/psychiatric**
    - Untreated depression
    - History of sexual abuse

System 2 differential diagnosis for chronic pain in a non-pregnant woman

- **Extrauterine causes**
  - Adhesions
  - Adnexal cysts
  - Chronic ectopic
  - Chronic PID
  - Endometriosis
  - Endosalpingiosis
  - Ovarian neoplasia
  - Ovarian remnant syndrome
  - Ovarian dystrophy
  - Pelvic congestion
  - Peritoneal cysts
  - Residual accessory ovary
  - Salpingitis
- **Intrauterine causes**
  - Adenomyosis
  - Dysmenorrhea
  - Cervical stenosis
  - Chronic endometritis
  - Uterine/cervical polyps
  - Leiomyomata
  - Pelvic organ prolapse
  - Embedded IUD
  - Congenital uterine abnormality (such as a non-communicating horn)
- **Urinary Tract**
  - Bladder neoplasm
  - Urinary retention
  - Chronic UTI
  - Interstitial cystitis
  - Radiation cystitis
  - Urolithiasis
  - Destrusor dyssynergia
  - Urethral diverticulum
  - Urethral syndrome
  - Urethral caruncle
- **Gastrointestinal**
  - Colon cancer
  - Chronic intermittent bowel obstruction
  - Colitis
  - Constipation
  - Diverticular disease
  - Hernia
  - Inflammatory bowel disease
  - Irritable bowel syndrome
  - Celiac disease
  - Gas colic
  - Lactose intolerance
- **Musculoskeletal**
  - Abdominal wall myofascial pain
  - Chronic coccygeal pain
  - Compression of lumbar vertebrae
  - Degenerative joint disease
  - Disk herniation/rupture
  - Fibromyositis
  - Muscle strains/sprains
  - Neuralgia of iliohypogastric, ilioinguinal, and/or genitofemoral nerves
  - Pelvic floor myalgia
  - Piriformis syndrome
  - Rectus tendon strain
  - Spondylosis
- **Other**
  - Abdominal epilepsy
  - Abdominal migraine
  - Bipolar disease
  - Depression
  - Malingering
  - Somatization
  - Conversion disorder
  - Familial Mediterranean fever
  - Porphyria
  - Shingles
  - Somatic referral
  - Diabetic ketoacidosis
  - Hemochromatosis
  - Lead poisoning
  - Hypercalcemic crisis

Differential diagnosis tools

It is probably obvious why the average physician only includes 3-5 diagnoses in his differential diagnosis for any given complaint that he evaluates: it's just too much work. That's unfortunate. My hope would be that doctors at least do this work for the diagnoses they see commonly. For me, an Ob/Gyn, that might include things like pelvic pain, abnormal uterine bleeding, vaginal discharge, dyspareunia, decreased libido, breast pain, etc. I should be an expert at these chief complaints and have detailed knowledge about each diagnosis that occurs on those differential diagnoses.

But what do I do when someone presents with a complaint I don't treat every day? Something like diarrhea for a week? Well, I look it up, and so should you. Where?

Many times, a simple **Google search** will suffice. A search for "differential diagnosis diarrhea" returns several excellent websites in the first 10 results.

For diagnoses like this that I don't treat every day, I will often search for the diagnosis plus "AAFP" on Google. In this case, A Google search for "AAFP diarrhea" results in two excellent articles, one on acute and one on chronic diarrhea. Neither are protected by a paywall and both include not only excellent differential diagnoses but also all the information needed to evaluate and treat the complaints. This helps me ask the right questions while gathering information and usually I can become an expert in just a couple of minutes.

There are many published indexes of differential diagnoses including many that are constructed for a specific specialty. As a medical student, I carried around my *French's Index of Differential Diagnosis* religiously.

More commonly, a patient presents with several symptoms at once. For example, maybe it is a 45-year-old patient who presents with diarrhea for one week, a fever, abdominal pain, nausea, and vomiting. You could type all of those into a Google search and you might find some meaningful information, but not usually. The classic way of dealing with this situation is to consider each complaint individually and look for diagnoses that occur on each differential diagnosis. Or, consider the differential diagnosis for the chief complaint and then scan the list for diagnoses that also are associated with the other symptoms, though this is not always obvious. If you are experienced with the subject matter, this may have already triggered a "diagnosis script" in your head, and that's how most clinicians would make the diagnosis. However, if you don't have that diagnosis script in your head, or you are uncertain, then this System 1 method doesn't help.

Fortunately, there are several computer-aided diagnostic aids that can help with this. I use an app called *MayaPro* that generates differential diagnoses for you based on any information you can provide. When I put in the symptoms listed above, it asks further qualifying questions that I might not have thought about. Finally, it lists a differential diagnosis for that constellation of symptoms, listed in order of probability, with "can't miss" diagnoses like appendicitis or pancreatitis marked with a red flag. For these symptoms, it gave me 20 diagnoses to consider, with acute gastroenteritis listed as most probable. I can click on each diagnosis and learn more information about it and I can add lab and other findings as they become available to refine the probabilities of each diagnosis. Technology like this is the future of the diagnostic process (**computer-aided diagnosis**).

Exercises

6.1 Pick a common complaint that you see regularly or that you are interested in. Make a differential diagnosis in a spreadsheet using the organ system versus VINDICATE'M methodology. Make a goal of having at least one diagnosis in every cell.

6.2 Use a Google search for "differential diagnosis AND" another complaint of interest. Look at three or more results and use the sites to construct a differential diagnosis.

6.3 Search for "AAFP AND pelvic pain" and look at the resultant article. Note the differential diagnosis listed as well as how other diagnostic information is presented in the article. Try this with another diagnosis you are interested in.

6.4 Consider a process disease differential diagnosis, such as infertility or blindness. First note the normal process by which fertility or sight occurs, for example. Then make a differential diagnosis by disrupting each step of the process in as many ways as you can.

7

Gathering Information

Remember Sherlock Holmes. First, he had a preliminary idea of what crime had been committed (the chief complaint). Next, he developed some theories of the crime based on his extensive knowledge of prior crimes (a differential diagnosis). Then, he started looking for clues (gathering information). He found clues others missed because he knew what he should be looking for when he talked to suspects (patient interview or history), examined the scene of the crime (physical exam), or did forensic tests (labs and imaging).

Thus, it's no accident that the chapter on differential diagnosis comes *before* the chapter on gathering information. I realize that this is a departure from traditional teaching, but it is a necessary one. Recall that in addition to avoiding some cognitive biases that are inevitable during the information gathering process, placing the differential diagnosis earlier in the diagnostic process allows us to gather better information. It helps us know which questions to ask and where to look.

It has been noted that both too little history and too much history are associated with incorrect diagnosis.[42] We need to take the *correct* amount of history, and we will see also that we need to perform the *correct* amount of physical exam. In the same way, we need to order the *correct* labs and imaging. The differential diagnosis, ordered by probability, is our guide to knowing how much is the correct amount.

History of present illness

The history of present illness (HPI) is the main source of narrative information that aids us in making a diagnosis. It has been said (and there is scientific evidence to support the claim) that 80% of diagnoses can be made by history alone. Ask any seasoned physician and she will agree. In fact, most diagnoses by experienced clinicians are made within just a minute or so of the patient interview. This isn't necessarily a good thing.

This happens for a couple of reasons. First, many diagnoses are self-evident, such as a viral illness, a urinary tract infection, or some other minor and self-limited complaint. Because they are self-limited (meaning they usually improve regardless of treatment), even if the diagnosis is wrong, then most physicians don't realize it, nor does the patient. Other times, the diagnosis may be slightly less obvious, but it is likely to be one of the 3-5 things that the physician has thought of in her differential diagnosis and she can quickly pick the right diagnosis with a few questions that distinguish those diagnoses from each other. But this relies on the fact the common things are common. For any given complaint, fewer than five diagnoses likely account for well over 80% of all cases.

But I'm interested in the remaining 20%. It's the rare and slightly less common diagnoses that are more likely to go undiagnosed. It's not uncommon that I see patients who have seen a dozen different doctors with the same complaint and never received a correct diagnosis. We know that at least 15% of diagnoses in emergency departments are incorrect and the real rate of misdiagnosis might be significantly higher.[43] Indeed, if physicians are correct 85% of the time, it's more because common and obvious diseases account for about 80% of cases. In other words, in the subset of patients for whom the diagnosis is less common or not obvious, physicians may be wrong more often than they are right.

I recently saw a patient who complained of pain during sex. She had been misdiagnosed with all the common things that might be associated with this complaint by a parade of primary care doctors, nurse practitioners, and even gynecologists: urinary tract infection (more than once), yeast infection (more than once), bacterial vaginosis, estrogen deficiency, endometriosis, interstitial cystitis, vaginismus, latex allergy, and lack of lubrication. Her actual diagnosis was vulvodynia and it was obvious after taking her history and patently obvious after performing an examination. She was not a patient who had a common cause of pain during sex, so providers who only focused on common things (thinking of five things at most on the differential diagnosis) had all failed to diagnosis it. They likely never thought of vulvodynia – not even the gynecologists. Because they never thought of it, they didn't ask the right questions, they didn't perform the correct physical exam, and they didn't find some clues that were there and misinterpreted others that they did find. They also too quickly settled on the most likely diagnosis that popped into their heads (anchoring) and they too easily accepted any evidence of that diagnosis once found, like leukocyte esterase on a urinalysis supporting a diagnosis of urinary tract infection (confirmation bias).

So, how can we do better? First, we need a better a differential diagnosis. We did that in the last chapter. Then we need to question the patient thoroughly.

Frequently, students are told that the patient will tell you what is wrong with them if you will simply listen. Active listening skills are emphasized, which is a good thing. But patients don't usually know what's wrong with them and their impression of what is wrong with them is often completely wrong. They believe that their pelvic pain is due to an ovarian cyst rather than constipation or that their viral respiratory infection needs an antibiotic. What the patient tells you leads to sometimes inescapable bias.

Patients also don't know what they should tell you. They will tell you about symptoms which are unrelated to the chief complaint while ignoring other factors which might be related. So, you should also learn to ask the right questions and to be thorough in this approach. Some of these questions are informed by the differential diagnosis and some are just part of a thorough history.

Let's consider the patient from the last chapter with pelvic pain. When interviewing her, let her speak. Ask her when she's done if she can think of anything else related to the complaint that she hasn't mentioned. Give her time. When's she done, it's your turn. You should have a systematic approach to gathering information and you'll need to ask the right questions to satisfy any unknowns.

For pelvic pain, I immediately want to know the following:

- What is the timing of the process? Divide symptoms into acute, subacute, and chronic. There will be some overlap in the lists, but there will be lots of differences too.
- Is the patient pregnant or not?
- What is the nature of the pain? Crampy, sharp, dull? Crampy pain for example implies the contraction of a viscus hollow (e.g., bowel, bladder, ureter, uterus). Learn what these different types of pain mean.
- What are associated or alleviating factors?
- Is the pain associated with menses?

I ask those questions related to pelvic pain, but a more general approach can be used for any complaint. A mnemonic like **SOCRATES** can be helpful for you to remember the types of questions to ask for each complaint:

- **S**ite: Where is it located?
- **O**nset: When did it start? Was the onset gradual or abrupt?
- **C**haracter: What does it feel like? Sharp, dull, crampy, throbbing, tearing, burning, etc.
- **R**adiation: Has it moved? Does it radiate to anywhere?
- **A**ssociations: Nausea, fever, pain during sex, loss of appetite, changes in bowel or bladder habits, etc.
- **T**iming: Has it gotten better or worse since onset? Does it come and go? How often does it happen? Etc.
- **E**xacerbating/Alleviating factors: What makes it better or worse? What remedies/medicines have been tried? What limitations does the pain cause?
- **S**everity: Rate the pain on a scale of 0-10.

These general questions are made more useful based on your differential diagnosis. Remember that for pelvic pain I use a differential diagnosis divided into five compartments (gastrointestinal, urinary, gynecologic, musculoskeletal, and nervous system/psychiatric). If you take a history with these five domains in mind, you can quickly focus on one or two of those specific domains by asking just a few specific questions; namely:

- What are your bowel habits like?
- How many times a day do you urinate? Do you urinate at night? If so, how many times?
- Does it hurt to have sex? If so, with insertion or penetration?
- Is the pain present only when you menstruate? Does it worsen with menses?
- Do you have a history of abuse? Have you felt depressed recently?
- Is the pain worse with exercise or other specific physical activities? Are there activities you avoid because of the pain?

If the answers to any of these questions are positive, then further questions will quickly help focus the work-up and narrow the differential diagnosis. Only by thinking of the differential diagnosis in advance can a process like this be designed. In the past, when I have given students oral exams with mock patient complaints, it is amazing how many students didn't consider appendicitis in their differential diagnosis for a patient presenting with pelvic pain. They didn't consider it because I was a gynecologist asking the question (triage cueing), but it changed the types of questions they asked. Most didn't ask questions about anorexia or nausea, for example. Even when I volunteered in the history all the classic symptoms of appendicitis, most still didn't consider the diagnosis because they had already narrowed their search fields to gynecologic diagnoses only.

A **diagnosis script** (or **illness script)** is a short constellation of symptoms, exam findings, risk factors, and perhaps lab or imaging results that call to mind a diagnosis. If a diagnosis is on your differential, be sure to ask the questions necessary to trigger the diagnosis script. In other words, if anorexia is part of your diagnosis script for appendicitis, you need to ask about anorexia specifically.

We will discuss diagnosis scripts in detail in Chapter 11. Another useful concept is the **management script**, which is a high-level, precompiled, conceptual knowledge structure of the courses of action that we may undertake to address a patient's problem.[44] Development of management scripts should be a secondary goal after development of diagnosis scripts.

Review of systems

Many times, physicians emphasize form over function in the practice of medicine. Physicians are increasingly overwhelmed with paperwork and documentation. They realize that most of it is unnecessary. These issues contribute to physician burnout and unhappiness. Some parts of the patient history have become relegated to something that we must do for billing purposes. In general, because of the way we are paid for our work, physicians feel like they must document both the review of systems (ROS) and the family history (FH) in order to be paid. Lost in the drudgery of "being required" to do this for billing is the actual purpose of these important parts of the history. In others words, physicians care more about the form of the documentation (to ensure payment) rather than the purpose (to ensure quality patient care).

The ROS is one of the most important parts of the history (that's why payers insist that it be done for payment). It typically serves one of two purposes. In an asymptomatic patient (e.g., a patient who has presented for an annual health exam), it serves to elicit potentially important symptoms that a patient might neglect to offer during the initial interview. For example, a middle-aged woman might not consider a constellation of minor symptoms important enough to mention, like increased abdominal bloating and girth, subtle pelvic pressure, and increased fatigue with occasional nausea. However, I am trained to recognize those symptoms as concerning for ovarian cancer. A proper ROS lists the symptoms of most diseases and can lead to the early detection of a latent disease process.

However, in a patient who presents with a complaint (like pelvic pain), the ROS serves to gather additional clues which may expand the differential diagnosis and to ask questions specifically related to a differential diagnosis. The ROS questions should be informed by the differential diagnosis but should also include general questions that might trigger new lines of inquiry that might broaden the differential diagnosis.

For example, if I were interviewing my patient's partner to discuss infertility, one ROS question that I might not normally ask would be whether he can smell properly; in this case, I would think to ask this additional ROS question (if not already pursued in the initial interview) because my differential diagnosis for male infertility includes Kallmann's Syndrome. In this way, the ROS is made more meaningful because of a good differential diagnosis.

A general ROS may provide additional clues which trigger thoughts of new diagnoses. For example, perhaps he reports headaches or visual changes and that reminds me to consider a pituitary tumor as a cause of infertility, or he reports frequent respiratory infections and I recall Young's or Kartagener's Syndrome. If I am using a computer-aided diagnostic tool, these additional symptoms may help the software recall and sort probabilistically a variety of diagnoses that I simply didn't think about.

Going system by system is akin to how we make differential diagnoses in the first place. Where we used the process of pathology (such as vascular, infectious, etc.) earlier to think about a specific organ, now we can apply different functional systems of the body and think about how they relate to our patient's complaint. It's not just about getting a list of questions answered for billing purposes – it's a part of the cognitive process of diagnosis. This also means that it is simply not as useful if the ROS is taken by your nurse assistant or by the receptionist who hands the patient a clipboard. The ROS is a part of diagnosis. It's the less-directed snooping around Sherlock Holmes does at the scene of the crime for clues he hasn't thought of yet.

A good ROS includes questions regarding the following systems:

- General/constitutional symptoms
- Eyes, ear, nose, and throat
- Cardiovascular
- Respiratory
- Gastrointestinal
- Genitourinary
- Musculoskeletal
- Skin/breast
- Neurological
- Psychiatric

- Endocrine
- Hematologic/lymphatic
- Allergic/immunologic

As an exercise for students and residents, think about how symptoms discovered from any of these systems might relate to the chief complaint. Remember our patient with pelvic pain. Let's consider just the general or constitutional ROS: fever might make us think of pelvic inflammatory disease, appendicitis, or pyelonephritis; fatigue or early satiety might suggest ovarian cancer; night sweats could remind us to consider pelvic tuberculosis; weight loss might lead us to think of new onset diabetes (which can present as abdominal pain); weakness in the lower extremities might suggest a spinal lesion; etc.

Each complaint should trigger a new round of questions using a focused method like SOCRATES as described above.

Medical history

No less important is the rest of the medical history, including past medical history (PMH), past surgical history (PSH), current and recent medications, allergies, social history (SH), family history (FH), as well as an obstetric and gynecologic history for women.

The purpose of this history is, again, not just to fill out a form. Each question is looking for clues. Gathering information is the process by which we refine our differential diagnosis and order it. We need information to assist us in determining pretest probabilities. In the medical history portion of the interview, information that helps us might include preexisting diagnoses that affect the complaint. For example, if she already has endometriosis, then that would seem to make that a more likely cause of her current complaint of pelvic pain. Likewise, if she has already had an appendectomy, then appendicitis becomes far less likely.

Inheriting someone else's thinking. We must be careful, though, to avoid inheriting someone else's thinking. For example, learning that the patient had endometriosis may adversely bias us. Delaying the medical history until after the HPI and ROS can help us not be overly biased by such information. You should question prior diagnoses. If you have learned anything yet, it hopefully is that misdiagnosis is common, but getting rid of incorrect diagnoses from patient histories is not.

Take endometriosis as an example. Roughly half of all patients who have been told that they have endometriosis, even after surgical exploration, do not. The gold standard for diagnosis is biopsy, but biopsy is rarely done, even if surgery is performed. Most patients who carry the diagnosis have not had a biopsy. The visual diagnosis at the time of surgery is incorrect about half the time. Among patients who do have endometriosis, roughly half have no symptoms from it. For many women, this means that symptoms of pelvic pain have been misattributed to endometriosis (premature closure) and the real cause of her pain remains undiagnosed. Thus, when a patient presents with a history of pain and endometriosis, there is roughly a 1 in 4 chance that her pain is due to endometriosis. Be careful about inheriting someone else's thinking.

Risk factors. Most of the past medical history serves to help us understand a patient's risk factors for various diseases. Risk factors are often misunderstood and misused in clinical medicine. Sometimes we are interested in identifying a risk factor so that we can modify it to reduce the risk of some disease. But most of the time we are looking for risk factors to help us predict the probability of a disease being present.

For example, which patient has the greatest risk of pulmonary embolism? A 25-year-old healthy woman who runs marathons and has no family history of stroke or thromboembolism with a complaint of shortness of breath, or a 65-year-old male smoker who has recently had an internal femur fixation after a car accident and who has a history of lung cancer who presents with shortness of breath? Risk factors help us understand the likelihood of a diagnosis. The medical history is littered with risk factors.

Protective factors. As opposed to a risk factor, a protective factor helps to reduce the probability of a disease. Receiving perioperative anticoagulants reduces the risk of thromboembolism compared to baseline; use of oral contraceptive pills reduces risk of ovarian cancer; etc.

Both risk and protective factors help us better organize our differential diagnosis according to the probability of each included disease by increasing or decreasing the likelihood of a diagnosis for a patient.

Medical records

I see patients who have transferred their care from other physicians or who were referred from other physicians nearly every day. Most of the time, I have old medical records to sort through. In a hospital setting, if I am called to consult on a patient, the computer is filled with old records from the current and past admissions. Many medical students and residents start their evaluation of a patient here, in the old chart, summarizing what is already "known" about a patient before they see the patient. As I said before, I know many doctors who essentially have their entire history and physical completed before they even see the patient. Usually, they have already decided what is wrong with the patient.

Clearly, this approach is wrong for two major reasons. First is the issue of inheriting someone else's thinking, which we just discussed. As a debiasing strategy, you should just assume that everyone who has seen the patient previously was wrong. Second, though, is that anything which leads you to a conclusion (a diagnosis) before gathering your own information is usually not good. Humans innately look for information that is coherent and supportive of their beliefs, not evidence that negates their beliefs.

We must deal with the medical record but we should do so with a skeptical perspective. If there is a previous diagnosis recorded, then how was that diagnosis made? Are the current medications appropriate or are they just being continued by each subsequent provider? What did previous physicians miss? Most importantly, interview the patient *before* reviewing the medical record.

Exercises

7.1 Consider a medical diagnosis such as chest pain secondary to coronary artery disease (CAD). What are some risk factors for CAD? What are some protective factors? How much relative risk does each factor confer? You should find the specific RRs from a source.

7.2 What questions are important to ask of a patient for whom CAD is a concern? How much risk is increased or decreased based on the answers to those questions? Again, you should find the specific answers from a source.

8

Physical Exam

So far, we have heard a complaint, developed a differential diagnosis, interviewed the patient, reviewed relevant records, and narrowed our differential diagnosis down to a few likely things, ordered by probability. Now we come to physical exam.

The first thing to be determined is whether we need to do any physical exam at all. Recall that we must consider three stopping points in the diagnostic process: **the diagnostic threshold** (are we certain enough to proceed with diagnosis?), **the test threshold** (is it worth ordering a test?), and the **treatment threshold** (should we start treatment without further testing?).

The usual reason to do a test, as we will discuss in the next chapter, is because we haven't reached the diagnostic or treatment threshold; we need more information because we are still uncertain about the diagnosis. However, as noted before, many diagnoses can be made based on history alone. In this context, doing further testing is usually inappropriate. Thus: **further testing is usually inappropriate if the diagnostic threshold has been met.**

Physical exam is a test, and you should think of it that way. We select tests based upon their ability to help determine something useful for the patient and we should select physical exam choices with this in mind. To do this, we need to know the value of each type of exam as it relates to the diagnosis we are considering, then select appropriately. Too much or too little physical exam, just like history, can have a negative effect on the accuracy of our diagnoses.

Obtaining a history prior to performing a physical examination improves the quality of the exam and it should dictate what examinations are performed. It also changes the predictive values of the examination and the rates of detection of abnormalities. Importantly, this is most true when a high-probability diagnosis is in the mind of the physician. Contrast this with a physician who does an exam just for the sake of doing the exam (or to enable coding at a higher level). In one study of stethoscopy, when the examiner had taken a good history and had a probable diagnosis in mind, the rate of correct diagnosis after examination was 87% versus 29.9% when an incorrect diagnosis was suspected or a poor clinical history was obtained.[45]

Physical exam is a way of testing hypotheses. Normal physical exam makes some diagnoses more or less probable. Abnormal physical exam makes some diagnoses more or less probable. Therefore, just like any test (see the next chapter), we must have already decided what our pretest probability for a given diagnosis was before we apply the test – in this case, physical exam. We do physical exams with intent: to make ourselves appropriately confident with our diagnosis.

Let's look at an example of this.

Recall our patient with pelvic pain. Let's say she presents with acute pelvic pain to the emergency department. We include pelvic inflammatory disease (PID) in our differential diagnosis. After interviewing her, we are uncertain of the diagnosis but consider PID a diagnosis worth pursuing with additional examination and testing (that is, we could not sufficiently exclude it as a diagnosis). We think that the patient has a 1 in 6 chance of having PID based upon our history, or a roughly 15% pretest probability. What physical exam findings can help us make this diagnosis more probable or less probable?

We discussed likelihood ratios (LR) in Chapter 5 and now we can start to learn how to use them clinically. A likelihood ratio of greater than 1 increases the odds of a disease and a likelihood ratio of less than 1 decreases the odds of a disease. **Tests with a LR of greater**

than 5 or less than 0.2 are most helpful in furthering the diagnostic process, since this magnitude of effect is usually necessary to make us sufficiently certain to make or eliminate a diagnosis about which we are uncertain.

Here is a table of LRs for physical exam features (and a couple of labs) for PID (each of these is an independent factor so each can add to the total post-test probability):[46]

			Post-test probabilities if 15% pretest probability	
Sign	**LR+**	**LR-**	**Present**	**Absent**
Purulent endocervical discharge	3.3	0.8	37	12
Abdominal rebound tenderness	2.5	0.8	31	12
Fever	1.3	0.8	19	12
WBC > 10,500 per mm3	1.3	0.9	19	14
ESR > 15 mm per hour	1.2	0.6	17	10
Cervical motion tenderness	1.1	0.9	16	14
Pelvic organ tenderness	1.0	1.0	15	15

Table 6. Likelihood ratios for common tests for pelvic inflammatory disease.

We shouldn't perform any test or exam without understanding its value for the patient, based upon that patient's pretest probabilities. What diagnostic and treatment thresholds are we comfortable with? We will talk extensively about this in the next chapters. For now, though, how confident do we need to be that a person has PID before we initiate treatment or stop looking for other diagnoses? This is a complicated question and its answer involves the cost for delayed or missed diagnosis as well as the risks of the treatment in question. In the case of PID, the interventions are relatively harmless (some antibiotics) and the cost of delayed or missed intervention might be infertility or worse. So, most physicians don't need to be too certain about the diagnosis before proceeding with treatment.

For the sake of the example, let's say a 30% probability was enough to initiate antibiotic therapy (though not enough to stop looking for another diagnosis). Look at the table above. If you are 15% certain based on history, then doing a bimanual pelvic exam to assess for pelvic organ tenderness is so useless that it shouldn't make you any more or less confident about the diagnosis. In other words, don't do the exam. Eliciting cervical motion tenderness isn't much more valuable and barely nudges your confidence in either direction. In fact, none of the findings in the table are that valuable except for testing for abdominal rebound tenderness and purulent endocervical discharge. Realistically, then, the decision to administer antibiotics (given our 30% treatment threshold) should be based upon those two findings alone and those are the only two bits of physical exam that should be performed.

Note that the post-test probability is not simply the pretest probability multiplied by the likelihood ratio; likelihood ratios multiply pretest *odds,* not probabilities. I use an online calculator to make these calculations easy. A good one can be found at getthediagnosis.org/calculator.htm.

Not appreciating the statistics that underlie the diagnostic process can easily lead to diagnostic and treatment errors. For example, a physician might be considering a diagnosis of PID. He may have been taught to treat anytime he discovers a patient with acute pelvic pain who has a purulent endocervical discharge and cervical motion tenderness. He sees a patient whom he believes has a reasonable chance of having PID after her interview (say a 1 in 6 to 1 in 4 chance). She does have some endocervical discharge but doesn't have cervical motion tenderness. Because she lacks the "classic" symptoms of PID, he stops pursuing that diagnosis and doesn't initiate treatment. In this case, the patient was harmed because he did a test that shouldn't have been done and which carried no diagnostic weight.

Likewise, a patient might present whom the physician believes has a 50% chance of PID after interviewing her. With a pretest probability this high, even if she didn't have a mucopurulent endocervical discharge, she would still have a 44% chance of having PID since that finding has a LR– of 0.8. Thus, he would be harming the patient by performing any of the physical exams shown in the table because none of them have the ability to lower the treatment threshold to below 30%, if that treatment threshold is desirable.

One quick way of implementing LRs is to use Fagan's nomogram, which is presented in Figure 17. In the figure, a dotted line is placed starting with a pretest probability of 30% and +LR of 5; the end of the line tells us that the post-test probability is around 65%. An exact calculation yields 68%, but the nomogram can provide a good enough calculation very quickly.

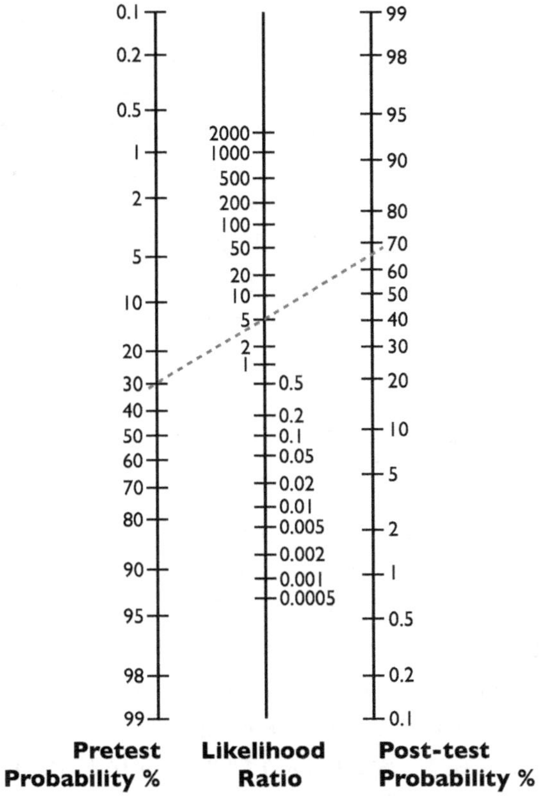

Figure 17. Fagan's nomogram.

Note that a –LR here is expressed as a fraction; so, a –LR of 5 would be 0.2. The online calculator mentioned on the previous page will let you start with a variety of inputs, including sensitivities and specificities.

Of course, the problem is that the physician may not be that good at determining pretest probability based on history alone for a given disease. This is, indeed, a significant problem. But just because a physician isn't good at doing that clinical work doesn't mean that the problem of inappropriate exam and testing goes away. We must approach clinical exams with pretest probabilities in mind. Accurate diagnosis scripts are the cornerstone of this, even when the prevalence of the suspected disease is not known.

Still, we can see that some exams should never be done. For example, in this case, testing for pelvic organ tenderness or cervical motion tenderness is just not valuable. This is much like Homan's sign for deep vein thrombosis (DVT); Homan's sign is completely invaluable for increasing or decreasing the probability of DVT and shouldn't be done.[47] Continuing to teach this exam to medical students and residents only serves to skew in their minds the predictive value of the exam. How many patients have been harmed because someone failed to do appropriate testing for DVT due to the absence of a Homan's sign?

The lesson is that **we need to know the predictive value** of any exam that we do and do exams with a specific and predefined intent. Always be able to answer the question: *What am I looking for with this exam that will help me make a clinical decision?*

Physical exam technique

Just like the ROS or other parts of the patient interview, much of physical exam has become "form over function," and for the same reason: billing. Physicians have been taught that to bill at certain levels, they need to "document" certain amounts of physical exam. However, this doesn't mean that a physician should do physical exam merely for the sake of being able to bill at certain level; that's unethical. It means that if a patient requires physical exam as part of the diagnostic or evaluative process, then physicians should get paid for their time. Lost in every billing and coding course that physicians attend is this simple admonition: do what's appropriate for the patient and then bill for it.

If physical exam is necessary, then it is important that it is done correctly. Poor technique in performing physical exam is pervasive. Many physicians have learned to do an exam as quickly as possible so that they can say they did it for billing and as a consequence there has been a decline in exam quality. As exam quality has declined, physicians have lost confidence in their abilities to gather meaningful data from the exam and have become over-reliant on labs or imaging to make a diagnosis, which in turn may lead to patient harm.

Recall that in one study of 208 patients who had a missed or incorrect diagnosis, 63% of the time this was related to a failure to perform a physical exam.[48] But even more common than this would be a failure to perform physical exam correctly. It is important that evidence-based techniques be utilized, and physicians should search the literature and look for opportunities to continue to improve physical exam techniques and to learn about their predictive values.

Another consequence of both the decline in exam technique and over-reliance or under-reliance on exam findings has been several new evidence-based recommendations that many screening exams should cease being performed, like clinical breast exam. At the same time, most clinicians who do breast exams do not employ evidence-based techniques. For example, clinical breast exam should be performed with the following principles in order to make the exam more beneficial than harmful:

- Travel across the breast in vertical stripes (lawnmower method)
- Use the pads of the middle three fingers with just one hand
- Palpate by moving the fingers in small circles
- Use three successive levels of pressure in each location
- Don't attempt to express discharge from the nipple

Surprisingly, few clinicians incorporate all these recommendations in their breast exam technique, making the exam less sensitive and less specific. This makes their exams more prone to false positives and false negatives, both of which are harmful to patients.

Unfortunately, what is true for breast exams is true for most exams. Cardiac exams have become little more than auscultation for three seconds at one point, which is inadequate even to determine normal rhythm. Thyroid exam too often involves palpating only the thyroid cartilage, not the thyroid gland. Physicians will state that cranial nerves are "grossly intact" without examining a single nerve.

There are many resources for improving physical exam skills. One of the best is the Stanford 25, a website which teaches the correct techniques and interpretations of 25 commonly used physical exams (stanfordmedicine25.stanford.edu).

Exercises

8.1 What is the +LR of spiral CT for pulmonary embolism?

8.2 Review two physical exam modules from the Stanford 25.

9

Selecting the Right Test

In 2010, the Institute of Medicine estimated that nearly 1/3 of US healthcare costs, about $750 billion, was wasted. This money is wasted in a variety of ways, but much of it is spent on unnecessary tests. Many of those unnecessary tests lead to other unnecessary interventions, including more tests, surgeries, and prescription medications. We have an epidemic of over-utilization, over-prescription, and over-diagnosis. This is equivalent to about $6,000 per household per year in the United States, and that's money that is not spent on food, clothing, or other essentials. The first step in stemming this tide of waste is to improve the diagnostic process, and a huge part of this is learning to select the right test.

Here are few clinical scenarios to orient us to the problem:

A patient presents to the emergency department with a complaint of a headache. She receives a "rainbow" panel of tests: a urinalysis, a complete blood count, a comprehensive metabolic panel, and a pregnancy test. The urinalysis revealed moderate leukocyte esterase and she was treated with an antibiotic; the complete blood count revealed a platelet count of 135,000 and she was referred to Hematology/Oncology for work-up; the metabolic panel revealed a potassium of 3.1 and she was started on potassium replacement.

A healthy, asymptomatic 37-year-old female school teacher receives "routine labs" as part of a yearly medical screen. She received a comprehensive metabolic panel, a uric acid level, a magnesium level, an iron level, a lipid panel, a thyroid stimulating hormone level, a free T4, and a complete blood count. Her iron is high (193) and her platelets are low (139,000) and she is referred to Hematology/Oncology. Her TSH is 3.55 and her free T4 is 0.7 (both normal) and it is recommended that she have the values repeated in 6 months. The hematologist ordered an extensive panel of labs and she is noted to have marginally high monoclonal gammaglobulins. Extensive testing, including bone marrow biopsy and repeated lab tests, are normal but the physician tells her he is worried about "pre-leukemia." He recommends IV IG therapy. A second opinion with another hematologist (suggested by her gynecologist) results in the patient deciding that no further intervention is appropriate.

A young woman presents to the emergency department with complaints of abdominal pain. She has not had a bowel movement in 6 days and has a history of constipation. She receives a CT scan which shows constipation. It also reveals a small, non-descript 2-mm nodule at the base of the left lung, a 2.8-cm right ovarian cyst, and a 2-cm simple cyst on the left kidney. She received a pelvic ultrasound which characterized the ovarian cyst as likely a hemorrhagic luteal cyst but also revealed a 1.3-cm uterine fibroid. She was told to follow-up immediately with gynecology. She also was referred to pulmonology and nephrology for follow-up of the other findings.

The pulmonologist recommended repeating the CT scan in 6 months and the nephrologist recommended renal biopsy. She underwent a renal biopsy and had a retroperitoneal hemorrhage requiring transfusion. The biopsy was benign and the repeat CT scan showed no lesion. The gynecologist recommended a diagnostic laparoscopy because she "might have endometriosis." No endometriosis was discovered at the time of her laparoscopy.

Do these scenarios sound crazy? These are all real patients, and unfortunately I see these sorts of examples almost every day. Unnecessary testing doesn't just waste money, it harms patients. This epidemic of over-testing in the US has resulted in the highest healthcare costs per capita while achieving *poorer* outcomes, because patients are harmed by over-testing.

Do I need a test?

Tests serve the following purposes:

- Screen for a disease in an at-risk population
- Decrease the likelihood of a "can't miss" diagnosis in a person at risk for that condition
- Increase or decrease the probability of a likely diagnosis on the differential diagnosis
- Inform the decision for a certain therapy or therapy change

Screening. Screening tests are not diagnostic tests. They are typically done in an asymptomatic patient population. Few screening tests apply to all populations; most are appropriate for a limited and predefined vulnerable population. The characteristics of that population increase the pretest probability of the disease and therefore the utility of the test. The principles of a good screening test include:

- The test should identify a clinically important disease that, if left untreated, will cause significant morbidity/mortality
- It should be inexpensive and noninvasive
- The disease being screened for should have a preclinical, presymptom phase for which there is a good intervention available
- The test should be highly sensitive

A great example of this is a Pap smear. Cervical cancer is a significant disease. Pap smears (or, really, cervical cytology) and HPV testing are inexpensive and sensitive tests that can detect precancerous lesions of the cervix which are highly treatable. Cervical cancer, with appropriate screening, can be almost entirely prevented. Cervical cancer screening is so effective that the screening tests apply to almost all women.

Still, the over-utilization of screening tests is a major contributor to adverse outcomes and waste. For example, under current guidelines at the time of writing (subject to change constantly), Pap smears should be performed starting at age 21 until age 30 every three years, and then Pap smears plus co-testing with HPV should be performed every five years until age 65. Yet, far too many providers perform Pap smears inappropriately: before age 21, yearly, over age 65, after hysterectomy, etc. Too much of a good thing is too much. It leads to over-diagnosis, unnecessary intervention, waste, and, ultimately, patient harm.

We might introduce the term **screening threshold** here. A patient needs to be beyond the screening threshold to warrant a screening test. For cervical cancer screening, the screening threshold is the criteria already listed; for lipid disorders, the screening threshold for a woman is if she is over age 45 or she has risk factors for heart disease; etc.

Screening tests should be performed at the right time and in the right patient population according to evidence-based recommendations. Physicians should also be careful to ask this question about screening tests:

> Does this test merely increase the rate of diagnosis or does it lead to improved patient outcomes?

We will talk later about picking the right outcomes rather than surrogate markers. A surrogate marker too often utilized when discussing tests is the rate of diagnosis. But merely looking at the rate of diagnosis isn't usually valuable because that doesn't allow us to know whether we are over-diagnosing or misdiagnosing. Instead, pay attention to patient outcomes. For example, we shouldn't care about the difference in the rate of breast cancer detection with routine mammography versus less or no mammography; rather, we should care whether the strategy results in fewer cancer deaths and better patient outcomes. Mammography in the fifth decade, as an example, does increase the rate of breast cancer diagnosis but doesn't decrease the risk of death from cancer overall.[49]

Can't-miss diagnoses. I have already introduced the concept of the "can't-miss" diagnosis earlier when discussing endometrial cancer. When I evaluate the endometrium with ultrasound or biopsy in a woman who presents with abnormal uterine bleeding and risk factors for endometrial cancer, I am not doing so because endometrial cancer is near the top of my differential diagnosis. I am doing so because, even though I am over 95% confident that the patient *doesn't* have endometrial cancer, I'm just not certain enough when it comes to a highly treatable cancer.

Of course, even when the endometrial biopsy comes back negative for cancer, that still doesn't mean that the patient doesn't have endometrial cancer; it simply means that I am confident enough that she doesn't have it to let me proceed with treatment for some other diagnosis. She might still have endometrial cancer. We never *exclude* or *rule-out* diagnoses, and those words should be stricken from our vocabulary. We simply make them less probable. Around 5% of women who present with an unknown invasive cervical cancer will have a normal Pap smear that is HPV negative. Depending on your clinical suspicion (pretest probability), a cervical biopsy might be needed, not just a Pap smear. In other words, you shouldn't feel like cancer is excluded due to a normal Pap and HPV test in a woman with a history of postcoital bleeding and a grossly abnormal appearing cervix on exam.

This begs the question: How confident must we be to not worry about can't-miss diagnoses? There is no easy answer to this question. We must weigh the risks of delayed or missed diagnosis versus the cost of increased certainty. Every patient who has ever presented with symptoms of acid reflux might have pancreatic cancer, but that doesn't mean we should do a CT scan on every patient who presents with reflux symptoms. The likelihood of pancreatic cancer is already sufficiently low that the risks of a CT scan (incidental findings, radiation exposure, cost, etc.) don't justify this testing.

On the other hand, up to 5% of women who present with abnormal vaginal bleeding after the age of 45 have a malignant or premalignant condition and endometrial biopsy or pelvic ultrasound is relatively inexpensive and associated with few false positives; thus, we use these tests to increase our confidence level that the patient doesn't have cancer. Treatment guidelines must weigh these factors when recommending an algorithm for testing for "can't-miss" diagnoses.

Unfortunately, these thought processes are heavily perverted in medicine, especially in areas of high litigation. What is a can't-miss diagnosis? What should trigger further testing? If a patient presents to the emergency department with typical abdominal pain associated with constipation, should he receive a CT scan to rule out mesenteric ischemia? No; but unnecessary CT scans happen every day. Did the patient in the vignette at the beginning of the chapter with slightly low platelets deserve a work-up for leukemia? No. Does every patient with a headache who presents to the ED deserve a CT scan? Do uncomplicated musculoskeletal complaints warrant imaging? Do women with stable fibroids need an MRI? We could go on for hours.

Next time you want to order a test to "cover your $%*," try stating the indication clearly: *Since x% of patients with this presentation have this serious disease, I need to order this test*. Most physicians don't know the 'x' in that statement. Typically, it should be well above 1% but it varies by disease process. This level is called the **test threshold**.

Diagnostic tests. We've been discussing the diagnostic process and most tests are ordered as part of that process. If, after gathering information and examining the patient, we are still unable to cross the **diagnostic threshold**, then we need more data. Sometimes this may mean asking questions that were unanswered previously, gathering information from outside records, performing previously unperformed physical exams, or even trying an intervention and seeing if the patient improves. But it also often means ordering a test.

Tests are not binary; they don't return 'yes' or 'no' answers. Each test must be appreciated for its predictive values. As with physical exam, we often express this in terms of likelihood ratios. Physicians should know the likelihood ratios of the tests they order. Ask yourself the following questions before ordering a test for diagnostic purposes:

- What disease am I testing for? How certain am I currently that the patient does (or doesn't) have the disease? How confident do I need to be before I make the diagnosis?
- How will this test change how I manage the patient?
- How good is this test? How will it affect my certainty about the diagnosis?

It is inappropriate to order a test without thinking about these questions in advance.

The first question requires us to have a differential diagnosis ordered according to probability and it requires us to have some idea of what the pretest probability is that the patient has the disease. It also requires us to think about how certain we want to be to make the diagnosis (diagnostic threshold).

The second question forces us to think how we will act in the presence of a positive or negative result. If we can't answer this question, we likely have either picked the wrong test or don't need the test.

The third question requires us to understand the positive and negative likelihood ratios of the tests we are considering.

We will see how to do this effectively in the next chapter.

Therapeutic test. The last reason to order a test is to determine the appropriateness of a certain therapy or to monitor or change a therapy. Common examples include a hemoglobin A1c for diabetes management, an INR level for Coumadin management, or perhaps genetic or hormonal status tests for malignancies to determine the most appropriate chemotherapy. Again, answer the question: *How will this test affect my management?*

Exercises

9.1 Think of three screening tests you perform regularly or have an interest in. What are the sensitivities and specificities of those tests? What is the prevalence of the diseases in the populations you screen? Calculate the positive predictive value of a positive result given these parameters.

9.2 Now consider a diagnostic test, such as troponin for myocardial infarction. What is the sensitivity and specificity of the test? What is the positive predictive value for the disease of interest over a range of pretest probabilities, such as 5%, 50% and 95%? How certain do you need to be to proceed with treatment for the disease (i.e., how high does the positive predictive value need to be)?

10

How to Interpret Tests

Let's recap: we saw a patient with a complaint; we made a differential diagnosis for that complaint; we gathered information from that patient; we refined our differential diagnosis and organized it from most probable to least probable; we performed any indicated physical exam and refined our list again; if we still couldn't make a diagnosis confidently or we needed to decrease the chance of a "can't-miss" diagnosis, then we selected a test that would help us and ordered it.

Now we need to make sense of what we've learned from the test result. The first thing we need to understand is what the test result means. Remember, tests aren't black and white. They don't indicate that a person does or does not have a disease. Some results *seem* binary, like a CT scan that either shows a lung nodule or it doesn't. But even these sorts of tests are limited by the sensitivity of the test and other testing parameters (such as the slice width of the CT scanner).

Most lab tests are returned as normal if the result falls within a certain range. This range is usually the average result plus or minus two standard deviations for a given population. In a few minutes, we will discuss how to use these results that are clearly "abnormal," that is, greater than two standard deviations different from the mean. But we also need to introduce another source of diagnostic error that arises in interpretation of tests: diagnostic (and therapeutic) drift.

Diagnostic and therapeutic drift

To illustrate the problem of diagnostic and therapeutic drift, let's conduct a thought experiment.

Let's imagine that we develop a test for an imaginary hormone called Mollynium (Molly is the name of one of my daughters). We study a "normal" population and we determine that normal levels of Mollynium range from 3-12 units, with an average of 7.5. Essentially, we define the normal range of Mollynium as the mean ± 2 standard deviations. In this fictional case, this is 3-12 units. We assume that patients who have Mollynium levels greater than 12 units have 'hypermollynemia' while those less than 3 units have 'hypomollynemia.' Let's say that the purpose of Mollynium is to make people happy. Therefore, people with low levels tend towards sadness and people with high levels are giddy all the time.

If a patient presents with normal levels of happiness, we would expect to find Mollynium levels between 3 and 12 units. Still, some patients who are completely "normal" clinically will have Mollynium levels above 12 or below 3. The mean ± 2 standard deviations doesn't capture everyone who is "normal," but we must draw a line somewhere. Statistically, plus or minus two standard deviations covers 95.45% of the population, so it is an unsupported leap to suggest that exactly 4.55% of all patients either have too much or too little of every substance we can measure.

What if a sad patient presents? If we check the level and discover that it is only 1 unit, then we might reasonably conclude that the patient needs more Mollynium. Similarly, someone who can't stop laughing and whose level is 15 might have too much. There are other reasons why a person might be too sad or too happy, though. Many sad or giddy patients will present with normal Mollynium levels and have a different problem - say, depression or a frontal lobe stroke. Clearly, though, it helps to have symptoms which raise the pretest probability that someone might have abnormal levels.

There are also patients who will present with abnormal levels of Mollynium (as compared to our "normal" population) but their "abnormal" Mollynium level is not actually the reason that they are too happy or too sad. A patient might have a level of 14 units and that might be normal for that individual; the real cause for her uncontrollable laughter might be a frontal lobe lesion. So, from a Bayesian perspective, the diagnosis is never certain, just more likely or less likely.

Diagnostic drift has been defined in various ways, but it is essentially when the parameters that define the diagnosis are expanded (for whatever reason), leading to more diagnoses. In the same way, I use the term "therapeutic drift" to indicate either over-treatment due to over-diagnosis or to indicate extra treatment with a given intervention (caused by *if some is good, more must be better* thinking).

Let's talk about some real examples. One method of determining amniotic fluid volume in pregnancy is to use the *amniotic fluid index* or AFI (it is probably not the best way but that's irrelevant to this example). The AFI is the sum of the maximum vertical depth of amniotic fluid measured in four quadrants and is reported in centimeters. We define "normal" as between 5-25 cm. In theory, these values represent the median ± 2 standard deviations. Outside of these values, we would expect to see an increase in abnormal outcomes or other associated problems, and, for the most part, that's true. Both too little fluid (less than 5 cm) and too much fluid (greater than 25 cm) are associated with adverse outcomes. Low fluid may result from ruptured membranes or uteroplacental insufficiency, while too much fluid might be associated with things like uncontrolled diabetes or fetal tracheal atresia.

So if a pregnant woman has an AFI of 6 cm, is it "normal" or not? Well, it is below the average, certainly, but it is *still normal*. She should be treated no differently than if her fluid were 12 cm. Similarly, 23 cm is above average, but it is still normal. The patient should be reassured of the normalcy and should be treated no differently than if her fluid were 12 cm.

Unfortunately, this isn't usually what happens. Instead, the patient is often treated as abnormal. She is told that her fluid is "low normal" or "high normal" and offered interventions, either in the form of extra testing or other things believed to affect the fluid level. This over-intervention can lead to harm with no evidence of benefit (if there were evidence of benefit, we would change the definition of "normal" to some range that would encompass those who benefit from diagnosis and treatment).

Does a normal AFI of 6 cm mean that a patient doesn't have ruptured membranes or uteroplacental insufficiency? Does an AFI of 24 cm mean that a patient doesn't have uncontrolled diabetes or a fetus with tracheal atresia? Of course not. But from a Bayesian perspective, there should be more evidence that would factor into assessing the probabilities of those diagnoses. Normal is normal. Labeling a person with "low" or "high" amniotic fluid in these situations is an example of diagnostic drift and it is harmful when it leads to unnecessary interventions without specific benefit to the patient.

What's more, where does drift end? How about an AFI of 7 cm? Or 8 cm? Where is the line in the sand? It's 5 cm. Another example might be ultrasound estimated fetal weight. If we decide to do extra testing or monitoring on fetal weights less than the 10th percentile or greater than 90th percentile, what should we do with the patient whose baby is in the 11th percentile or 89th percentile? Nothing. Tell her that her baby is normal (unless you have other data that would indicate a problem).

What about the margin of error of our tests? Ultrasound just isn't that accurate, so that 11th percentile fetus might easily be 8th percentile given the margin of error of the test. But this error is already reflected in our testing parameters. For example, we don't care about small babies until they are less than 5th percentile (or even 3rd percentile). By including fetuses less than 10th percentile (and not the 11th percentile), we capture this whole population, even given the errors in the test. In this case, normal is normal and everything else is drift.

Drift happens when we treat a temperature of 100.3 as a fever (though, arguably, we need to change our understanding of what a fever is given recent discoveries). It happens when we treat fatigued patients with a normal TSH for hypothyroidism anyway because her TSH is "high normal." It happens in all fields of medicine and leads to over-diagnosis of everything from myocardial infarctions to cancer, with all the unnecessary interventions and harms that one might expect.

Therapeutic drift is a natural consequence of diagnostic drift. The more we relax our standards for diagnosis, the more we apply treatments. When we treat healthy people, we usually have good outcomes. This, in turn, reinforces the behavior because the treatments seem to be associated with positive results (until a dangerous side effect occurs).

Therapeutic drift happens in other ways, too. If 600 IU of Vitamin D per day is good for you (I'm not saying it is or isn't), then 1200 or 1800 IU must be even better. If a blood level of at least 20 ng/ml of Vitamin D is healthy, then having a level of at least 30 ng/ml must be even healthier (though there is evidence that Vitamin D levels above 50 ng/ml is associated with increased mortality). If a Pap smear every 3 years is good, then one every year must be amazing. You can probably think of dozens of other examples. More is not always better, particularly when there are harms associated with more.

Less isn't necessarily better either, of course. If a Cesarean delivery rate of 15% is better than one of 35%, that doesn't necessarily mean that a rate of 5% is better than 15%. This is where we need evidence-based medicine: it allows us to study large populations and determine the parameters for diagnosis and treatment that maximize benefits while minimizing risks. One practitioner in the scope of one practice cannot see and interpret enough data to make these decisions. So, we must rely on large populations to do so and trust the parameters that the data indicate.

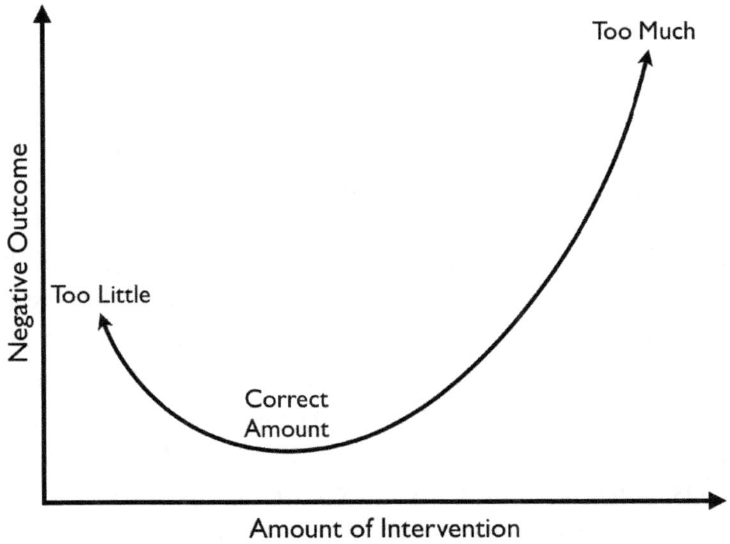

Figure 18. J-shaped curve of outcomes versus intervention.

We can see this concept in Figure 18. Too little and too much of an intervention, test, or treatment can be both good and bad. A certain number of Cesarean deliveries are necessary to reduce neonatal mortality and morbidity; but, when too many are done, the side effects and complications of Cesarean delivery (infection, bleeding, thromboembolism, etc.) soon outweigh the benefits. **All tests and treatments have risks.** Even rare risks become meaningful when tests or treatments are used in patients who receive no clear benefit.

Tests are harmful when they lead to over-diagnosis. At the same time, none of this means that we should blindly accept the results of our tests, either. Bayesian reasoning frees the practitioner to realize that sometimes a patient has the disease, even though the test says that the patient does not; it helps us avoid the pitfalls of diagnostic and therapeutic drift.

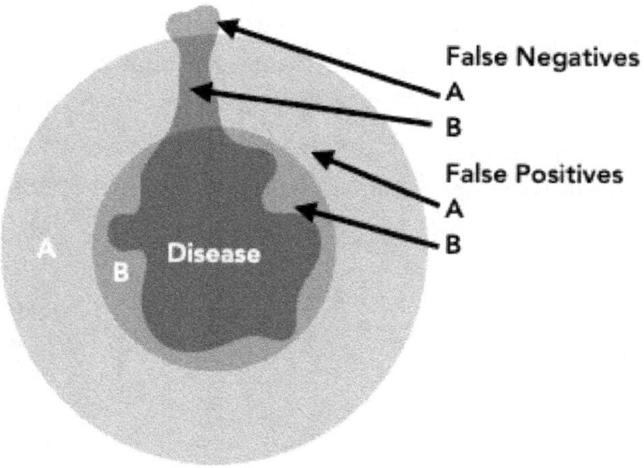

Figure 19. Tests A and B with differing sensitivities for a disease.

Figure 19 demonstrates the problems with selecting tests that are *too* sensitive for a disease (or creating disease criteria that lead to excess over-diagnosis). Imagine that A and B are two different tests (or two different disease criteria). Neither is perfect. Each misses some patients with the disease. However, B is associated with relatively few false positives compared to A. At the same time, A is associated with fewer false negatives but many more false positives. So, we must decide how much over-diagnosis is permissible given the consequences of over-diagnosis (and missed diagnosis). Drift occurs when we do this in an undisciplined manner without a full analysis of the risks, benefits, and costs. Imagine circle B slowly getting bigger. That's drift.

I could catch every ectopic pregnancy very early if my only criteria were a positive pregnancy test; it would be a very, very sensitive test, but it would be the wrong test.

Bayesian statistics in diagnosis

The diagnosis of ruptured membranes in pregnancy is clinically very important. Decisions about delivering a pregnancy early, hospitalization, and even termination of pregnancies, often depend on being correct about this diagnosis. Understanding how to diagnose ruptured membranes is fundamental to the basic practice of obstetrics; but understanding the clinical reasoning and statistics that underlie diagnosing ruptured membranes is fundamental to all clinical medicine, **and this skill may be the most important skill in clinical medicine!**

We use Bayes' theorem to integrate test results into our clinical diagnostic processes. Let's look at how we diagnose ruptured membranes to bring together what we have learned so far and learn how to properly use test results.

We have discussed the numerous cognitive errors and distortions that can lead to mistakes in diagnosis and management of patients. But there is no bigger problem than the non-intuitive nature of statistics and a poor understanding of Bayesian probability and the subsequent misapplication of clinical data.

First, let's look at the tests that are commonly available for the diagnosis of ruptured membranes:

- **History.** Yes, history is a test in a sense. We gather information (we interrogate) and we determine the probability that the patient has ruptured membranes. Our sense of the likelihood that she has ruptured membranes becomes the pretest probability for our other clinical exams and tests.
- **Nitrazine or pH test.** The pH of amniotic fluid is more basic than vaginal fluids, so the nitrazine test is a quick screen of the vaginal pH. Amniotic fluid will turn the test positive, but so might blood, semen, urine, and some vaginal infections, all of which also are more basic than normal vaginal fluid.
- **Ferning test.** This is a microscopic test of amniotic fluid which

has been allowed to dry on a slide (for around 8 minutes). The fluid will develop an arborization pattern (like a fern) due to NaCl crystals in amniotic fluid. False positives are commonly due to fingerprints on the slide.

- **Vaginal pooling.** Placing a speculum in the vagina (or sometimes just seeing fluid overflow out of the vagina) is fairly specific for ruptured membranes, but it might be due to other things, like vaginal infections, urine, or vaginal transudate.

To use these tests, we first need to know how good each one is. As an Obstetrician, it is my duty to understand in detail each test I utilize. We can start to understand how good these tests are by looking at the sensitivity and the specificity of each test. Recall that sensitivity (or the *true positive rate*) is the percent of patients who have the disease who will test positive; while specificity (or the *true negative rate*) is the percent of patients who don't have the disease, whose test will be negative. Here are the numbers for these tests:

	Sensitivity	Specificity
Nitrazine[50]	93	83
Ferning (in labor)[51]	98.0	88.2
Ferning (not in labor)	51.4	70.8
Pooling	Low	High

Table 7. Clinical performance of common tests used to diagnose ruptured membranes.

Knowing these numbers, let's ask a simple question: If a woman presents with questionable leakage of fluid and you perform a ferning test that is positive, what is the chance that she has ruptured membranes? Unfortunately, most clinicians are tempted to answer either 88.2% or 70.8% depending on whether she is in labor or not, since this is the "specificity." In other words, most clinicians, in their thinking, understand the specificity of a test as being roughly equivalent to the test's accuracy or predictive value. But nothing could be farther from the truth. Recall that specificity is the true negative rate.

The actual answer to the question is: ***We need more information.*** Neither sensitivity nor specificity considers the prevalence of the disease. The question we want an answer to is this: *What is the post-test likelihood that the patient's membranes are ruptured?* This is also known as the positive predictive value (PPV). The corollary of this is the negative predictive value (NPV), which would answer the question, *How sure am I that her membranes aren't ruptured since this test is negative?* We don't know enough yet to determine this. Why?

Recall how we defined PPV in Chapter 5:

$$PPV = \frac{True\ Positives}{True\ Positives + False\ Positives}$$

Let's actually derive the formula for PPV that we learned about in Chapter 5. To use this equation, we need to know what a true positive and false positive are.

Recall that sensitivity is also called the true positive rate. Since it is a rate, it is dependent on the prevalence or likelihood of the disease. Thus,

$$True\ positives = Sensitivity\ \times\ prevalence$$

Since specificity is the true negative rate, then by the same logic,

$$False\ positives = (1 - Specificity)\ \times\ (1 - prevalence)$$

By substitution then, we see that

$$PPV = \frac{Sensitivity\ x\ prevalence}{sensitivity\ \times\ prevalence + ((1 - specificity)\ \times\ (1 - prevalence))}$$

Thus, the reason we can't answer the original question is because we don't know the prevalence of the disease (ruptured membranes).

Determining the prevalence is easy if we are talking about a well-defined population for a screening test like a mammogram; for example, the prevalence of breast cancer in asymptomatic white women in their 40s. We can look that prevalence up in a study if we are screening an asymptomatic white woman in her 40s.

But, in clinical medicine, we are not usually talking about a well-defined population. We are instead considering a patient who presents with symptoms and has a unique combination of risk factors and complaints. The prevalence of ruptured membranes depends on the patient's history and her risk factors! It would be silly to use the prevalence of ruptured membranes in pregnant women in a general population just as it would be silly to use the general prevalence of HIV when determining the pretest probability of a sex worker who presents with pneumocystis pneumonia.

Thus, we must determine the **pretest probability** of ruptured membranes in the individual patient before us, after the gathering information stage.

Imagine that **Patient #1** presents at 22-weeks' gestation after she noted a small 3 cm wet spot in her underwear after getting home from exercising. There has been no subsequent leakage. She has no contractions, bleeding, or cramping, and a history of two healthy pregnancies with term deliveries.

Patient #2 presents at 41-weeks' gestation and reports a large gush of fluid while shopping for pickles, with fluid running down into her shoes. She reports continued leakage and she is now contracting every 3 minutes, painfully. She also reports that some of the fluid she has seen is green in color, looking very much like meconium. She is 8 cm dilated.

Is there a difference in the pretest probability of ruptured membranes in these two patients? Absolutely. **This is the whole point of clinical medicine.** Determining pretest probability **is** the purpose of the history (and, in many cases, the physical). We don't know the exact pretest probability of ruptured membranes in these two patients; but, we know that the probability of ruptured membranes in Patient #1 is very low, say 5% or less, and we know that the probability of ruptured membranes in the second patient is very high, at least 95%.

Now we can plug this information into the formula and determine the positive (and negative) predictive values of one of our tests when they are performed for each individual patient. Let's say that we do a nitrazine test on both patients – the same test, performed in the same way, by the same doctor, on the same day. Each test has the *exact same* performance characteristics, determined by its sensitivity and specificity. If the test comes back positive in both cases, what is the positive predictive value in each case? (There are calculators to do this for you, so don't worry about the math):

- Patient #1: PPV = 22.4%

- Patient #2: PPV = 99.05%

The false positive rates are simply calculated with the formula *100%-PPV*, or 77.6% and 0.95% respectively in this case.

What if the test were to come back negative? What would be the negative predictive value (NPV)?

- Patient #1: NPV = 99.56%
- Patient #2: NPV = 38.4%

The false negative rates are then *100%-NPV*, or 0.44% and 61.6% respectively for these patients.

All of this makes sense intuitively. In fact, Bayesian inference is how we naturally interpret the world. In the case of Patient #1, if the test were positive, we would assume it was likely a false positive and not trust the result. In the case of Patient #2, if the test were negative, we would intuitively assume that it was a false negative. The statistics bear this reasoning out, showing that our assumptions are statistically, or probabilistically, correct.

It should be clear that you cannot interpret tests – any test – without knowing the prevalence of the disease or the pretest probability of the disease based on clinical information, including risk factors. For screening tests, we use the prevalence of the disease in the population being screened. For symptomatic patients, we use our clinical knowledge and the likelihood of things on our differential diagnosis, to inform the pretest probability. This, in turn, combined with the performance characteristics of the test (the sensitivity and specificity), determines the positive and negative predictive values of the test, which is the knowledge we need to effectively utilize the test in clinical medicine.[52]

In the diagnostic process, many tests are often used; as we learn new information we change our understanding of the probability of the disease in which we are interested. It's a continuous, dynamic process that is changing with each new fact that we learn or each new test result that becomes available. One test's posterior probability (the post-test probability) becomes the next test's prior probability (the pretest probability).

Another important point to observe is that patients can be harmed by false negatives and false positive tests. Patient #1 might be harmed by a false positive exam and Patient #2 might be harmed by a false negative exam. Intuitively, you would not trust either of those false results and would treat the patient without respect to the test result. This begs the question asked in the last chapter, *Why order the test in the first place?* For these two patients, you should not. Tests are meant for patients with indeterminate probabilities, not for patients with extremely low or extremely high probabilities. If you are going to treat the patient regardless of the test outcome, don't order the test in the first place unless it is important to gain confirmation or to inform treatment.

Tests are not *positive* or *negative*. They are only useful for statistical modeling and at some point we have to gain the confidence, statistically, to either treat for a condition or not treat (the diagnostic or treatment threshold). ***But we are never certain.***

In the next chapter, we will also look at an example that uses likelihood ratios instead of the true positive and negative rates. First, here's another clinical example that I use commonly to illustrate this point to learners.

Cell-free DNA (cfDNA) analysis is a revolutionary new type of testing that can be used to detect fetal aneuploidies, like Down Syndrome. It directly tests circulating fetal DNA in the maternal blood, and has remarkable performance characteristics. The exact sensitivities and specificities vary by company and by study, but for Down Syndrome, the sensitivity is around 99.9% and the specificity is around 99%. This is not only an excellent test, but these performance characteristics are extraordinary. Almost no test in any field of medicine functions this well. So how does this extraordinary test perform in the clinic when used to screen for Down Syndrome? Let's again imagine two patients. The first is 20-years-old and the second is 40-years-old. We can look up the risk of Down Syndrome for those ages on a table (they are about 1/1000 and 1/70 respectively).

Now, each patient is sitting in front of you with a positive test result for Down Syndrome; again, the same test, performed on the same day, analyzed by the same lab. What do you tell them? Remember, this is about the best test in medicine in terms of accuracy and has both high sensitivity and specificity. Each patient is considering terminating the pregnancy in the event that the fetus has Down Syndrome. So, what's the PPV? Here are the results using the assumptions stated above:

- 20-year-old patient: PPV=9%
- 40-year-old patient: PPV=59%

These results are very non-intuitive. [53] Most obstetricians are shocked by these results. I have given this little quiz to hundreds of students, residents, attending physicians, maternal-fetal medicine specialists, and even statisticians. Nearly everyone has stated that the chance the fetus has Down Syndrome in both cases is over 99%. Most obstetricians would recommend termination for the 40-year-old patient without further testing (like CVS or amniocentesis) and many, in fact, would recommend termination for the 20-year-old as well. If this is one of the best tests in medicine, imagine how poor the performance for other tests are when the pretest probability for the disease is low, like CT scans, CBCs, TSHs, X-rays, and even biopsies. Let's look at this using a probability tree, which is sometimes more intuitive. First, let's look at the 40-year-old woman:

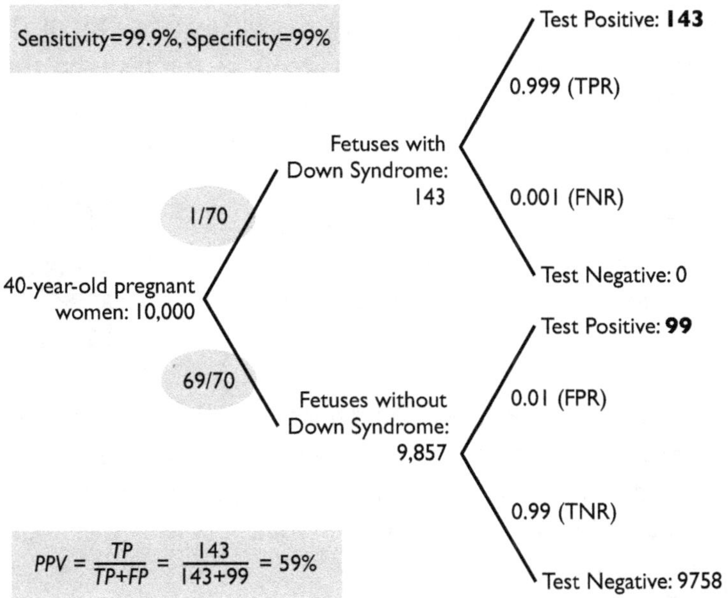

Figure 20. Positive predictive value (PPV) of a cfDNA test in a 40-year-old woman.

And now the 20-year-old woman:

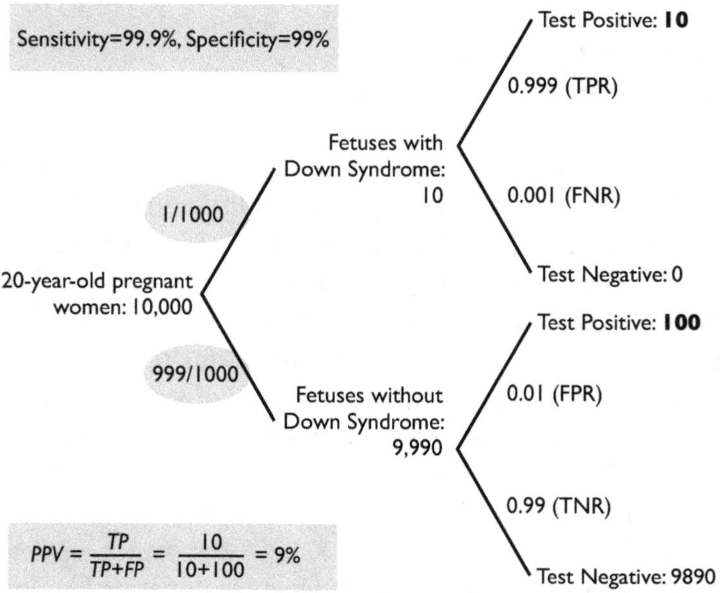

Figure 21. Positive predictive value (PPV) of a cfDNA test in a 40-year-old woman.

These probability trees are a quick and intuitive way of understanding and applying pretest probability to test results. Recall that the True Positive Rate (TPR) is the sensitivity of the test and the False Positive Rate (FPR) is 1-specificity. The rest is easy. The bottom line: false positive results are not equally distributed among different populations.

Most importantly, **tests are not binary!**

If you would like to play around with these ideas for other tests, check out getthediagnosis.org.

Exercises

10.1 Go to the website above. Search for "community-acquired pneumonia." Study the reported sensitivities and specificities for the various reported tests and exam findings. Draw probability trees (a blank tree is found at Figure 14 in Chapter 5) for a very sensitive finding and for a very specific finding.

10.2 Search for at least three other diseases or tests you are interested in and review the reported sensitivities and specificities.

10.3 Go to choosingwisely.org and search for recommendations from your specialty. Are there any things on the list you are doing wrong?

11

Making the Diagnosis

We've learned a lot of important concepts and many of them are extremely non-intuitive, at least at first. We need to integrate them all together and make a diagnosis. The process we are learning uses Bayesian updating (or probabilistic reasoning) to move us from chief complaint to diagnosis in the most accurate and efficient way possible. Let's implement our new skills by putting them to work in order to make a common diagnosis: a urinary tract infection (UTI).

How do we diagnose a UTI?

Urinary tract infections (UTIs) are commonly over-diagnosed in emergency departments and ambulatory settings. As many as half of all women presumptively treated with a UTI do not, in fact, have one.[54 55] In some settings, the rate of over-diagnosis may be dramatically higher. Some over-diagnosis is necessary (it's better to treat a few healthy people than to leave sick patients untreated); but in ambulatory or acute-care settings, a pattern of over-diagnosis may lead to several problems. For this diagnosis, consider these issues:

- The patient's actual problem may go untreated (such as an STI or other causes of the presenting symptoms).
- Unnecessary antibiotic exposure may later harm the patient, even leading to *C. difficile* infections or other severe complications.
- The clinician, who incorrectly diagnosed the UTI, may learn to misattribute some symptoms to UTIs that should not be attributed, leading to more over-diagnosis and misdiagnosis in the future with other patients.

Why are UTIs over-diagnosed? The leading cause is inappropriate utilization of urinalyses (UA). Many patient-care settings order routine urinalyses as if they were screening tests. Instead, they should be ordering them after considering the patient's symptoms and her pretest probability for having a UTI. One study found that more than half of UAs ordered in the hospital were inappropriate.[56] The rates may be even higher in ambulatory settings.

A better understanding of how to appropriately diagnose a UTI will help us solidify our understanding of the cognitive tools necessary to correctly diagnose any pathology. Let's look at what knowledge and skills are necessary.

- **We need a knowledge of how the disease presents.** The more we understand about the actual symptoms and risk factors of a UTI, the better we will be able to formulate an idea of the pretest probability of a UTI after we gather information.
- **We need to understand how well the tests for the disease work, and how they compare to the gold-standard test.** In the case of a UTI, we need to know which parts of a UA are informative (nitrites and leukocyte esterase) and what the positive predictive values of each part of the UA are compared to the gold standard of a urine culture with more than a 100k colony-forming units (CFUs).
- **We at least need to understand how well the test works in patients with a low, intermediate, or high pretest probability.** This allows us to integrate the pretest probability with the test results and arrive at a clinically useful post-test probability in a real-world setting.

- **We need to understand what level of certainty is necessary before initiating treatment (diagnostic or treatment threshold).** This level of confidence varies with the disease and the risks of the treatment; it varies with how important it is to be right or how dangerous it is to be wrong. This last step balances the risk-benefit equation which dictates all therapeutics.

These four elements are necessary to diagnose or treat any disease and it should be the goal of medical education to explicitly answer each question for the learner.

How do UTIs present? UTIs present as dysuria with or without other symptoms like urgency, frequency, hematuria, or suprapubic pain. These symptoms would be explained by the most common types of UTI, including urethritis and cystitis. In the case of pyelonephritis, additional symptoms of fever, chills, flank pain, nausea, vomiting, and costovertebral-angle (CVA) tenderness might be present. Also, the timing of these symptoms, in most cases, should be acute and progressive.

What symptoms are commonly misattributed to UTIs? In ambulatory settings, a variety of pain symptoms, including generalized pelvic pain and lower back pain, are often falsely attributed to UTIs. Also, fevers of unknown origin may be falsely attributed to UTIs. Note that flank pain and CVA tenderness should only be attributed to a UTI in the case of a patient who also has fevers and chills (that is, patients with pyelonephritis or a renal abscess). In other words, flank pain without fever is very unlikely to be any sort of UTI.

Risk factors for UTIs include sexual intercourse, new sexual partners, use of spermicides, history of a prior UTI, or history of UTI in a first-degree family relative. The following are **not** risk factors for recurrent UTI: precoital or postcoital voiding patterns, daily beverage consumption, frequency of urination, delayed voiding habits, wiping patterns, tampon use, douching, use of hot tubs, type of underwear, or body-mass index.[57]

The differential diagnosis for UTI symptoms in women includes: vaginitis, urethritis, interstitial cystitis, PID, herpes, chlamydia, bladder cancer, post-menopausal atrophic vaginitis, etc. This differential diagnosis assumes that symptoms like low back pain, generalized abdominal, or pelvic pain are not falsely attributed to UTI.

More than 50% of women who acutely have one or more symptoms of a UTI will have a UTI on urine culture, and more than 90% of women who have one or more symptoms of a UTI without symptoms of vaginal irritation will have a UTI on urine culture.[58] This means that the history is extremely important for diagnosing a UTI and most UTIs can be diagnosed from history alone.

How good are the tests for UTIs? A urine culture revealing more than 100k colony-forming-units is the gold standard for the diagnosis of a UTI. The disadvantage of this test is time and cost. Point-of-care urinalysis (UA) is the most commonly employed diagnostic tool. Only three parts the UA have any potential importance: Leukocyte esterase, nitrites, and blood. Of these, leukocyte esterase and nitrites are the most important. How do they perform?

In the last chapter, we showed how to determine this information from the published sensitivities and specificities. Here I would like to use positive and negative likelihood ratios so that we can demonstrate both methods for determining post-test probability.

- Nitrites alone: LR+ = 6.5 and a negative likelihood ratio or LR– = 0.58.
- Leukocyte esterase alone: LR+ = 1.4 and a LR– = 0.44.[59]

How well does this test work in a patient with low, intermediate, and high pretest probability for a UTI? What we really want to know when we order a diagnostic test is, *What is the post-test probability of the disease if the test comes back positive given what I already know about the patient?* To calculate the positive predictive value (PPV) using likelihood ratios, we use this formula:

$$PPV = \frac{\left(\frac{pretest\ probability}{1 - pretest\ probability} \times LR+\right)}{\left(\frac{pretest\ probability}{1 - pretest\ probability} \times LR+ + 1\right)}$$

Don't worry about the formula; use an online calculator to do the math (easycalculation.com/statistics/post-test-probability.php). But take note of the role of pretest probability in determining the positive predictive value.

Let's consider three ranges of pretest probabilities:

- *Low pretest probability,* such as an asymptomatic pregnant woman or a patient who receives a "screening UA." The prevalence of asymptomatic bacteriuria in pregnancy is about 6%, so we can use this as our pretest probability.
- *Intermediate pretest probability,* such as woman presenting with acute onset of one or more (real) symptoms of UTI. As we saw above, this history alone gives a 50% pretest probability.
- *High pretest probability,* such as the woman presenting with one or more symptoms of UTI and no symptoms of vaginal irritation, giving a pretest probability of 90%, also as noted above.

A lot of clinicians are freaked out by all this math. The thought of having to consciously assign a number for a precise pretest probability after seeing a patient is daunting. But don't get bogged down. Simply thinking in terms of *low, intermediate,* and *high (10% or less, around 50%,* and *greater than 90%)* is usually enough.

Using the above online calculator, here are the positive post-test probabilities or positive predictive values (PPV) for each scenario:

Low pretest probability (6%):
- Nitrites, PPV = 29.3%
- Leukocyte esterase, PPV = 8.2%

Intermediate pretest probability (50%):
- Nitrites, PPV = 86.67%
- Leukocyte esterase, PPV = 58.34%

High pretest probability (90%):
- Nitrites, PPV = 98.32%
- Leukocyte esterase, PPV = 92.65%

Thus, in a patient with a high pretest probability, performing a UA doesn't do much to update the pretest probability, which is so high to begin with that we might harm the patient in the event of a false negative. The presence of nitrites, in the case of the low and intermediate pretest probabilities groups, does significantly change our confidence about the diagnosis, but the presence of leukocyte esterase does not.

Tests are also sometimes important to help exclude a disease. But if you do the math using the negative likelihood ratios, you will find that a UA is of little value to exclude UTI; a urine culture is necessary for this.

What level of certainty is necessary to begin treatment? This question is the most complex question we must consider, and there are many variables for determining the diagnostic threshold or treatment threshold (which are often, but not always, the same thing). Overall, the interventions we use to treat UTI (antibiotics) carry extraordinarily low risks of unintended harm; still, they can cause allergic reactions and even death, as well as antibiotic resistance, alteration of GI flora, or simply the risk of missing the correct diagnosis due to misdiagnosis.

What are the risks of not treating, or delaying treatment while awaiting the results of a urine culture? Perhaps more pain and suffering for the patient, as well as the potential for the development of a worse infection (like pyelonephritis or sepsis). In the case of pregnant women or immune-compromised patients, the risks of not treating can be devastating.

Given these considerations, it makes sense that we do not need an extraordinarily high level of confidence to begin treatment, and how much confidence we need varies with the patient. For an immune-competent, non-pregnant young woman who is symptomatic, treatment based on symptoms alone (with a predictive value of at least 50-90%) is sufficient. We would likely do more harm than good if the UA came back normal (false negative) and we withheld treatment (especially since the test has poor NPV).

In the case of an asymptomatic pregnant woman, given the potential negative sequelae of untreated asymptomatic bacteriuria (like preterm labor or pyelonephritis), it seems reasonable to treat any patient who has nitrites present on UA (PPV of 29.3%) but probably not leukocyte esterase alone (PPV of 8.2%). Furthermore, ordering a urine culture for the asymptomatic pregnant woman with nitrites makes sense because there is a 70% chance that she is being treated mistakenly, and knowing this may influence the rest of her prenatal care (such as the decision to give continuous antibiotic prophylaxis for recurrent UTIs).

The most common location for over-diagnosis or misdiagnosis of UTI is in the ambulatory care or acute care setting, where a UA may be performed without sufficient pretest probability (often as part of "standard" or "rainbow" labs), and leukocyte esterase is discovered. The patient's pretest probability may be even lower than the 6% we have considered here in our calculations. The patient is then treated for a UTI, even though there is a 90-95% chance that she does not have one, with her presenting symptoms falsely attributed to a UTI. Her real condition often goes undiagnosed.

The most important skill necessary to prevent this from happening is a correct appreciation of the symptoms of a UTI in a non-febrile patient: new-onset dysuria, urgency, frequency, hematuria, or suprapubic pain (this is a *diagnosis script*). Don't order a UA unless one or more of these symptoms are present. If they are present, consider not ordering a UA anyway and treating empirically unless something complicates the picture (like vaginitis or chronic symptoms).

Diagnosis scripts

We have been utilizing the idea of diagnosis scripts throughout this book, but let's look at them more formally. Recall that a **diagnosis script** is a short constellation of symptoms, exam findings, risk factors, and perhaps lab or imaging results that call to mind a diagnosis. These scripts are important because they form the backbone of the information our brains (intuitively) search for as we gather information about a patient's complaint.

A faulty diagnosis script can destroy the diagnostic process before it even gets started. Creating accurate and discriminating diagnosis scripts should be at the very core of clinical medical education. If we misattribute certain symptoms to the wrong diagnoses or have a false impression of the risk factors or frequency of the disease (base-rate neglect), then we may still happen upon the wrong diagnosis even though we might have done everything else correctly.

So, what are the components of a good diagnosis script? Let's think about ectopic pregnancy for a moment. Here's an example of two diagnosis scripts representing different extremes of the clinical spectrum:

1. A woman presents with signs and symptoms of pregnancy or a

positive pregnancy test. A quantitative HCG is 7000. Ultrasound reveals no evidence of an intrauterine pregnancy.
2. A woman presents with severe abdominal pain and signs of hypotension. She has a positive pregnancy test. She has a history of pelvic inflammatory disease. On exam, she has evidence of peritoneal irritation and a Fast Scan reveals a large amount of free fluid in the pelvis and abdomen.

Diseases may present with different levels of severity and with vastly different symptoms and predisposing factors. This spectrum must be accounted for when generating diagnosis scripts. Students tend to learn "typical" presentations, but those typical presentations may not even represent a simple plurality of the various ways a diagnosis might present. Activation of an incorrect diagnosis script or failure to activate the correct diagnosis script will often follow.

What information must we know then to generate quality scripts? I would recommend considering the following issues for each disease process:

1. Frequency of the disease.
2. Symptoms of the disease and how often patients with the disease have each symptom.
3. Risk factors for the disease and how much each risk factor increases the probability of the disease.
4. Diagnostic criteria for the disease.
5. Special cases that may occur that may short-circuit activation of the diagnosis script for the typical presentation.

These five points are heavy on statistics, but as we have learned, this sort of information is essential for the type of reasoning we are learning in this book. These numbers needn't be precise; it is enough usually to think in terms of common and rare or usually and unusually.

For ectopic pregnancy, my diagnosis script is something like this: Ectopic pregnancies occur in about 1.4% of pregnancies. Patients may be asymptomatic or may present with mild to severe pain. If there is rupture, they might range from having mild peritoneal irritation from blood to having severe hemorrhagic shock. Most patients are asymptomatic or mildly symptomatic when diagnosed. Patients with previous pelvic infections or previous tubal disease or tubal surgery have an increased risk. Women using birth control are at an increased risk. A woman with a previous tubal ligation has a 1 in 3 risk of an ectopic if she becomes pregnant, while a woman with an IUD has a 1 in 6 risk of an ectopic with pregnancy. The diagnosis is clinical and should be presumed in any woman with severe pain and shock and a positive pregnancy test or when there is evidence of no intrauterine pregnancy and an HCG level greater than 3500. Some women can have both an intrauterine pregnancy and an ectopic pregnancy concurrently (a heterotopic pregnancy)

That's a lot of information for a "diagnosis script." The truth is, I know much more about the subject than those brief highlights but that information represents the minimum stuff I need to know to even begin to appropriately consider the diagnosis (remember, I can always look up more information about the disease or treatments, but I can't diagnose something I didn't think of first). The information in that paragraph covers both presentations I listed earlier and it prevents me from making fatal errors, like ruling out an ectopic pregnancy just because there is an intrauterine pregnancy. The information in the diagnosis script for a disease should prompt me to think of the disease and consider it in my differential on every appropriate occasion. So, for example, I will consider ectopic pregnancy in my differential diagnosis for every woman with a positive pregnancy test as well as every woman who presents in hemorrhagic shock or with acute pelvic pain, even if they have a known intrauterine pregnancy.

Generating these sorts of informed diagnosis scripts should be the goal of every student, resident, and physician for each diagnosis they might encounter. Students are hopefully used to seeing diagnosis scripts in clinical vignettes that form the basis for many standardized exams.

Accurate diagnosis scripts are essential to diagnosis. Remember, the mind is like a parachute: it only works when properly packed. Diagnosis is the chief clinical function of a physician. Every other function (therapy, surgery, medication, etc.) is secondary to accurate diagnosis.

Errors in the diagnostic process come in three main flavors: delayed diagnosis, missed diagnosis, and overdiagnosis. All three can cause severe patient harm.

Misdiagnosis

Misdiagnosis is an easy concept to understand and we have hopefully explored many of the reasons that misdiagnosis might occur. As we previously discussed, misdiagnosis occurs far too often in clinical medicine. The consequences of misdiagnosis include:

- Failure to provide treatment for the actual diagnosis.
- Harm and cost encountered from exposure to potentially wrong treatments.
- Erosion of the patient's trust in the medical system and harm in future illnesses due to decreased likelihood to access care from a physician or trust her recommendations.

Most misdiagnosis likely goes undetected; the patient may heal despite an incorrect diagnosis or the treatment provided for the incorrect diagnosis might also benefit the true diagnosis. Unfortunately, these cases often are a source of feedback sanction, meaning that the physician never learns the lesson of the misdiagnosis and may even increase the rate of misdiagnosis in the future.

When a misdiagnosis becomes apparent, particularly if the patient is negatively affected by it, then far too often there is also a second victim: the physician. This sequence of events is often a source of guilt and shame for the physician. While guilt can be a positive learning experience, leading to education and work to improve, shame can be very destructive. Shame is associated with disengagement, a loss of confidence, and feelings of the *imposter syndrome*.[60] This emotional environment, in turn, may create an unfavorable cognitive disposition to respond on the part of the physician that unfortunately leads to more mistakes and potential patient harm. *Error begets error.*

Naturally, there may also be medicolegal, privileging, or licensure consequences for the physician.

A useful way of communicating the impact a screening test can have on the rate of misdiagnosis is the **number needed to misdiagnosis** or **NNM**, where,

$$NNM = \frac{1}{1 - accuracy} = \frac{1}{1 - Specificity - Prevelance(Sensitivity - Specificity)}$$

So, a test that has an accuracy of 95% would have an NNM = 20, meaning that one out of every 20 persons tested would suffer a misdiagnosis (either a false positive or false negative). Tests with higher NNMs are more useful clinically.[61]

Delayed diagnosis

For diseases that are not self-limited, one of three things might happen: the correct diagnosis is eventually made, the patient recovers with supportive therapy even though a correct diagnosis was never made, or the patient dies. Recall the autopsy study that found that the cause of death in nearly half of the patients was different than the diagnosis that they were being treated for at the time of death.

Delayed diagnosis is often better than misdiagnosis, especially if the correct diagnosis is made soon enough to avoid permanent harm. Delayed diagnosis is almost inevitable with many rare diseases. For cost and efficiency reasons, it should not be a goal of the diagnostic process to eliminate delayed diagnosis. If it were, we would have to test for every disease (horses, zebras, AND Tasmanian tigers) every time we encountered any complaint. But we should seek to eliminate the delayed diagnosis of **"can't miss diagnoses."** Can't miss diagnoses are those diagnoses which, if not diagnosed in a timely manner, might lead to permanent disability or death, like ruptured ectopic pregnancy or appendicitis or pulmonary embolism.

Overdiagnosis

In practice, the concept of overdiagnosis may be somewhat harder to appreciate. When a disease is overdiagnosed, it is not a misdiagnosis, but it is diagnosis that shouldn't be made because the disease is indolent, non-progressive, or just irrelevant. It's not a false positive either; treating a false positive result would be an example of misdiagnosis. The patient has the disease – it's just that it doesn't matter and the patient may be harmed by potential treatments.[62]

Slow growing and relatively indolent cancers are often overdiagnosed. Examples might include papillary thyroid carcinoma, early stage prostate tumors, and even ductal carcinoma *in situ* of the breast.

Early diagnosis and treatment likely doesn't change patient outcomes compared to treating patients later whose cancers become progressive or symptomatic. Yet, many patients undergo unnecessary surgeries and other interventions as well as suffer sometimes life-altering anxieties for cancers that would have been of no consequence to them if not diagnosed.

Overdiagnosis is common with all screening tests and the frequency and limits of screening tests must be selected to help prevent overdiagnosis. For example, many screening tests are no longer offered after a certain age because the disease being screened for isn't likely to be the cause of death for the older patient. Understanding this, it is justified to extend some screenings for patients with an above-average life expectancy, but also to limit some screenings for those with below-average life expectancies. Imagine a 55-year-old patient who has a terminal pancreatic cancer with a life expectancy of less than a year: what sense does it make to do lipid screening, colorectal cancer screenings, pap smears, or any other test? Is it worthwhile to detect DCIS in the breasts of an 82-year-old with a five-year life expectancy? Or early stage prostate cancer in a 74-year-old patient with end stage emphysema?

In short, overdiagnosis occurs whenever a diagnosis is made of a condition that is not likely to cause symptoms, disability, or death for the individual patient. Harm occurs when the patients undergo interventions that are unneeded.

Many authors believe that overdiagnosis is not related to a failure of the diagnostic process itself. However, I would disagree with this assessment. The diagnostic process has not failed in the sense that a true and correct diagnosis has been made; but it is up to the clinician to not misuse this diagnosis in a way that might harm the patient. Few patients who are told that they have papillary thyroid cancer refuse surgery and few surgeons would feel comfortable recommending against surgery for it. Part of the diagnostic process is the judicious use of screening and diagnostic tests. Less overdiagnosis would occur if tests were only ordered in patient populations more likely to benefit from them, which means that the physician must account for the incidence of the disease or pretest probability of the disease for the individual patient as well as understand and communicate to the patient what harms or benefits might come from the testing before ordering the test.

We also must focus on picking the right outcome when trying to understand the risks and benefits of a test. If you want to increase the diagnosis and detection of papillary thyroid cancers, then you should perform thyroid ultrasounds on every patient you meet. But if your goal is to reduce the rate of death and disability associated with papillary thyroid cancers, you will find that those ultrasounds are the exact wrong strategy. Our goal shouldn't be to diagnose more cancers with screening tests; rather, it should be to reduce the overall risk of morbidity and mortality. We'll explore this concept in the next chapter.

Exercises

11.1 Pick three diagnoses that you commonly treat or are interested in. Write diagnosis scripts for each one. Don't do this based on memory; instead, use standard textbooks or review articles to write an informed diagnosis script for each one that includes an awareness of how common certain signs and symptoms are.

11.2 For each of these three diagnoses, which symptoms, exam findings, or tests might you commonly employ to make the diagnoses? Determine the sensitivities and specificities for these diagnostic clues and then calculate with an online calculator for PPV how well each finding predicts the disease given a low, intermediate, and high pretest probability.

11.3 Pick a diagnosis from your field of interest that is commonly overdiagnosed. What are some specific implications of this overdiagnosis? Consider financial, psychological, and medical implications.

12

Picking the Right Outcome

Now that we have done our best to accurately diagnose the patient, we need to think about how we select appropriate treatments. Many treatments in everyday use in medicine provide no benefit to patients. Companies extensively – even aggressively – market treatments to physicians and patients, and the products they sell are always presented in the best possible light. Physicians must be able to clearly and cogently decide which outcomes are important, and then search for and interpret the scientific literature with that aim in mind. In the next few chapters, we will discuss these skills. To do so, we must think about how to pick the right outcome.

What matters to your patient?

For many years, the American Academy of Family Physicians has promoted a summary of evidence-based medicine called a POEM or "patient-oriented evidence that matters."[63] The concept is simple and powerful. A POEM must address these three questions:

- Did the research focus on an outcome that patients care about

(e.g., morbidity, mortality, quality of life)?
- Is the problem that has been studied common and is the intervention feasible?
- Does the information have the potential to change the practice of many physicians?

What sorts of outcomes truly matter to your patients? They want to live longer, healthier lives; they want to hurt less; they want to have a higher quality of life; they want to be independent and functional; they want to feel well and energetic; they want to be able to work and pursue their interests; they want to be able to take care of themselves or their families; etc.

Illness trajectories are useful graphic depictions of the natural course of diseases with and without interventions. We used an illness trajectory earlier in Figure 10 in Chapter 4 while discussing regression to the mean. On the Y-axis is the state of function or health and on the X-axis is time. I like to apply this concept to life as a whole. In the graph below, we see that the ideal life would be to have excellent health and function for a long time and then die suddenly, without years of decline.

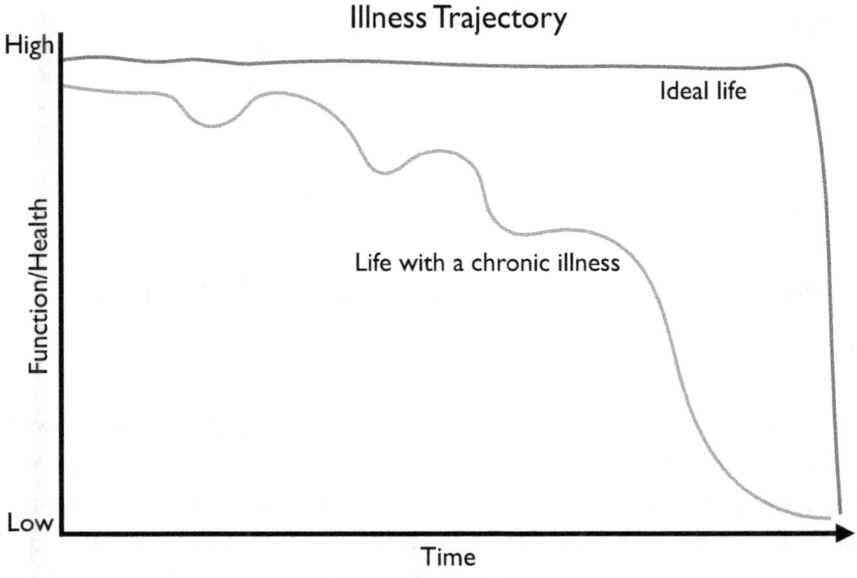

Figure 22. Illness trajectory.

Most lives are more like the second line, which depicts life with a chronic illness. There are ups and downs in health and function but a great number of years are spent in gradual and then rapid decline before death. Many (if not most) patients would rather have fewer years but with greater function and health than more years with poorer function and health. What would your patient choose?

Let's think about two interventions available for the same patient and compare them on an illness trajectory (Figure 23). If the patient receives no intervention, she declines rapidly and dies soon. If she receives Intervention 1, she maintains her quality of life for several more years before declining rapidly and dying. If she receives Intervention 2, her quality of life declines rapidly but she lives for many more years; in fact, she lives for several years longer than with Intervention 1. Which intervention would you choose? Which would your patient choose?

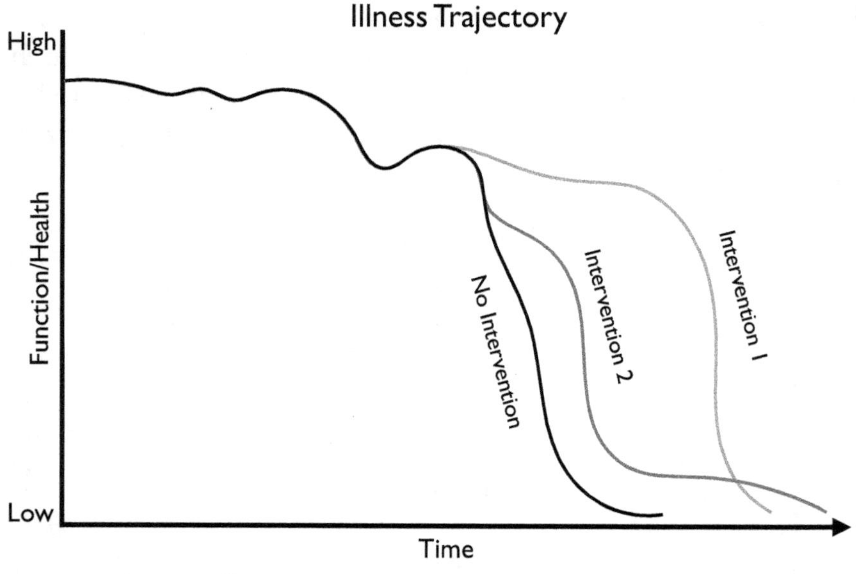

Figure 23. Illness trajectory comparing two interventions to placebo.

These sorts of trade-offs are common but are rarely discussed with patients. For example, perhaps a blood pressure medication will prolong a patient's life by a couple of years, but it comes at the expense of feeling fatigued, mentally foggy, and with diminished sexual function. The patient needs to understand the benefits clearly and the downsides clearly to make a choice. How important is it to take a particular medication? How does it benefit (or harm) the patient's life expectancy and his quality of life? A company marketing Intervention 2 would likely do well in the marketplace because it would tout its mortality benefit, though most patients, if fully informed, would pick Intervention 1.

Surrogate markers

Most physicians (and patients) can pick the outcome that matters, but many are led astray by surrogate markers. A surrogate marker is something other than the outcome that matters but it is viewed as something that corresponds to the outcome that matters. For example, you may care about decreased risk of death from stroke but you study blood pressure levels. The blood pressure is a surrogate marker.

Surrogate markers are easier to study. Following mortality benefit may take many decades, but studying whether a drug affects blood pressure, cholesterol levels, or blood sugar levels can be done quickly. But use of surrogate markers is responsible for most of the ill-advised treatment choices in medicine. Let's think about a few examples.

We already mentioned lipids. Most medications that treat lipids were studied and marketed for their ability to lower LDL or triglycerides or raise HDL with the assumption that this would translate into reduced cardiac mortality and morbidity. But this assumption has proven mostly incorrect. Niacin, for example, lowers blood cholesterol effectively but it does not reduce the risk of heart attack or stroke or prolong life.[64] Cardiovascular drugs are ripe with examples of heavily-marketed use based upon a surrogate marker but long-term medical reversal when definitive studies emerge that show no true mortality or morbidity benefit.[65]

We also mentioned blood sugar. Nearly all diabetes medications have been marketed for their ability to lower HgA1c (the average blood glucose over three months). Yet, there is strong and consistent evidence that HgA1c that is too low leads to an increased risk of death. So, again, merely looking at blood sugars isn't enough. In addition, we must consider adverse events related to the drug. Let's say that a diabetes drug was shown to lower the risk of heart attack in diabetics and death related from heart disease; that's still not enough. Heart attacks are only one thing that kills diabetics. We need to look at all-cause mortality. The drug may end up leading to the death of more patients than it prevents due to unintended, adverse effects.

These ideas matter both for diagnostic tests and screening tests. For example, what outcome matters when considering the benefits of mammography? Is it number of breast cancers diagnosed? Is it survival rate for breast cancer diagnoses? Is it mortality from breast cancer? None of those are correct. The number of breast cancers diagnosed isn't a good marker because all tests inherently over-diagnose and misdiagnose illnesses; many of the new breast cancers detected may not be clinically relevant or might not be cancers at all.

If we over-detect lesions that don't need treatment or diagnose healthy women with breast cancer, then "breast cancer" survival statistics will improve; women who don't have breast cancer or have other indolent lesions tend to do very, very well with treatment. What about breast cancer mortality? Mammography may lead to a decline in the number of women who die each year from breast cancer, but at what expense? If women develop other cancers (like lung cancer) from years of chest radiation, then total cancer deaths may be unaffected (and there's good evidence that they aren't). Also, how many healthy women suffered or even died due to therapies and interventions that they didn't need? The only fair outcome to consider is total mortality and morbidity with and without the intervention; that's the only way to capture both intended and unintended consequences.

How about if we were to design a study to look at the effectiveness of a drug designed to treat preterm labor, like a tocolytic drug? What would be the primary outcome of the study? Since the drug is designed to stop contractions, many researchers might pick frequency or number of contractions. Or, perhaps, whether the drug leads to fewer deliveries in 48 hours, one week, or before 37-weeks' gestation. Yet, again, these are all surrogate markers that may not predict the real outcome of interest: whether there is a reduction in neonatal morbidity and mortality. Even when considering neonatal morbidity, we must consider consequential outcomes. One recent study recommended a new intervention with potentially adverse side-effects because it led to fewer newborns needing extra oxygen by mask in the first two hours of life. Such a small benefit might easily be outweighed by other factors (like cost or increased risk of neonatal hypoglycemia).[66]

The use of surrogate markers may lead to false conclusions in many other ways. For example, among those who question the safety and efficacy of vaccines, or claim that vaccines cause autism or other serious health issues, there is a heavy focus on surrogate markers rather than the outcome that matters. For example, if one wanted to determine whether vaccines cause autism, it is a simple matter to compare the rates of autism (and other neurodevelopmental disorders) among children who have been vaccinated versus those who have not been vaccinated. Since the rates are not different (in fact, there are fewer neurodevelopmental disorders among vaccinated children), then opponents of vaccines focus on surrogates like aluminum or mercury or vaccine schedules. They demand evidence that those individual elements do not cause autism, but the studies rarely focus on individual elements rather than vaccines as a whole. In this context, the surrogate marker is a form of the *straw man argument*.

When we discuss the critical analysis of research studies in Chapter 15, an important part of the analysis will be to determine what outcome truly matters, and if the researchers used a surrogate marker, then what evidence exists that the surrogate marker is meaningful.

Unintended consequences

Just as important as picking the right outcome is appreciating other outcomes caused by the intervention. Medical decisions are made in the context of weighing risk versus benefit. To do this well, we must understand the real benefits and the real risks, and be able to communicate these to our patients effectively.

Medicine is rife with unintended consequences, partly because we tend to focus only on benefits when studying interventions. Many risks develop over a longer period than is usually captured in the context of a single study, and may not become obvious until years later. Worse, many unintended consequences aren't obvious because we simply aren't looking for them.

A drug may lead to lower blood sugars in the short term, but higher cardiac mortality in the long term. A tocolytic drug might prolong a pregnancy by a couple of days but lead to higher neonatal death due to higher rates of sepsis overall. In pregnancy, we screen all pregnant women for Group B Streptococcus (GBS) between 36 and 38 weeks' and treat colonized women with intravenous antibiotics during labor to reduce the risk of GBS-associated neonatal sepsis and meningitis. The strategy works and does reduce (but not eliminate) the risk. Yet, unintended consequences might include things like harm caused to the fetus or mother due to maternal antibiotic anaphylaxis, changes in GBS resistance patterns that might make treating affected newborns more difficult, changes in the resistance pattern of other pathogens that affect newborns such as *E. coli*, and changes in gut flora for the mother and child that might increase the risk of future serious bacterial infections. Unintended consequences might also include longer hospitalizations for children born to GBS-positive mothers that might increase the risk of iatrogenic exposure to other pathogens, increased cost associated with maternal and pediatric care that might force cuts elsewhere, etc. The 2002 study that led to our current treatment strategy did not consider most of these adverse outcomes.[67]

Some unintended consequences are worse than others.

Thalidomide was brought to market in 1956 and was used in pregnant women for treatment of nausea before any human safety studies were performed. More than 10,000 children were born with serious birth defects and more than 2,000 died before the drug was finally taken off the market in 1961. The companies that marketed the drug around the world had declared it safe during pregnancy even though no such data existed. Doctors, unfortunately, were more than willing to recommend it to thousands of pregnant women without asking critical questions such as *Does it work?* or *Is it safe?* Thalidomide was withdrawn from the market not because doctors were outraged after observing birth defects in the babies they were delivering but because of public outcry.

From 1940-1971, diethylstilbestrol, or DES, was given to pregnant women to prevent miscarriages. In 1971, a report in the New England Journal of Medicine linked DES to clear cell adenocarcinoma of the vagina in the children of the women who had taken DES while pregnant.[68] It took several more years for doctors to finally stop using the drug; off-label uses as an emergency contraceptive and to suppress lactation continued for nearly a decade after this report. Like thalidomide, no safety studies confirmed that the drug was safe to use during pregnancy. But, unlike thalidomide, doctors should have known early on that DES was not effective at preventing miscarriage. In fact, in

1953 in the American Journal of Obstetrics and Gynecology, Dieckmann et al. published a double-blinded clinical trial from the University of Chicago which showed that DES had no effect on the miscarriage rate.[69] There had never been any data that showed that it worked. Yet doctors embraced it as an acceptable therapy based on anecdote and marketing and despite a lack of efficacy or safety data. Then, even after a large, well-designed study showed that it didn't work, doctors continued to prescribe it for another 20 years, exposing almost 2 million women to the drug.

These examples show us how important it is to both define specifically what outcome you would like to achieve and to think through all the potential adverse outcomes that are possible. In the case of DES, the outcome should have been an observable decrease in miscarriage. But this clear outcome was muddied with surrogate markers ("it raises the patient's hormone levels and we know that low hormones are associated with an increased risk of miscarriage"). Worse, even the simplest unintended consequences weren't considered: if you give a drug to a pregnant woman, you must consider birth defects as an adverse outcome. In the case of thalidomide, the drug was effective and the adverse outcome should have been observed immediately; in the case of DES, the adverse outcome wasn't obvious, but the drug didn't do what it was intended to do and shouldn't have been used for that reason alone. The story of DES is a potent reminder that we should never do things that don't work, even if they seem harmless or they "make sense." The real harm may be unnoticeable or might even have its effect decades later. Thus, do nothing unless it clearly is beneficial, even if it appears harmless.

Population versus individual

It is often difficult for physicians and their patients to grasp small risks and benefits when appreciated in terms of the scale of a single patient. We will talk more about effective ways of communicating these risks in Chapter 17. But here I would like to focus on how we practice medicine and how our thoughts become biased. We are often taught to focus on the patient before us; we are told to think about what's best for her as an individual. Of course, it is true that we should make decisions with an appreciation of what she prioritizes, as we have already discussed. But when thinking about the best test or intervention, we must think in terms of a population, not an individual. Imagine that you have 1,000 or 10,000 patients like her, not just one. What treatment pathways would lead to the best aggregate outcomes for all 10,000 women?

Too often, physicians get caught up in *but what if?* sort of thinking when assessing treatment plans. For example, based on current evidence, a 40-year-old woman who has no risk factors for breast cancer should not get a screening mammogram. We have made that analysis based upon population level data that considers the pros and cons of performing mammography on a generic 40-year-old woman with no risk factors for breast cancer. But for the individual woman sitting in front of you, what if? What if she has a malignant lump? Mammography could save her life! She doesn't want to die of breast cancer, and her insurance will pay for the test. The risks of mammography are small and are only apparent when examined over a large population of patients. So, for the individual patient, it might seem like a worthy intervention. But the small risks taken by each patient exposed to the risks of mammography soon add up. Some women, somewhere, will be harmed by the test, and the number of women harmed will soon outweigh the number of women who receive benefit. Recall the "aggregate bias" which refers to this inappropriate belief that what is best for a whole population does not apply to an individual. Thinking on a population level can help keep risks and benefits in perspective.

The Quadruple Aim

What are the ultimate goals or outcomes we would like to see in medicine? This idea is important to think about. Everything we do in medicine, ideally, would lead towards these ultimate goals. One frequently cited set of goals for medicine is called The Quadruple Aim. These four goals are:

1. Improved population health
2. Lower cost of care
3. Better patient experience
4. Improved physician experience

Note carefully that the first aim is to improve *population health,* not individual health. We do not dedicate all our resources to make one individual life as good as possible. This would be a violation of the ethical principle of *justice*. Rather, we use our scarce resources to make the health of a population as good as it can be with the resources we have available. Again, to guard against risk/benefit distortion, you must think in terms of 1,000 patients like the one sitting in front of you, not just the individual.

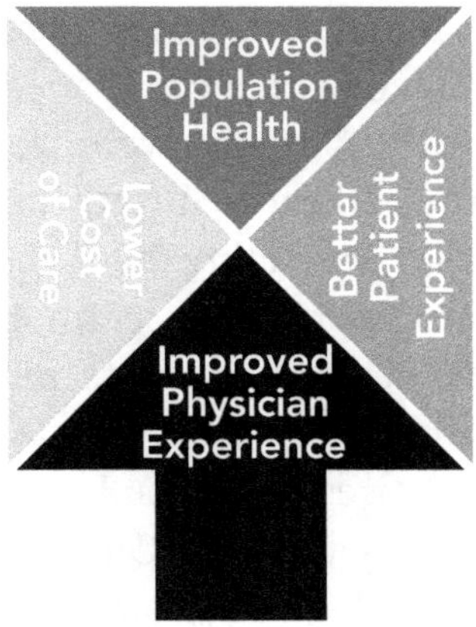

Figure 24. The Quadruple Aim.

Lower cost of care follows adoption of evidence-based practices. With so much over-testing, misdiagnosis, over-diagnosis, unnecessary and incorrect interventions, we end up wasting a huge amount of resources. The more strictly we follow quality, evidence-based guidelines, the less money we will spend per capita while also achieving better outcomes.

The patient experience is improved when doctors provide the right diagnosis at the right time, and consequently recommend the right treatments. When these first three aims are achieved (formerly called The Triple Aim), then the result is an improved physician experience. Physicians are increasingly burnt out because they don't feel like they can take care of patients in the way they should and provide high-quality care to them. Fears of litigation, increasing paperwork burdens, frustrations with employers and payers, all contribute to high levels of frustration among physicians. Frustrated physicians make more mistakes, retire early, leave medicine, develop substance abuse disorders, and even commit suicide. None of those things are good for patients.

Ultimately, the quality of healthcare must be measured by outcomes. If a patient presents with a problem, the only real question is whether we make her life better, worse, or no different by our actions. It is, therefore, vitally important to understand the desired outcome and to measure it.

Exercises

12.1 Go to tripdatabase.com. Perform a '**PICO search**.' Specify the patient population, intervention, comparison, and outcome that your patient desires. Examples are provided on the website. This type of search can also be performed directly in PubMed and is the process for generating POEMs.

12.2 Think of three surrogate markers commonly studied in your area of interest. What unintended consequences might be masked by using each of those surrogate markers?

12.3 Pick the most recent treatment you have recommended to a patient. Does the treatment satisfy the Quadruple Aim? Was there a lower cost option? Does high-quality data show that it improves population health? Does it improve an outcome important to your patient? Does it have any undesirable trade-offs for your patient?

13

Selecting Treatments

The goal of the medical processes we have been describing is to accurately diagnose problems and then offer effective solutions that improve the quality and quantity of life for our patients. Sometimes we cure diseases, sometimes we keep them stabilized on a chronic basis, sometimes we offer supportive care while the body cures itself, and sometimes we offer supportive care while the body loses its battle with the disease. But each time we offer a treatment we should be explicitly clear about its risks and benefits, as well as how effective it might be towards achieving our desired goal.

Treatments come in a variety of forms:

- Advice or counsel about coping with the disease
- Recommendations about how to modify issues under the patient's control that affect her health or disease process (e.g., habit modification, diet, exercise, avoidance of stressors, etc.)
- Statements of encouragement, affirmation, and support
- Therapy (e.g., physical, cognitive, occupational, rehabilitative, etc.)
- Medications
- Surgery

Ethically, no treatment should be selected or offered to a patient unless the physician can specifically answer these questions:

- What outcome does my patient desire?
- What are the risks of the treatment and how common are they?
- What are the benefits of the treatment and what is the magnitude of effect (i.e., how much will this improve my patient's disease)?
- Is there scientific evidence that this treatment will help my patient achieve her goals?
- Does my patient have a compelling reason to avoid the treatment?
- Are there unintended consequences of the treatment?
- Are there less expensive, less invasive, less dangerous, or less noxious treatments which provide the same or nearly the same outcomes?

Many, if not most, interventions offered to patients by physicians offer little to no specific benefit to the patient. Too much intervention is rampant in the healthcare industry. Many surgeries are performed unnecessarily: 70% of hysterectomies, 12% of angioplasties, 22% of pacemakers, at least 17% of spinal fusions, 26% of hip replacements, 38% of knee surgeries, and probably half of cesarean deliveries are unnecessary.[70] Overall, 10-20% of surgeries are performed for no benefit. Even when benefit is present, it may not be worth undergoing the surgery to the patient given how small the magnitude of benefit is versus the risks or costs. The same issue exists for prescription drugs; it is estimated that at least 1 in 3 drugs are unnecessary and many more provide such a marginal benefit that the cost/benefit ratio or the benefit/side effect ratio is not justified.

The **therapeutic imperative** leads to too much intervention and has led to an epidemic of polypharmacy. Physicians feel that they must offer *some* treatment for a patient's complaint, even if the treatment is of little or no value. Advice and health recommendations, too, are doled out routinely with very little scientific evidence of their efficacy. Sometimes these recommendations are innocuous; other times they can be deadly. Recall the example of bed rest for pregnancy complications or threatened preterm labor or miscarriage: the practice offers no benefit but can lead to serious harm, such as pulmonary embolism. Even when physical harm doesn't occur, psychological harm may. **Guilt sanctioning** may occur when, for example, a woman is told to be on strict bed rest due to a threatened miscarriage. She cannot comply with the requirement because she has two other small children to care for and she does. When the miscarriage happens, she blames herself for noncompliance with the physician's (nonscientific) recommendation. The physician sanctions her guilt, sometimes explicitly, sometimes implicitly. Seemingly benign phrases like *take it easy and let's see what happens* can and often do lead to guilt sanctioning.

Our treatments for patients are selected in the context of medical ethics. A good understanding of medical ethics makes obvious the previously mentioned questions that the physician needs to answer about interventions before recommending them. Let's quickly review medical ethics.

Medical ethics

Medical ethics are quite simple:[71]

- **Beneficence:** Always act in the best interest of the patient (not in your best interest or your employer's best interest);
- **Nonmaleficence:** Don't be the source of harm to your patient (or if you are, make sure it's worth it – make sure the benefit of what you're doing outweighs the risk);
- **Autonomy:** Let the patient decide the best course of action given her wishes, personal values, and goals (and remember, you have autonomy, too, in the physician-patient dyad);
- **Justice:** Treat all patients equitably and consider the economic impact of how you treat an individual patient in terms of how it impacts your ability to treat others (when you overspend resources on one patient you are harming other patients by making those resources unavailable).

Again, consider the questions about interventions that I posited above: if you can answer these questions, you will select and apply interventions in an ethical manner. Unfortunately, there are many perverse incentives that seem to dominate medical care in the United States. Combined with a lack of knowledge about the risks, benefits, costs, and alternatives to various treatments, these perverse incentives have created an environment of over-utilization and poor treatment choices that often don't satisfy these ethical principles. Some of these perverse incentives include a fee-for-service environment where over-utilization is rewarded; fear of litigation for inaction; a desire to please employers who financially benefit from utilization; etc. Physicians are also manipulated into inappropriate utilization from companies that stand to benefit, including pharmaceutical companies, medical and surgical device manufacturers, and medical services providers.

To illustrate the thought processes needed to pick appropriate interventions, let focus on a task that all physicians must do at some point in their careers: prescribing medications.

Which drug should I prescribe?

The first key to correctly picking a drug is making the correct diagnosis. If we treat the wrong disease, even with the right drug, we aren't likely to get the desired outcome. Selection of prescription drugs is one of the most important (and costly) things that physicians do. How should we decide which drug to prescribe?

First, let's consider a patient who represents an almost typical patient for many primary care providers: a diabetic who has neuropathy, hypertension, hyperlipidemia, gastroesophageal reflux disease, hypothyroidism, bipolar depression, and an overactive bladder. By the way, the average American between ages 19 and 64 takes 12 prescription drugs (many of which are unnecessary).

For each issue on the problem list, let's select two common treatments. Doctor A tries to use lower cost drugs and generics when available. Doctor B favors newer drugs and believes that he is providing world-class medicine (at least that's what the drug salespeople keep telling him). Let's see what it costs Doctor A and Doctor B to treat this patient each year. Also, we'll see what the ten-year cost of treatment is and multiply that for 500 patients (to represent a typical treatment panel of patients that a primary care doctor might have like this).

Comparison of Two Prescribing Habits

Disease	Doctor A	Cost A	Doctor B	Cost B
Hypertension	Lisinopril	$4	Edarbi	$209
OAB	Oxybutynin	$4	Enablex	$328
Diabetes	Metformin	$3	Victoza	$1126
Diabetes (2nd)	Glyburide	$12	Januvia	$565
Depression	Escitalopram	$4	Viibryd	$342
Hyperlipidemia	Simvastatin	$6	Livalo	$372
Hypothyroidism	Levothyroxine	$10	Synthroid	$53
GERD	Omeprazole	$8	Dexilant	$202
Neuropathy	Gabapentin	$12	Lyrica CR	$376
Bipolar	Quetiapine	$9	Latuda	$1219

Costs above are for one month prescription (October 2019 prices)

Monthly cost per patient	$72	$4,792
Yearly cost per patient	$864	$57504
10-year cost (at 6% interest)	$11,755	$782,362
10-year cost for 500 patients	$5,877,500	$391,181,000

Table 8. Cost comparison of a doctor who prescribes generic drugs (Doctor A) versus branded drugs (Doctor B).

Table 8 above shows this comparison, along with an adjustment for interest of a 10-year period. The difference in the two approaches is dramatic: Doctor B has spent over one third of a billion dollars more than Doctor A in just 10 years, treating just 500 patients. Add to this the increased costs of poorly selected short-term medications (such as choosing Fondaparinux versus Lovenox), unnecessarily prescribed medications (like antibiotics for earaches, sore throats, and sinus infections), unnecessary tests (like CT scans for headaches or imaging for low back pain), and the cost differences between Doctor A and Doctor B soar, all based upon decisions that the two doctors made in different ways.

In a career, Doctor B may easily cost the healthcare system $1 billion more than Doctor A. The cost of prescription drugs (and more importantly, the way doctors decide to use them) more than explains the run-away expense and relatively low value of the US healthcare system. The top 100 brand-name prescription drugs in the US (by dollar amounts) netted $194 billion in sales per year as of 2015 (plus pharmacy mark-ups, etc.).[72] The list of drugs that year included several of the drugs listed in Table 8. Some of those drugs are now available as generics, and others have seen their prices double between 2015 and the time of writing this chapter. So, Doctor B's choice of drugs is representative for the choices far too many doctors are making in the US.

Now in fairness, the most prescribed drugs by number of prescriptions are generics, but because they are so inexpensive, they do not make the top money list. Doctor A's choice of Metformin, Levothyroxine, and Omeprazole are all among the Top 10 most prescribed generics. Unfortunately, the most prescribed drug of all (with 123 million prescriptions in 2015) was generic Lortab.

So, who's right, Dr. A or Dr. B? Let's go through a process to determine what drug we should prescribe. To do this we need consider a series of questions.

Is a drug treatment necessary?

Just because a patient has a diagnosis doesn't mean she necessarily needs a drug to match it. Physicians tend to overestimate the benefits of most treatments and do a poor job of conveying the potential benefits (and risks) to patients. Non-medical treatments (like lifestyle changes, counseling, etc.) are undervalued and perceived as less effective or just too much work. Providers often feel like they have to write a prescription or they will have an unsatisfied patient (in other words, they feel that if a patient presents with a complaint, it must be addressed with an intervention rather than just education and reassurance). Even worse, many clinicians believe that if they prescribe a medication during a visit, they can bill for the visit at a higher level, so they feel financially rewarded for prescribing. Here are some common issues that serve as examples of such over-prescribing:

- Psychotropic medications are notoriously over-prescribed (rather than utilizing more time-consuming interventions like psychotherapy), particularly antipsychotics, which are heavily marketed and often prescribed by non-psychiatrists. Stimulant medications, too, have become heavily abused and over-prescribed.[73]
- The elderly tends to accumulate diagnoses over time and are vulnerable to polypharmacy, often treating the side effects of one

drug with another drug. An Austrian study found that 36% of elderly internal medicine patients were receiving at least one unnecessary drug.[74]

- Antibiotics are incredibly over-prescribed. For example, a 2016 study revealed that at least 1 in 3 antibiotic prescriptions was unnecessary.[75]

In general, overuse of prescription medications happens for the following reasons:

- A drug is used for a condition that should first be treated with lifestyle changes or other non-pharmacological interventions (e.g., mild hypertension, obesity, mild hyperlipidemia, insomnia, anxiety, mild depression, etc.).
- A drug is used for a condition that is actually a side-effect of another medication (e.g., constipation due to anticholinergics, sexual dysfunction due to an antihypertensive or an SSRI, or insomnia due to a variety of drugs).
- A drug is used for a condition that would get better without treatment (e.g., viral URIs, distal urethritis, musculoskeletal sprains and pains, earaches, etc.).
- A second drug is used for a condition when another drug is already being used. Many times, this is necessary, but all too often, two drugs are used when only one would do (perhaps at a higher dose).
- A drug is used for a condition but offers almost no benefit for the condition (e.g., using SSRIs to treat premature ejaculation, even though the time from penetration to ejaculation is only lengthened by a few seconds).
- A drug is prescribed with a proven benefit but the measured benefit is not considered worthwhile by the patient (e.g., a healthy 45-year-old woman taking a statin drug to reduce her risk of a nonfatal heart attack by 1 percentage point over ten years).

Once it is decided that a drug is necessary for treatment, the next question is, Which one?

Which drug is appropriate?

To determine this, we must first decide what the goal of treatment is in order to select a medication that satisfies that goal. This sounds obvious, but it isn't always done appropriately. The next part is trickier. In general, we need to select the lowest-cost medication that fulfills this treatment goal. Dr. A has done this in the previous example quite well. Dr. B would rationalize his choices by adding that he has also picked either the most effective drug or the one with the least side effects. This rationale sounds attractive, and drug companies prey on the desire of physicians to use the best drug with the fewest side effects (and the ego-boosting idea that they are being innovative). But it is the wrong strategy. Here's why.

Let's say that Drug A is 90% effective at treating the desired condition and carries with it a 5% risk of an undesirable side effect. Meanwhile, Drug B is 95% effective at treating the condition and carries only a 2.5% risk of the side effect. Drug B is then marketed as having half the number of treatment failures as Drug A and half the number of side effects as Drug A. Drug A costs $4 while Drug B costs $350. So, which drug should we use? If 100 people use Drug B, then approximately 92 will be treated successfully without the side effect at a cost of $386,400 per year. Eight people will remain untreated. If 100 people use Drug A, then approximately 85 people will be treated successfully without the side effect at a cost of $4,080 per year. After failing Drug A, an additional 7 people will use Drug B successfully at a cost of $29,400 per year, with 8 people still untreated. This try-and-fail or stepwise approach, rather than the "Gold Standard" approach, treated just as many people successfully but saved $352,920 per 100 people per year.

This scenario assumes that such dramatic differences between the two drugs even exist. The truth is that this is rarely the case. Such dramatic differences in side effects and efficacy do not exist for the drugs listed in Table 8. What's worse, in many cases the more expensive drug is inferior. HCTZ has long had superior mortality benefit data compared to ACE-Is and ARBs, but due to clever marketing, it has never been utilized as much as it should be in non-diabetic hypertensive patients. Metformin is simply one of the best (and cheapest) drugs for diabetes, yet no one markets it and there are no samples in the supply closet. Fondaparinux is an important drug for the rare person who has heparin-induced thrombocytopenia (HIT), but the incidence of HIT among users of prophylactic Lovenox is less than 1/1000, hardly justifying the 60-fold increased price of fondaparinux. The clinical differences between Cozaar and Benicar can hardly begin to justify the 31-fold price difference (before Benicar became available as a generic). To decide by policy that every patient with overactive bladder should receive Vesicare ($374), Detrol LA ($235), or Enablex ($328) as a first line drug when most patients are satisfied with Oxybutynin ($4) is the attitude that is bankrupting US healthcare. In some cases, physicians prescribe the exact same drug at a costlier price (e.g., Sarafem for $486 instead of the chemically identical fluoxetine for $6).

A word of caution about interpreting drug comparison trials: drug companies have tens of billions of dollars at stake when marketing new drugs. The trials that are quoted to you are usually produced and funded by those drug companies. They only publish the studies that show positive results. Drug studies are, by far, the most biased of all publications so be incredibly careful in believing the bottom line from the studies. We will talk about this issue in a few chapters. Here's another of my axioms: **If the drug were really that good, it wouldn't need to be marketed so aggressively.**

Even when one drug is substantially better than another, it still may not be appropriate to prescribe it. Many times, the prohibitive cost of the perceived "better" drug prevents patients from getting and maintaining use of the drug, leading to significantly poorer outcomes than would have existed with the "inferior" drug. The best drug is the one the patient can afford to take.

Thus, we should pick the drug that is designed to achieve the desired goal that is least expensive (or at least pick a drug that is available as a generic), then specifically decide why that drug is not an option for a given patient before picking a more expensive one (e.g., the patient is allergic to the drug, the patient previously failed the drug, etc.). To aid in this cognitive work, physicians must know the cost of the drugs they prescribe. A good source is goodrx.com, which lists the retail prices of drugs and also provides coupons for patients to get the best price at local pharmacies.

Often during marketing of drugs, the conversation about drug costs is steered toward the cost to the patient. Many makers of very expensive drugs will help pay for the patients' copays or even help meet the patients' deductibles so that the financial burden to the patient is lessened. But the actual cost of the drug (to our healthcare system) is just as important. The ethical principle of justice demands that we consider the societal impact of the cost of our interventions, not just the personal impact. Patients are harmed by the total cost, just in a less direct way. Thousands of dollars per year of medications result in higher insurance premiums, higher deductibles, and higher taxes to finance government-funded insurance plans.

Of course, drugs are not the only thing we spend money on; the same thought processes must be applied to surgeries and procedures as well as labs and imaging, etc. The healthcare industry actively seeks to influence in any way possible the decisions we make about interventions and medications and this influence weighs heavy in physicians' thought processes.

What influences our prescribing habits?

There are two important facts about drug company marketing.[76] The first is that physicians do not believe that they are influenced by drug company advertisements, representatives, free lunches and dinners, pens, sampling, CME-events, etc. The second is that drug company marketing is the number one influence on clinicians' prescribing habits. In 2012, US drug companies spent $3 billion marketing to consumers while spending $24 billion marketing directly to doctors.[77] They spend that money because they know it yields returns.[78]

Remember that just one Dr. B is worth over one-third of a billion dollars in a ten-year period to pharmaceutical companies. They are more than happy to buy as many meals as it takes. But doesn't this money support research? It's true that most drug companies spend billions per year on research and development, but the total R&D expenditures are usually about half of marketing expenses alone, and obviously, a fraction of the $350+ billion revenues. But don't samples help me take care of my poor patients? No. They help influence your prescribing habits and they give you the impression that the drug is a highly desirable miracle that would be wonderful to give away to the needy because it's just that good. What can help you take care of your patients? Four dollar generic drugs.

These problems would be much worse than they are if insurance companies ("the evil insurance companies") didn't have prescribing tiers that force patients to ask their doctors for cheaper alternatives and if they didn't deny coverage entirely for many brand-name drugs. But it is still an enormous problem, resulting in hundreds of billions of dollars of excess cost to the US healthcare system each year.

Simply by being more like Dr. A and less like Dr. B, physicians could rapidly and dramatically reduce the cost of US healthcare.

Value-based care

In the last few pages, we have discussed the Quadruple Aim and medical ethics. I would like to take a moment here to discuss value-based care and to unify these concepts into one framework.

Value-based care is popular today as an alternative to the fee-for-service models that have dominated medicine for the last several decades. In a fee-for-service environment, physicians (and hospitals and everyone else) are rewarded financially for doing something – whether that something is valuable or not. This has created a system of perverse incentives because the healthcare industry is remunerated not for excellent outcomes but simply for doing something, and this necessarily had led to over-utilization of services. In a fee-for-service environment, it doesn't matter if my patient needs a hysterectomy or even if I am good at doing the procedure, it just matters that I do the hysterectomy. In a perverse twist, if I don't do unnecessary things, I am financially harmed. In a fee-for-service model, I almost daily choose between doing what's best for the patient and what's best for my income. This must end.

Value-based care seeks to change these perverse incentives. It's complicated to implement and the healthcare industry is struggling to find a fair method. But the idea is simple and it follows a simple formula:

$$Value = \frac{Quality}{Cost} = \frac{Outcomes}{Direct + indirect\ costs}$$

To be a faithful practitioner of value-based care, then, you must understand two things about the care you deliver:

- How much does it cost? This includes both direct and indirect costs, like lost economic productivity while recovering from a surgery, for example.
- What is the magnitude of effect? In other words, how beneficial is the intervention compared to its potential harms?

For a drug that costs $400 per month compared to a generic drug that costs $4 per month, the 100-fold price difference might be worth it in a value-based environment, but the difference in magnitude of effect would have to be very significant. It's not enough that it might be a *little better* or have a *slightly better* side-effect profile. This is all the more true when you realize that the excess money being spent might be better spent elsewhere (i.e., *justice*). In value-based care, it's all about bang for the buck. **Just because something has value doesn't mean it's valuable.** An expensive drug might provide slightly better outcomes for the patient, but there isn't an infinite amount of money to spend on each patient, so we must spend it wisely.

This way of thinking is uncomfortable to many physicians and patients. It forces us to put a dollar amount on life or quality of life. But only by doing that can we make sure that we are providing the most we can with our scarce resources. It must be quantified.

The concept of value-based care is really not different than the Quadruple Aim or the four principles of medical ethics. They are different ways of saying the same thing.

Let's use a clinical example to illustrate this.

Let's say a woman needs a hysterectomy. The physician must choose which route of hysterectomy is best for her. In almost all cases, the answer is a vaginal hysterectomy. From a value-based perspective, it is the cheapest method of hysterectomy compared to other methods like open, laparoscopic, or robotic, considering both direct costs (the actual cost of doing the procedure) and indirect costs (since it has the quickest return to normal activities). In fact, it is the only method of hysterectomy that is routinely profitable for hospitals since the cost is several thousand dollars less compared to other methods. It is the method of hysterectomy with the best outcomes (less bleeding, less pain, less infection, fewer complications, least invasive, etc.). It is a value-based win.

From the perspective of the Quadruple Aim, it lowers the cost of healthcare, improves population health compared to other methods, provides a better patient experience (quicker return to activities, less pain, better cosmesis, etc.), and improves the physician experience since she will be rewarded for not losing money for the hospital and insurance companies (less hassle, fewer prior authorizations, etc.). It is a Quadruple Aim win.

From the perspective of medical ethics, it satisfies the principle of justice for the reasons just discussed. Since it provides the best outcomes, it satisfies the principle of beneficence. It doesn't expose the patient to unnecessary harms (like punching holes in patients' abdomens that might damage their intestines) and therefore satisfies nonmaleficence. A properly-informed patient would almost always choose vaginal hysterectomy compared to other routes and therefore satisfies autonomy. Yet, vaginal hysterectomy is almost never performed.

These three frameworks all lead to similar conclusions when applied to a given medical question. The difference is how that conclusion is communicated. Physicians are offended when their ethics are challenged. They want to make a distinction between intent and outcomes (e.g., their intentions were ethical even if it didn't turn out well). The Quadruple Aim is a lofty aspiration that asks the healthcare system to keep its eye on the ultimate goals, but provides no incentive (moral or economic) to do so. But value-based care, appropriately implemented, will hopefully provide economic incentive to do the right thing.

Whatever framework you choose to use, every action, from prescribing a drug to ordering a test or performing a surgery, must be considered from this perspective.

Exercises

13.1 Go to goodrx.com. Look up the last five drugs you prescribed. Are there less expensive alternatives?

13.2 If you have recently received an industry-sponsored drug or product presentation, did you choose to use the product? Did you do any independent research? Did you look for literature that compared it to less expensive alternatives? If not, perform a PICO search that does so now.

14

When to Ask for Help

No physician can know everything. Medicine is so complex that, by necessity, we live in an age of specialization. But there must be a balance: asking for help too often encourages over-utilization and diminishes the patient experience; asking for help too rarely may lead to bad outcomes for patients. This idea may not seem worthy of its own chapter, but there are some important issues to consider.

What are your limitations?

In a broad sense, this question is answered by your training in medical school and, more importantly, in residency. It's not just what you know but what you can do.

If you've gone to medical school, you should know how to treat hypertension, but even a simple diagnosis like this could occur in such a complicated patient or be so refractory to care that the patient will be better treated by a subspecialist, such as a cardiologist (even if you are an internist). Where these lines occur are blurry, but it necessitates the physician knowing her limitations and putting the patient first. This sometimes even means referring to another physician who shares your specialty. Let's say you are a gynecologist and your patient needs a hysterectomy but you don't have the skillset to perform a minimally-invasive hysterectomy (such as a vaginal hysterectomy) but your competition across town does; you should refer the patient. If your patient has a diagnosis that you never treat, particularly of a chronic condition, you should refer the patient to someone who treats that diagnosis more often. Asking for such help occurs both too often and too rarely. Inappropriate incentives affect both cases and both can lead to patient harm.

Some of the factors that can lead to not asking for help when it is needed include:

- Fear of lost income (particularly with surgeries)
- Fear of appearing incompetent in front of the patient (particularly when referring to another physician within same specialty)
- Over-confidence

But the opposite issue occurs more frequently; physicians and other providers tend to consult far too often and refer patients with little impunity. In the hospital setting, some of this over-consultation is encouraged by economic demands to support subspecialists. If a hospital has a subspecialist on staff, they will be utilized or they will be lost. An internist might suddenly find herself consulting the new cardiologist for a patient with hypertension that she might have previously treated on her own. She does this out of a spirit of collegiality and because she wants the new physician to be utilized enough to stay, but this sort of over-utilization causes an explosion of healthcare costs.

Others over-consult purely out of laziness or perceived medicolegal implications. Far too frequently, a busy hospitalist may consult four or five subspecialists during a simple admission with a prime incentive being reduction in work for the hospitalist, not because the patient's diagnoses are beyond the training of the hospitalist. These sorts of fruitless consultations are expensive, frustrating to the patient, and often lead to disjointed care.

Some subspecialists will promote a sense of medicolegal fear to gain routine referrals from primary care physicians. In Obstetrics, for example, it is common to see every patient over age 35 in a practice referred to a Maternal-Fetal Medicine (MFM) subspecialist for consultation regarding "advanced maternal age." But this is counseling that every practicing obstetrician should be able to perform and only in very rare cases is there a need for such consultation. But, the referral saves the obstetrician time, provides a false sense of medicolegal protection, and perhaps helps keep the MFM in business. Still, it is inappropriate. Such referrals must be thought of in terms of the Quadruple Aim. In this case, the referral increases the cost of care, doesn't improve population health, and worsens the patient experience (who now must pay more for her care, miss more work, wait longer for answers, etc.). There is no virtue in sending a patient to another doctor to do what you are already trained to do, no matter how one might rationalize it.

What help is available?

Asking for help doesn't necessarily mean referring a patient or requesting a consultation; in fact, it shouldn't normally. The most basic way to get help is to read. As we discussed earlier in this book, overreliance on our memories is a major contributor to medical error. Physicians should be adept at scouring the literature for answers to clinical questions. We will discuss this activity in detail in the coming chapters.

Another form of help might be mentorship from a colleague or partner. This might be a more experienced colleague or it might even be a junior colleague. Someone recently out of training may offer new ideas and skills about a topic and she might be invaluable even to the most veteran of physicians.

Help also comes from our allied health partners. Physical and occupational therapy, nursing, social work, case management, etc. offer skillsets different than those directly in the bailiwick of the physician. Some of the most important aspects of treating disease involve these allied professionals who can provide the needed day-to-day rehabilitative therapies and also address negative social determinants and other obstacles to treatment.

Finally, help comes from our physician colleagues, both informally (the curbside consultation) and formally through consultation and referral. Note that consultation and referral are not the same thing. Referral generally implies that all aspects of diagnosis and treatment will be handled by the new physician while consultation implies that the new physician will answer some specific questions while the overall care will still be managed by the original physician.

One important note: in the ethical framework being described, a supervised midlevel should neither make nor receive consultation from a physician. In other words, a supervised midlevel should ask for help from his supervising physician before deciding to consult another physician. If the supervising physician decides that such outside consultation is appropriate, then that's okay. In the same way, when one physician asks another physician for consultation, it is generally inappropriate for that consultation to come from a supervised midlevel provider without direct input in the consultation from his supervising physician. I am speaking here strictly of consultation, not referral.

When seeking help, you should be able to answer these specific questions:

- What questions/issues do I need addressed?
- Are these questions/issues beyond my scope of practice and training?
- How quickly do I need this information to take care of the patient?

What help do I want?

The answers to the above questions will help ensure that consultation is appropriate but, importantly, will help the physician use the consultation in an appropriate manner. Many complex diagnostic challenges are referred out to a cadre of specialists each of whom, in turn, declares that their organ system isn't responsible for the present symptoms. This too often results in frustration for the patient.

The referring physician's job is to drive the care of the patient and to put into context the many sometimes disparate pieces of advice she receives. Subspecialists provide a myopic view of the case and this disjointed care can harm the patient. I recall treating a patient many years ago who was pregnant with markedly elevated liver enzymes; she was not hypertensive. A consultation was requested by the gastrointestinal service and the subspecialist offered in his consult that he felt that the patient had acute fatty liver of pregnancy (AFLP) and recommended delivery. The patient was in her second trimester and had normal glucose and fibrinogen levels, all factors that would tend to go against AFLP. I told him this. He then told me that he had diagnosed AFLP over a dozen times in fellowship training in the second trimester. I reminded him that the incidence of AFLP is only about 1 in 10,000 and that at least 95% of it occurs in the third trimester and then wondered if his hospital had delivered 2.4 million women while he was in fellowship training. Okay, so I'm not known for suffering fools. But, the point is that I am the expert on pregnancy and AFLP, not him. My job is to consider his advice and use it, if appropriate, to help the patient; not to sheepishly follow it (and in this case, harm the patient) because he went to a fellowship or claims some experience (appeal to experience). I discounted his advice and pursued my own diagnostic evaluation and I was right: she had herpetic hepatitis (which isn't treated by delivering a baby at 25 weeks-gestation).

Consultants are not smarter than you, they just have a different area of expertise than you might and you have expertise (and perspective) that they do not. **Consultation is not abdication** of your duties as the primary caregiver for the patient.

The consultative process, then, is made much more useful by the following:

- Tell the consultant specifically what you would like help with and what work-up and considerations you have already pursued. If this is a hospital consultation, speak directly to the physician from whom you are asking for help.
- Discuss the consultant's findings and recommendations with her when she is done, especially if this occurs in the hospital setting.
- Consider the consultant's recommendations in view of your own expertise and in the context of information and opinions from other consultants, etc.
- Follow-up with the consultant if you have information or perspective that might change her opinion.

Shooting the elephant

In 1936, British writer George Orwell published an essay called *Shooting An Elephant,* in which he describes the story of a young British police offer in Burma (presumably Orwell himself as he was a police officer in Burma) who was called to shoot an elephant.[79] In the story, a normally tame elephant goes on a rampage, destroying property and even killing a man. The locals call on the police officer to track down and kill the elephant. As he tracks down the elephant, he finds it peacefully eating and obviously no longer a threat. The right thing to do would have been to allow the elephant's mahout (its handler) to be reunited with the valuable animal. In the brilliant essay, Orwell describes his internal conflict between two demands: earning the respect of the Burmans by killing the elephant or doing what he feels is right by letting the animal live. He finally shoots the elephant and the garish description of the more than half-hour long execution will leave you feeling sad that man could treat beast so poorly. For Orwell, it was an awakening.

In medicine, *shooting the elephant* has become a euphemism for a doctor providing care just because someone asked him to. This happens any time that a physician orders tests or interventions simply because he feels it is expected of him to do so. Perhaps a pediatrician orders an unnecessary antibiotic for a child because he feels like the parent expects it or a primary care physician orders unindicated labs due to a feeling that the patient will be dissatisfied if she does not.

For consultants, however, shooting the elephant has become an Olympic sport. When one physician refers a patient to another physician, the feeling of the subspecialist is that he must do *something* to make the referring physician and the patient feel appreciated and satisfied. Unfortunately, this often means ordering unnecessary tests and procedures, which have the dual incentive of providing more income to the subspecialist as well as appeasing the referring physician.[80]

Imagine that a primary care physician hears a murmur on an asymptomatic patient during a health screening and she refers the patient to a cardiologist. The cardiologist listens to the heart and appreciates no murmur and takes a history and learns nothing of concern. That should be the end of the visit. But the patient might feel like her time has been wasted or that her primary care physician is incompetent; the primary care physician might feel like her clinical skills have been questioned or that the cardiologist isn't capable. These angsts create a scenario ripe for elephant hunting. So rather than create these stresses, the cardiologist happily orders an echocardiogram (and makes a larger fee in the process).

Don't kill elephants without a good reason.

Exercises

14.1 What are some examples of inappropriate consultation you have seen? Why was it inappropriate?
14.2 In your specific clinical arena, what are some practical ways that the consultative process could be made more valuable?

15 How to Interpret Literature

So far, we have been talking about clinical skills and thought processes. But just as important is the physician's ability to read and interpret the scientific and medical literature. Scientific studies form the basis of what we know about diseases, how they are diagnosed, and how they are most effectively treated. However, physicians tend to be poor at this task and tend to interpret new literature (and old) in a very biased way: if it roughly agrees with what they already believe, then it is a good study, and if it doesn't, then it must be a bad study.

Fortunately for us, we have already learned most of the basic cognitive skills necessary to avoid this trap. Let's dive in.

Lies, damned lies, and statistics

Imagine that you've just read a study in the prestigious *British Medical Journal* that concludes the following:[81]

> Remote, retroactive intercessory prayer said for a group is associated with a shorter stay in hospital and shorter duration of fever in patients with a bloodstream infection and should be considered for use

in clinical practice.

Specifically, the author took a group of 3,393 patients who had been hospitalized for sepsis up to ten years earlier and randomized them to one of two groups. He prayed for one group and did not pray for the other group. He found that the group which he prayed for was more likely to have had shorter hospital stays and a shorter duration of fever. Both findings were statistically significant, with a P-value of 0.01 and 0.04, respectively.

So, are you currently praying for patients who were hospitalized 10 years ago? If you aren't, some "evidence-based medicine" practitioners (of the Frequentist statistical school) might conclude that you are a nihilist, choosing to ignore the latest science. Yet, intuitively most physicians just don't believe this study. I don't know any who make it a practice to go to medical records and pull 10-year-old charts and pray for them. Doctors just ignore this study and studies like it.

But why? How do we know if a study is valid and useful? How can we justify ignoring one study while demanding universal adoption of another when both have similar P-values?

I would challenge you to answer these five questions when you read a study to determine its validity and relevance to your practice:

1. How good is the study?
2. What is the probability that the discovered association is true?
3. What other hypotheses explain the data?
4. Is the magnitude of the discovered effect clinically significant?
5. What cost is there for adopting the intervention into routine practice?

We will discuss the details of each of these questions in depth; but first, let's review the relevant concepts from introductory statistics that we'll be using. Most published medical studies report the results of some form of hypothesis test, although some report other kinds of findings (e.g., an estimate, such as a confidence interval for the mean survival time of a disease in a certain population.) A hypothesis test involves two competing hypotheses. The **alternative hypothesis** is generally a statement of what the researcher wants to substantiate through evidence, such as "Medication X is associated with lower cholesterol levels," or "the new drug has lower side effect rates," or "the average life expectancy increased after treatment." On the other hand, the **null hypothesis** is a default statement of no effect: "there was no change in cholesterol level" or "the groups had equal rates of side effects" or "there was no change in average life expectancy." It is important to understand that the null hypothesis is *presumed* to be true unless there is strong enough evidence found to refute it, much the same way that a criminal defendant in the US judicial system is presumed to be innocent unless proven guilty.

At the conclusion of a criminal trial, the defendant is either pronounced guilty (if there IS sufficient evidence to convict) or not guilty (if there is NOT sufficient evidence to convict). This is a subtle but important conceptual distinction: the defendant is never *proven innocent,* at least in the United States. An acquittal means that there was not sufficient evidence to convict; it does NOT mean there was sufficient evidence to prove anything. In much the same way, we never prove H_0 (the null hypothesis); we either conclude that there is sufficient evidence to reject the null, or there is not sufficient evidence to reject the null. Of course, the researcher undertaking the study wants to show that there is sufficient evidence to reject the null, lending credence to the alternative hypothesis under investigation. (There are many subtleties and ways that things can go wrong, sometimes drastically wrong, with this process – these issues are the theme of this chapter).

A P-value is the probability of finding data as extreme as the data we've found, assuming that the null hypothesis is true. If the P-value is very small – smaller than an arbitrarily selected value called *alpha,* often chosen to be 0.05 – our conclusion is to reject the null hypothesis. On the other hand, if the P-value is bigger than *alpha,* we fail to reject the null hypothesis. We will refer to these results as a positive test result and negative test result, respectively.

There are two types of error, due to chance, that can occur in the process of conducting a hypothesis test. A **Type I error** means rejecting the null hypothesis when it is actually true; the Type I error rate is given by α (*alpha*). (For $\alpha = 0.05$, this is equivalent to saying that if the null hypothesis is true, in 1 out of 20 studies we will obtain a P-value less than 0.05 just by chance and erroneously reject the null hypothesis; hence *alpha* is the Type I error rate). A **Type II error** means failing to reject the null hypothesis when it is actually false; the Type II error rate is called β (*beta*). The **power** of a study is defined as 1-β, and is the probability of a true positive result.

	Null Hypothesis True	Null Hypothesis False
Reject Null Hypothesis	Type I Error	Correct
Fail to Reject Null Hypothesis	Correct	Type II Error

Figure 25. Matrix of Type I and Type II errors.

In terms of study design, there is inherently a trade-off in establishing our Type I and Type II error rates. In general, a larger Type I error rate will mean a smaller Type II error rate, and vice versa. (Analogously, in our criminal justice system, if our standards of evidence are too low we will erroneously convict too many innocent people, like a Type I error; if our standards of evidence are too high we will erroneously acquit too many guilty people, like a Type II error). If patient enrollment in a research study is increased, this will increase the Type I error rate; if enrollment is decreased, this will increase the Type II error rate. If study enrollment is too high or low to maintain reasonable error rates, we call this an **overpowered** or **underpowered** study, respectively.

Of course, as we have discussed before, **Bayesian inference** is a vital tool in overcoming many of the shortcomings inherent in classical hypothesis testing, and we will see later that asking "What is the pretest probability that the alternative hypothesis is true?" is a key to meaningful analysis of our study results. We will also consider the role that **bias** plays in research, and consider specific ways to estimate the effect that researcher bias has on the probability that a research finding is actually true.

Now let's turn to the five questions listed above, and examine the myriad considerations that should be made when evaluating the validity of research conclusions.

1. How good is the study?

Most published studies suffer from significant methodological problems, poor design, bias, or other problems that may make the study fundamentally flawed. The ability to read scientific literature critically is one of the foundational skills of physicians.

The most common way (and perhaps the most incorrect way) that physicians read literature is to read the abstract of the paper (and sometimes just the conclusion of the abstract of the paper) and get a "take-away" message, while assuming that all of the other work and vetting has been done for them by the editorial team. Keep in mind, however, that, for a variety of reasons, most published conclusions in literature are untrue.[82] As many as 80% of papers published this year will contain false conclusions. Peer review is incredibly ineffective at findings mistakes or important errors in papers.[83] Therefore, recognizing these two points, let's look at one approach to analyzing a scientific paper.

Scientific papers are generally organized in the ***IMRD* format**: Introduction, Methods, Results, and Discussion. We will follow this common outline and look at an approach for evaluating the technical aspects of a paper. This is the sort of analysis frequently done at "Journal Clubs," but it is only the beginning of the necessary analysis.

Introduction

- **What is the quality of the journal in which the article is published?** Not all journals are created equal. Consider the quality and reputation of the journal in which the article appears and the quality of that journal's peer review and editorship.
- **Do the authors have any conflicts of interest or other biases?** Do the authors have commercial bias (such as working for a drug company, lab company, as a plaintiff's expert witness, etc.) which would incentivize them to be biased in their research? Bias also comes in other forms, such as a need for publications, fame, or promotion of a pet theory. Are the conflicts

of interest clearly stated? An Internet search of the author(s) is also helpful to discover potentially unstated conflicts of interest.

- **What question is being asked by the researchers?** Rephrase the central point of research succinctly and clearly in your mind.
- **What is the null hypothesis?** This should be clearly defined and stated. The entire research design is built around a null hypothesis. Remember, we are seeking to disprove – not prove – so we need to know what we are seeking to disprove. The authors likely have an *alternative hypothesis*, but the study should be designed to disprove the null hypothesis in most cases.
- **What outcomes are of interest? What is the primary outcome being observed? What are the secondary outcomes?** The primary outcomes are usually the only ones the study is adequately powered to assess; make sure you understand which outcomes are primary and which are secondary, as well as which were studied *a priori* and which were not. The appropriate statistical analysis changes based upon the answers to these questions and the study design may need to be different for these different outcomes.
- **What is already known about this topic?** It is hard to know what confounders (and other important factors which might affect the design of the study) exist without first reviewing what is already known about the subject. A good paper will do this in the introduction. A bad paper will skip over relevant, already-published data, ignoring previous studies which tend to conflict with the new findings. This knowledge is also necessary to assess the prestudy bias and odds, which we will discuss in detail soon.
- **Do the stated alternative hypotheses make sense based on what we already know?** What is the prestudy probability that the hypotheses are true? As stated above, even if the data in the paper turns out to be statistically significant, it still has a meaningful chance of being false based on the prestudy likelihood of the hypothesis (which we will discuss formally in the next section).
- **Restate and summarize, in your own words, what is known about the topic and what the investigators are seeking to learn.** Stating this in your mind will help you not be distracted

by clever writing.

Methods

- **What would be the best way to design a study to answer the research question?** In general, a placebo-controlled, triple-blinded, randomized trial is the ideal, but there are sometimes ethical, financial, or time constraints that prohibit this. If an optimal study design has not been used, are there specific reasons why not?
- **How did the authors design the study?** If they did not do the best type of study, is the method they used capable of answering the question asked, and with what degree of certainty? In considering study design, try to determine if the authors were honestly searching for evidence against their theory (in accordance with the scientific method), or making design decisions calculated to give favorable results.
- **What are the weaknesses of the study design?** Common examples include poorly described or too narrow a patient population, poorly matched controls, unclear study endpoints, confounding issues, differences in measurement, recall bias, loss of patients to follow-up, etc. A summary of common study designs and some strengths and weaknesses can be found at the end of this section.
- **What were the inclusion and exclusion criteria?** Did these criteria make sense? How might the study results have been affected by different choices? Choices about inclusion and exclusion criteria can have an enormous impact on the outcome of the study. Review each one and decide if it made sense to include or exclude the group. Are there other criteria the authors didn't consider?
- **What variables were held constant? Was there a control? Was the control appropriate?** As with exclusion and inclusion criteria, control of variables is of enormous importance. Do the variables controlled for make sense? Are there variables the authors didn't consider? Matching groups based on demographics alone is common but is a very poor way of deciding that two groups are alike. Of course, prospective randomization goes a long way to account for lurking variables.

- **Does the study method address the most important sources of bias for the question?** Biases come in many, many forms. The ideal study is free of bias (though no studies truly are). What did the authors do to compensate for bias? What biases did they not consider?
- **Was a power analysis done?** A power analysis is fundamental to assessing a trial for Type I and Type II errors. Claims of statistical significance have no meaning without a proper power analysis. If a power analysis was not done, why not? A study without a power analysis is a preliminary study. The goal of a study should be to enroll the proper number of patients based upon an *a priori* power analysis. Under-enrollment may lead to a Type II error (a false negative finding) but over-enrollment can lead to a Type I error (a false positive finding). If enrollment didn't match the power analysis, was an *a posteriori* power analysis done? If not, why not? We will talk about P-hacking in the next chapter.
- **Did the study follow the predetermined study design**? Was there over-enrollment, under-enrollment, or other deviations from the protocol? The authors should account for deviations from the protocol; if they did not, think about how the outcomes might have been affected.
- **Was the study Institutional Review Board (IRB) approved? Did the IRB controls adversely affect the study design?** Sometimes a study may be restricted, ended, or otherwise affected by the IRB in a way that significantly alters the protocol and outcomes of the study.
- **Were the statistical methods used appropriate?** This is one of the most difficult areas for the average physician to opine about. Was the correct type of hypothesis test performed? Were the theoretical criteria for that hypothesis test satisfied? Were the assumptions made about data distribution justified? It may take very advanced training to answer this question. Keep learning.

Results

- **How closely-matched were the patient populations in the different groups of the study?** Does this affect the conclusions offered? Even in well-designed studies, the compared groups may not match perfectly (e.g., one group is older than another). Could this affect the outcome?

- **Did the authors account for all the relevant confounders?** Which important ones were left out and how does that affect the conclusions?
- **What did the study find?** What did it *actually* find? State it in your own words. Usually, it was that the null hypothesis was rejected. That doesn't necessarily mean that the alternative hypothesis is true. Remember that a single null hypothesis generally has more than one possible alternative hypothesis associated with it, and in rejecting the null, the authors have not technically proven their theory. Don't jump to conclusions.
- **Were the findings statistically significant?** This assumes a lot, not just a P-value < 0.05. Was there a power analysis? Did it fit the results? Were appropriate statistical methods used? Were different methodologies used for multiple comparisons analysis? We will talk about these issues in detail in the next chapter.
- **How certain is the measured effect?** How wide is the confidence interval or other measure of the scatter of the data? What is the magnitude of effect? The size of the magnitude of the effect is very important. Clearly beneficial interventions will have a large magnitude of effect. Imagine a study that compared insulin to placebo for treatment of Type 1 diabetes. Clearly, insulin is effective. A study of two patients will demonstrate this. Now contrast that to a study which needs 10,000 patients to show a subtle difference between the intervention and placebo; this difference might be real or it might just be noise in data that has been overanalyzed.
- **Were appropriate statistical methods used in the analyses of subset or secondary outcomes? Was the study powered sufficiently to address all the secondary outcomes?** Remember that different statistical methods need to be used for subset analyses and multiple comparisons, and many papers get this part completely wrong (more about that later.) Often an *a priori* power analysis is not performed for secondary outcomes, and so

a P-value < 0.05 for the main outcome does not necessarily signify statistical significance for secondary conclusions.

- **If the data suggests a correlation, is there any evidence of causation?** Apply the Bradford Hill criteria we learned in Chapter 4 to understand how likely causation might be.
- If it is an epidemiological study, were any of the **seven deadly sins** committed? Raj Bhopal[84] warns us to look for these "sins:"

    1. Failing to provide the context and definitions of study populations.
    2. Insufficient attention to evaluation of error.
    3. Not demonstrating that comparisons are like-for-like.
    4. Either overstatement or understatement of the case for causality.
    5. Not providing both absolute and relative summary measures.
    6. In interventional studies, not demonstrating general health benefits.
    7. Failure to utilize study data to benefit populations.

Discussion

- **Were all the appropriate outcomes addressed?** Studies tend to focus on one type of outcome but often unintended harms from the intervention (for example, antimicrobial resistance in an antibiotic study) are not explicitly studied. Not vaccinating children may reduce the risk of death or injury from the vaccine but it also increases the overall risk of death (from the disease); we must consider all relevant outcomes.
- **Was a surrogate outcome used?** Just because a drug lowers cholesterol doesn't mean it reduces heart disease risks. Or, just because low Vitamin D levels are found in patients with fatigue doesn't mean that giving a patient Vitamin D will improve her fatigue. Beware of surrogate outcomes and the conclusions drawn from them.
- **What are other alternative hypotheses that fit with the data?** As we discussed earlier, there is almost always more than one alternative hypothesis that could potentially explain the data.

- **Are all the authors' conclusions supported by their data?** Authors like to draw inferences and conclusions that have nothing to do with disproving the null hypothesis; be careful what the take-away message is.
- **Are the conclusions novel and how do they add to what we know/do?** The more novel the conclusion (e.g., disagrees with previous ten trials), the more skeptical one should be as we will discuss shortly.
- **Is the outcome studied clinically relevant?** What is the magnitude of the effect, the number needed to treat (NNT), the number needed to harm (NNH), etc.? Many things are statistically significant but not clinically significant, either because the problem addressed is so unusual or insubstantial that intervention is inconsequential, or because the magnitude of effect is so small that the intervention may not be worthwhile.
- **Is the result of the study broadly applicable, or specifically applicable to your patient population?** Make sure that the patient population in the study is similar to the patient population to which you would like to apply the data.
- **What do others think about the study?** An internet or PubMed search about a study will often reveal important commentaries, letters-to-the-editor, and criticisms of many published papers that you might not have considered, as well as other studies that contradict its findings.

We will dig deeper into specific statistical problems and other issues that need to be addressed when looking at pitfalls and biases in interpreting the literature in the next chapter. For now, Table 9 provides a review of the most common types of studies we will encounter in assessing the scientific literature. Obviously, whole books have been dedicated to each type of study design mentioned in the Table and this information is meant only to orient us for our discussion and to remind us of some basic strengths and weaknesses associated with each type of study design.

There are many special subtypes of study designs within each category and dozens of advanced statistical methods sometimes used; each of these are subject to unique strengths and weaknesses. An excellent resource for evaluating study types and statistical methods in detail is available from the *Journal of the American Medical Association's* "Guide to Statistics and Medicine." This regular series in *JAMA* provides high level detail about almost any method you might need or encounter and you can access it with the QR code to the right.

Study Type		Design	Strengths	Weaknesses
Evidence Summaries	Meta-analysis	Attempts to combine data from RCTs	• Can provide more precise confidence in/estimates of outcomes compared to single RCT • Often more generalizable than single RCT	• Usually faulty with heterogeneous data • Subject to publication bias • Fails to correct for bias of original studies • May negate results of a quality RCT with several inferior RCTs
	Systematic review	Provides an expert review of all available evidence	• Can assess several types of evidence (not just RCTs)	• Subject to selection bias of author
Experimental Studies	Randomized controlled trial (RCT)	Patients are randomly assigned to an intervention or control group	• Gold standard for establishing causality • Can study more than one intervention	• Expensive • Takes time • Not always broadly applicable • Ethical limitations • Not useful for rare conditions
	Nonrandomized controlled trial	Patients are assigned to an intervention or control group without randomization	• Fewer ethical considerations than RCT	• Not randomized • Takes time • Lurking variables
Observational Studies	Cross sectional study	Snapshot of data about diseases at a given time	• Inexpensive, quick • Useful for determining prevalence	• Survivor bias
	Cohort study	Prospectively follows a group of matched patients over time	• Good for rare events • Can study multiple variables • Hypothesis generating	• Lurking variables • Takes time • Selection bias
	Case-controlled study	Retrospectively compares patients in controlled groups	• Inexpensive, quick • Hypothesis generating	• Recall bias • Lurking variables
	Case series	Reports details about several similar patients	• Same as case report below	• Same as case report below
	Case report	Reports details about a single patient	• Helpful for rare diseases or presentations • Hypothesis generating	• Very often contain false conclusions • Selection bias • Not generalizable

Table 9. Strengths and weaknesses of different study types.

In addition to the study designs in Table 9, there are other studies which involve non-human subjects, like animal studies and *in vitro* lab studies. Understanding levels of evidence and considering the evidential value of each type of study is essential to maintaining perspective about the value of the published literature. Table 10 below shows various types of experimental designs categorized into the traditional five levels of evidence. Notice that animal and bench research are among the lowest level of evidence – beneath even a case report involving a human subject. As a thought experiment, let's show what these levels of evidence would mean if we were considering the possibility of the existence of UFOs (start at the bottom of Table 10 and work your way up, from the lowest level of evidence to the highest level).

Higher

Level	Example of Evidence	Comment
Level 1	Meta-analysis of Homogenous RCTs Randomized Control Trial	Multiple, repeated, high-quality RCTs are best Single RCTs still are incorrect ~80% of the time
Level 2	Meta-analysis of Level 2 or Heterogenous Level 1 Evidence Prospective Comparative Study	Prospective data collection to remove bias (prospective collection of UFO observations)
Level 3	Review of Level 3 Evidence Case-control Study Retrospective Cohort Study	Systematic review of anecdotes (attempt to explain several UFO observations)
Level 4	Uncontrolled Cohort Studies Case Series	Collection of anecdotes (several UFO observations)
Level 5	Expert Opinion Case Report Personal Observation	An anecdote (like a UFO observation)
Foundational Evidence	Animal Research *In Vitro* Research Ideas, Speculation	"Makes sense" based upon basic principles (Surely, there must be life somewhere else in the Universe)

Lower

Table 10. Levels of evidence of clinical evidence (applied to evidence of UFOs).

There are several important insights from this illustration. The first is that Levels 3-5 evidence is of very poor quality, relatively speaking. Most conclusions made by retrospective data analysis are wrong due to things like recall bias, lurking variables, the regression to the mean problem, incorrect assumptions of causation, etc. It should be unusual that conclusions from these levels of data affect clinical practice. Most research of this sort is meant to influence the design of prospective trials, not to lead to robust conclusions.

Second, notice that foundational studies based on animal research or *in vitro,* bench-based research is of the lowest evidential value. Paradoxically, however, many consumers of science assume that these sorts of studies are of exceptionally high value. For example, those who believe that vaccines cause autism tend to ignore Level 1 evidence that there is no association between the two factors while pointing to animal or *in vitro* studies that claim some impact of vaccine ingredients like aluminum or formaldehyde on brain tissues. They make the mistake of thinking that these small-scale, lab-based observations are "hard-science" while epidemiologic studies are "soft-science." People want to draw broader conclusions from these small observations: *If aluminum causes brain inflammation, then vaccines which contain aluminum must cause autism and therefore the epidemiologic studies are wrong or rigged.*

In a science like physics, bench-based research can generally lead to generalizable conclusions; but not in biology. The difference is that biology is a **complex system** which we will discuss in more detail in Chapter 18. In short, though, the outcomes of a complex system cannot be judged by any component part of that system. Thus, large studies that look at outcomes with two or more compared interventions are the only fair way to judge complex systems.

Finally, notice the UFO analogy in the illustration. The anecdotal "evidence" for UFOs doesn't break down until we conduct a prospective, controlled trial. This observation leads us to rely on prospective studies with blinding to avoid problems with bias, lurking variables, and regression to the mean. Even then, there can be serious problems with Level 1 and Level 2 evidence and, indeed, most studies like this eventually fail replication.

In the next chapter, we will spend more time looking at how even Level 1 and Level 2 studies can be faulty. These concepts are key to answering our first question, How good is a study? But even if we decide that a study was conducted in an appropriate way, with controls for various biases, and that the statistical methods were appropriate and correctly executed, and that there was a statistically significant finding, then that still is not enough. We have four more questions to answer. Note, however, that most studies fail this first question. Still, if the study passes the first question, then we must address how likely it is that the research discovery is true.

2. What is the probability that the discovered association (or lack of an association) is true?

I have stated more than once that most published research is wrong, including randomized controlled trials. This statement dumbfounds many physicians and other scientists. We will explore many reasons why this occurs in the next chapter, but here I would like to demonstrate evidence for this claim and introduce a tool that we have seen numerous times already in this book to aid in answering this second question.

In 2005, John Ioannidis published a now famous paper called *Why Most Published Research Findings are False.*[85] He elaborates on many of the ideas that we will discuss in this section, and specifically he shows how to measure the impact of study design, bias, and the ratio of true to false findings in a given field (e.g., *genetics* or *nutrition*) on the positive or negative predictive values of a study. His paper has been very influential and has inspired many researchers to attempt to replicate previously published trials or somehow measure just how significant this problem might be. Here is a sampling:

- Begley and Ellis, in *Nature* in 2012, found that only 6 of 53 high-profile cancer studies could be replicated in repeat experiments.[86]
- Jager and Leek, in 2014, attempted to discredit Ioannidis' methods by analyzing 5,322 papers published in the five most prestigious medical journals.[87] They assumed that all these published papers were free of bias and that there was no manipulation of P-values or other malfeasance, but they still estimated that 14% of these papers contained false findings. These represent the very best papers in the best journals and assumes the authors didn't fudge any statistics. In other words, 14% wrong is the absolute minimum rate of false discoveries. The rate goes up with lower-quality journals and P-hacking.
- Ioannidis himself analyzed 49 highly cited articles in 2005 in the *Journal of the American Medical Association* and found that only 20 of them had results that could be replicated; 7 had been explicitly contradicted by other studies and most showed profoundly

different magnitudes of effect in subsequent studies.[88]

- The Reproducibility Project reported in 2015 in *Nature* that only 39 of 100 psychology studies could be reproduced. When effects were replicated, the magnitude of effect was generally about one-half the size that had been observed in the original studies.[89]
- A study published in 2010 in *Nature Genetics* found that 10 of 18 genetic microarray studies could not be replicated.[90]

In medicine, the **replication crisis** often takes the form of **medical reversal.** If a new study is given enough credibility that it changes standard medical practice, but then later that study is found to be faulty and the practice changes again, then we call this *medical reversal*. Most studies don't lead to a change in medical practice, so medical reversal occurs usually after the very best papers or multiple papers lead to a new consensus of thought; yet, even then, many of these studies are later falsified or fail replication. We previously discussed some examples of this including cardiac stents, HgA1c goals in diabetes management, and vertebroplasty or arthroscopic knee surgery. We might include aspirin for prevention of stroke in elderly patients, 17-hydroxyprogesterone caproate for prevention of recurrent preterm labor, mammography for low risk women in the fifth decade, and hundreds of other examples. Grab a medical textbook from 30 years ago in your specialty and you will find examples of medical reversal on every page. Practices then were based on studies and evidence, but over time new evidence developed. Some of this new evidence adds to or improves what was understood, but much of it frankly contradicts what was once believed.

Vinay Prasad et al., in a 2013 study, analyzed 1344 articles regarding medical practice published in high impact journals between 2001-2010.[91] Of these, 363 articles examined a current standard of care and the current practice was affirmed 138 times (38.0%) while the practice was reversed 146 times (40.2%). Another group published an analysis in 2019 of 3017 randomized controlled trials published in three of our top journals in the last 15 years: *NEJM*, *The Lancet*, and *JAMA*.[92] They found that 396 of the RCTs were examples of medical reversal. This doesn't reflect the percentage of those RCTs that contained false findings, but it does show how commonly new RCTs disagree with previous evidence that was considered good enough to constitute a standard of care.

In other words, even good papers (not counting all the bad ones that never made it into a good journal or had a lasting impact) are very often reversed in time. So, what's going on?

- Some of these eventually reversed papers had significant statistical problems (whether deliberate or unintentional). This includes things like P-hacking and other experimental design problems that were destined to yield the wrong conclusions. The next chapter will detail these issues.
- Some of these papers were just chance findings (false positives or false negatives) and needed to undergo replication. We will talk about Type I and Type II errors in a moment.
- Some contained deliberately falsified data (and many papers are eventually retracted when this dishonesty is discovered).

Fortunately, we have already learned about the tool that can help us more accurately determine whether the conclusion of a paper is true: Bayes' theorem. Recall the equation:

$$P(H|E) = \frac{P(E|H)P(H)}{P(E)}$$

Here we can let E represent an *experiment* (we called this *evidence* before, but in this specific case it can be the experimental results of a study) and H is the hypothesis being suggested. When we read a scientific paper, what we want to know is the probability that the hypothesis, H, is true, whatever that hypothesis might be. We test the hypothesis with the experiment, E. Before we do an experiment, we should have some idea of how likely it is that our hypothesis is true; this is the *prior probability* or *pretest probability*. Here, I will call this the *prestudy probability* of the hypothesis: How confident are we that the hypothesis is true before we see the experimental results? After we do the experiment, we update that prior probability and determine the posterior probability which becomes the new probability that the hypothesis is true. Let's label the terms in this equation in Figure 26:

Likelihood
The likelihood of observing the evidence (E) if the hypothesis (H) were true

Prior Probability
The probability that the hypothesis (H) is true before the new evidence (E)

$$P(H|E) = \frac{P(E|H)P(H)}{P(E)}$$

Posterior Probability
The probability that the hypothesis (H) is true given the new evidence (E)

Probability of Evidence
The probability that the evidence (E) exists for any hypothesis

Figure 26. Bayes' theorem.

What we normally see reported in scientific papers is the P-value; many naively assume that a significant P-value indicates that the alternative hypothesis is true, but it does not. The P-value corresponds to the *Likelihood* term in Figure 26; recall that the P-value is the likelihood of seeing the observed data if the null hypothesis were true. So, a P-value of 0.03 indicates that there is a 3% chance of finding the observed data if the null hypothesis were true (that there is no difference in the two groups being compared). This is a tool of Frequentist statistics that helps us understand if the differences in our data sets are due to chance or because the groups are different.

However, when we read a paper about some novel intervention, what we want to know is whether the hypothesis the paper advocates for is true; does the intervention make a difference? If we read a paper that indicates that a new blood pressure drug is associated with less cardiac mortality compared to placebo with a P-value < 0.01, then the only thing we know from that paper (even if it is otherwise a perfect study and paper) is that there is less than a 1% chance that the data observed would exist if the two groups were the same. Therefore, it seems as if the observed difference between the intervention and placebo would be due to the intervention. But our confidence in whether the new drug is as effective as the study seems to indicate is related by the *posterior probability* (or post-study probability) which incorporates the *likelihood* of observing the data as well as the *prior probability* that the drug was effective based upon what we already knew (e.g., from prior evidence about the subject).

None of this is any different than what we have already discussed regarding the interpretation of laboratory tests in the diagnostic process. In that case, we had a prior probability that the patient had the disease based upon the frequency of the disease or based upon our clinical history and physical exam; then we integrated the results of the test (and its specific sensitivity and specificity) to arrive at a new posterior probability that the patient has the disease.

Whereas laboratory tests and imaging studies have specified sensitivities and specificities, experimental studies have analogous performance characteristics that are informed by the Type I (*alpha*) and Type 2 (*beta*) error rates. Recall the definitions mentioned at the beginning of the chapter: A Type I error occurs when a study falsely rejects the null hypothesis (which generally states that there is no association between two variables). This is a "false positive" result, because it falsely gives credence to the alternative hypothesis. A Type II error occurs when a study fails to detect a true association. This is a "false negative" result, since it incorrectly discounts the alternative hypothesis.

We generally accept that it's okay to be falsely positive about 5% of the time in the biological sciences; thus, many research papers in medicine use a threshold value (*alpha*, the Type 1 error rate) of 5%. This leads us to say that anything with a P-value less than 5% is "statistically significant." To say this another way: if the null hypothesis is true, 5% of studies will produce a P-value less than 0.05, leading to a false positive result.

On the other hand, studies commonly use a Type II (false negative) error rate of 20%, meaning that 80% of all true associations will be found and 20% will be missed. "Power" is defined as 1-*beta*, or the "true positive" rate. A study should be appropriately "powered" to detect true associations at a reasonably high rate, while keeping the false positive rate reasonably low. Too few or too many patients in the study creates a problem; so, a power analysis should be performed before a study is conducted to make sure that the number of enrolled subjects (*n*) is correct. We will talk about this in detail in the next chapter.

Note the analogies between Type I and Type II error rates in hypothesis testing, and sensitivity and specificity in clinical tests:

- False positive rate is the Type I error rate (*alpha*), or 1-specificity; so, the more specific the test, the lower the false positive rate/*alpha*.
- False negative rate is the Type 2 error rate (*beta*), or 1-sensitivity; so, the more sensitive the test, the lower the false negative

rate/*beta*.

Notice how all of this is analogous to what we have already learned in previous chapters and we are simply implementing Bayes' theorem for a different type of test (in this case, a clinical experiment). Instead of testing whether a patient has a disease, we are testing whether a proposed hypothesis is true. Just as the pretest probability of the disease affected our answer, so too the pretest probability that a hypothesis is true affects our answer (the prior probability).

Not all hypotheses that we can test are equally likely, and that matters.

For example, compare these two hypotheses: 'Cigarette smoking is associated with the development of lung cancer' versus 'Listening to Elvis Presley music is associated with the development of brain tumors.' Which of these two hypotheses seems to be more likely, based on what we already know, either from prior studies on the subject or studies related to similar subject matter? Or if there are no studies, based on any knowledge we have, including our knowledge of basic sciences?

Let's say that we create a well-designed study to test our two hypotheses (the cigarette hypothesis and the Elvis hypothesis). It is most likely that the cigarette hypothesis will show a positive link and that the Elvis hypothesis will not. But what if the cigarette hypothesis doesn't find a link for some reason? Or what if the Elvis hypothesis does show a positive link? Do we allow these two studies to upend and overturn everything we currently know about the subjects? Not if we use Bayesian updating.

We call the prestudy probability π. I'm not sure what the prestudy probabilities that these two hypotheses are true should be, but I would guess that there is at least a 99% chance ($\pi = 0.99$), based on everything we know, that cigarette smoking causes lung cancer, and that there's no more than a 1% chance that listening to Elvis causes brain tumors ($\pi = 0.01$). Don't get caught up on precise numbers; many scientists have had difficulty embracing Bayes' theorem because they find difficulty in assigning specific numbers at this step. But just like in diagnostic testing, it is often enough just to say *high, low,* or *intermediate*. The actual probabilities are more like 99.99% and 0.01% but that detail isn't hugely important.

Let's use these two examples to see how to use Bayes' theorem to deal with these two hypotheses. At the end of the chapter, I will provide more details about the formulas used here and a link to an online calculator to make the calculations easier. For now, here is one of the main formulas we will use, which is based on Bayes' Theorem and makes use of prestudy probability (π), *alpha*, and *beta*:

$$Post\text{-}study\ probability\ (or\ PPV) = \frac{\pi(1-\beta)}{\pi(1-\beta)+(1-\pi)\alpha}$$

Notice that this formula is identical to the formula for Post-test Probability we presented in Chapter 5, replacing "1-specificity" with its analog α, and "sensitivity" with its analog 1-β.

We can calculate how these prestudy probabilities affect our study results and how our study results affect our assumptions that the hypotheses are true. Assuming a P-value (or α) of 0.05 and a power of 90% (or β of 0.10), then the positive predictive value (PPV or post-study probability) that the cigarette hypothesis is true (if our study found a P-value $\leq$ 0.05) is 99.94%. On the other hand, the PPV of the Elvis hypothesis, with the same conditions, is only 15.38%. These probabilities assume that the studies were done well and that the resultant findings were not manipulated in any way. The only difference is the prestudy probabilities (π) that we assigned to each hypothesis.

If we were to introduce some bias into the Elvis study (referred to as u), then the results change dramatically. The parameter u is defined as the proportion of what would be negative study results in an unbiased system that are instead reported as positive results due to the authors' biases and resulting data manipulation. For example, with u=0.3 (the author never did like Elvis anyway), the PPV of the study becomes only 2.73%.

Table 11 below shows the calculation results for a study testing whether smoking causes lung cancer. Notice that even if the study were biased, it makes little difference if we have a 99% confidence that this hypothesis is true before conducting the experiment. Conversely, if the study somehow found that smoking *didn't* cause lung cancer – even if this were an otherwise perfect study – we would only believe that result to be true with 8.76% confidence. In other words, we shouldn't change our beliefs or practice patterns based upon a single study that claimed no link between smoking and lung cancer, regardless of how good the study might otherwise be.

P-value = 0.05, β = 0.1, π = 0.99, u = 0.1	***No bias, 1 study***	***Biased, 1 study***
Positive Predictive Value	99.94%	99.84%
Probability of false positive belief in hypothesis	0.06%	0.16%
Negative Predictive Value	8.76%	

Table 11. Outcome probabilities of a study testing whether smoking causes lung cancer.

Remember, the relevant equations and link to an online calculator are at the end of the chapter, if you'd like to play with the numbers yourself. Let's look at the same chart for our Elvis hypothesis.

P-value = 0.05, β = 0.1, π = 0.01, u = 0.3	***No bias, 1 study***	***Biased, 1 study***
Positive Predictive Value	15.38%	2.73%
Probability of false positive belief in hypothesis	84.62%	97.27%
Negative Predictive Value	99.89%	

Table 12. Outcome probabilities of a study testing whether Elvis causes brain cancer.

In this case, with only a 1% prestudy probability that listening to Elvis causes brain cancer, even positive results from a well-done study translate to only a 15% chance that the hypothesis is true. If we introduce bias into the study, that chance quickly drops to 2.73%.

Bias comes in many forms and affects many parts of the production sequence of a scientific study, from design to implementation, analysis, and publication (or lack of publication). It is not always intentional on the part of the authors. Remember, intentionally or not, when people design studies, they often design the study in a way to show the effect they are expecting to find rather than to disprove the effect (which is the more scientifically rigorous approach). We will discuss the types of bias that affect scientific studies in the next chapter.

If we conducted five Elvis-type studies until we found a positive result, then even without bias the PPV of the study showing an association would only be 4.27% (and add a bit of bias and that number drops dramatically, well below 1%). How about 10 studies? Then the PPV would fall to 2.46%. This multiple studies issue can occur without us even knowing about it due to **publication bias**, wherein studies showing negative results are often not published, meaning it is hard to know how many negative studies were performed before someone published a paper with a positive result. Both examples assume a power of 90%, which is very good. Most studies are underpowered, with estimates ranging in the 20-40% range, allowing for a lot more Type II errors.

What if our cigarette study didn't show a positive result? What is the chance that a Type II error occurred? Well the negative predictive value (NPV) would be only 8.76%, leaving a 91.24% chance that a Type II error occurred (a false negative). But if the power were only 30%, an estimate closer to many inadequately designed RCTs, then the chance of a Type II Error would be 98.65%. Conversely, if the Elvis study says that Elvis doesn't cause brain tumors, its negative predictive value would be 99.89%.

Thus, the risk of a false positive or false negative study result is dependent both on study design (the *alpha* and *beta* boundaries) and upon the pretest probability of the hypothesis being tested.

As an aside, I will note here it was R.A. Fisher in the 1930s who suggested the arbitrary threshold of 5% as an acceptable limit for "rejecting" the null hypothesis (the probability of a Type I error), but it was Jerzy Neyman and Egon Pearson who realized that two mistakes were possible with this approach (Type I *and* Type II errors). In Fisher's view, if the P-value was 0.05, there was a 5% chance that the study result was wrong; but, Neyman and Pearson realized that this wasn't true because Fisher had ignored the Type II errors. Fisher dealt only with the null hypothesis, but Neyman and Pearson realized that there must also be an alternative hypothesis and that the possibility of error in a study was more than just finding a false positive study result.

Let's return to implementing what we have learned with two more examples. Let's apply Bayes' theorem to the retrospective prayer study I mentioned at the beginning of this chapter. The author found a P-value of 0.04 related to the *shorter duration of fever* outcome. We can use this P-value for the *alpha* in the calculation. There was no power analysis, so we might assume the study was underpowered, with a *beta* of 0.5. We might also assume significant bias in the construction of the study (u=0.5). Lastly, how likely do we think that prayer 10 years in the future will affect present-day outcomes? Let's pretend that this has a 1% chance of being true based on available data, which gives us $\pi = 0.01$. If we run the numbers then we find that the positive predictive value of this study is only 1.44%, if our bias assumption is correct. Even without any bias at all, and no P-hacking or other manipulations of any kind, the PPV is only 11.21%.

P-value = 0.04, β = 0.5, π = 0.01, u = 0.5	**No bias, 1 study**	**Biased, 1 study**
Post-study probability of hypothesis	11.21%	1.44%
Probability of false positive belief in hypothesis	88.79%	98.56%

Table 13. Outcome probabilities of a study of remote intercessory prayer for ICU patients.

When you read about this study, you knew already in your gut what Bayes tells you clearly: There is an incredibly low chance, despite a well-designed and statistically significant trial published in a prestigious journal, that the hypothesis is much more likely than the 1% probability we guessed in the first place. Note that the posterior probability is slightly higher than the pretest probability since the study is evidence of a positive effect, but still so unlikely that we should not change our practices as the author suggested.

If all of this sounds sketchy, it isn't. Remember, these are the exact same tools we've been using throughout the book, just applied to hypothesis testing. Not every hypothesis is of equal value, that is, they don't have the same pretest probability. The false positive results are more likely to be distributed among the low probability hypotheses and the false negative results are more likely to be found among high probability hypotheses.

Here's the general concept: Figure 27 compares a high probability hypothesis (a 50% prestudy probability) to a lower probability hypothesis (a 5% prestudy probability), using both an average test (one with lower power and a marginal P-value or some degree of bias present in the study) and a good study (one with a statistically significant P-value and greater power and less bias).

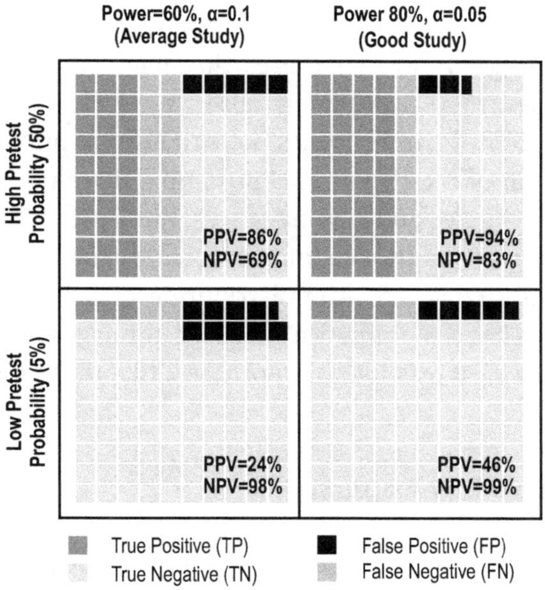

Figure 27. Comparison of likely vs unlikely hypotheses in 'average' vs 'good' studies.

Notice the positive and negative predictive values; they are clearly influenced by the likelihood of the hypothesis. Many hypotheses found in literature have a far lower prestudy probability than 5%; the hypotheses of most epidemiological studies, nutritional studies, and other "data-mining" type studies may carry a prestudy probability closer to 1 in 1,000 or even 1 in 10,000.

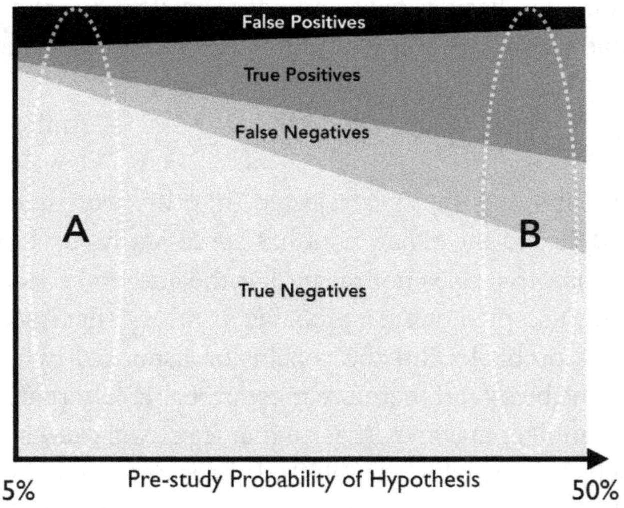

Figure 28. Distribution of false and true results across a spectrum of prestudy probability.

Figure 28 shows this same data in a different way. If you conduct a study among a group of hypotheses that have only a 5% chance of being true, then there will be few false negatives and true positives because only 5% of the hypotheses fit this description (sample A). Conversely, half of the hypotheses in sample B are true so there will be more true positives and false negatives, but far fewer false positives and true negatives. As the prestudy probability increases from 5% (π = 0.05) to 50% (π = 0.5), the distribution of potential Type I and Type II errors changes dramatically. You can imagine how this graph extrapolates to even less likely hypotheses, like those found in nutritional studies.

Pretest probability matters.

If we have decided that a study is high quality (question 1) and that the post-study probability of a true positive finding is reasonably high (question 2), then we can reject the null hypothesis with some high degree of confidence. But does this mean that the alternate hypothesis is true?

3. What other hypotheses explain the data?

> The casual view of the P value as posterior probability of the truth of the null hypothesis is false and not even close to valid under any reasonable model, yet this misunderstanding persists even in high-stakes settings.
>
> – Andrew Gellman

When the null hypothesis is rejected (in other words, when the P-value is < 0.05 and the experiment has successfully survived our first two questions), that does not mean that the authors' conclusions are automatically accepted or even probable. It means that *AN* alternative hypothesis is probable. But the conclusion espoused by the authors may not be the best (and therefore most probable) alternative hypothesis. For example, I may hypothesize that ice cream causes polio. This was a widely-held belief in the 1940s. If I design a study in which the null hypothesis is that the incidence of polio is not correlated to the incidence of ice cream sales, and then I find that it is, then I reject the null hypothesis. But this does not then mean that ice cream causes polio. That is one of many alternative hypotheses, and much more data is necessary to establish that any given alternative hypothesis is probable.

The polio versus ice cream sales correlation is a famous example that **correlation doesn't equal causation,** and it is used in many statistical books and has become an internet meme. Figure 29 shows how closely-correlated ice cream sales in 1949 were with the rates of diagnosis of new cases of Polio.

This is a mistake of "correlation equals causation" and this is the most likely mistake with this type of data. Comical examples of this sort are almost infinite, though many public health officials took the polio and ice cream connection quite seriously.

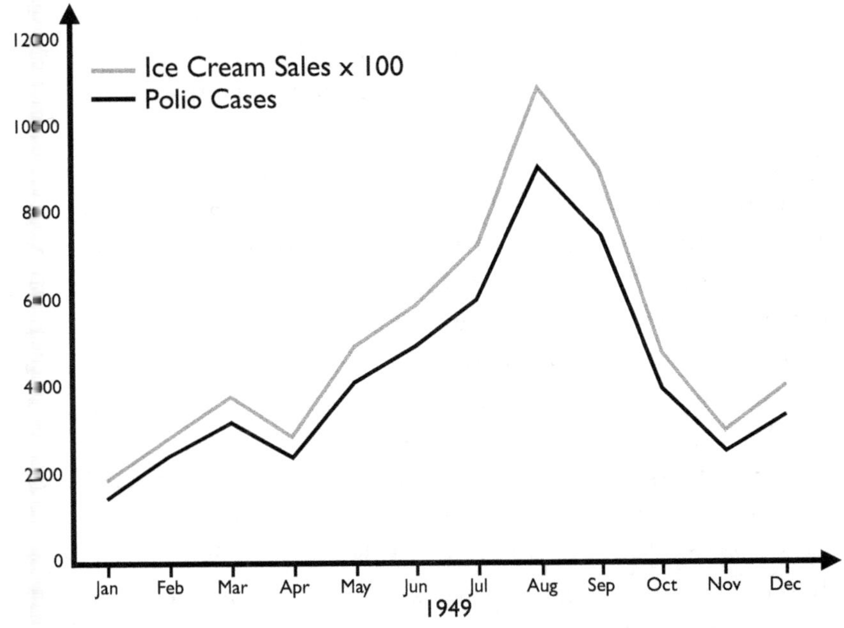

Figure 29. Ice cream sales versus cases of Polio in 1949.

Most people understand clearly that correlation doesn't equal causation. Tyler Vigen has written a book called *Spurious Correlations,* and his website uses public data sources to data-mine absurd correlations, like strongly linking the divorce rate in Maine to the per capita consumption of margarine.

You know that margarine isn't linked to divorce rates intuitively, and that's why you chuckled when you read that. But a hypothesis could be formed: perhaps husbands, angry that their wives have switched to margarine instead of butter, are quick to file for divorce. Still, you reject the causal relationship of the two events, almost without thought, just as today you reject the Polio and ice cream correlation. But that's simply because you understand the pathophysiology of Polio better than scientists did in 1949.

You might have deduced that the correlation between ice cream sales and Polio was absolutely true and reproducible. It was and is strong science. In the summer months, ice cream trucks went through neighborhoods; kids gathered at the truck and bought different flavors of ice cream. While gathered together in a tight pack and while sharing licks of other kids' flavors, the poliovirus could spread. But that doesn't mean that ice cream causes polio, which was the belief at the time (through a process by which pasteurization of milk didn't kill the virus).

But notice, too, that what I just said isn't directly supported by the data in the graph! You liked it and it makes sense. I created a narrative fallacy for you. We humans want to understand the world around us, and I just gave you a construct or narrative to understand the data. But is it the only one? Not at all. Perhaps the virus, like ice cream, thrives in summer months. Perhaps it uses water as a vector, and kids were more likely to go swimming or play with local fire hydrants when it was hot. Perhaps the correlation is purely accidental.

Be careful making any assumption of causation, regardless of how much "sense" it might make.

As I have said before, your brain is a Bayesian engine, so it rejects as silly the idea that margarine sales are linked to divorce rates or that the marriage rate in Kentucky is related to how many people drown each year after falling off a fishing boat (another of Vigen's spurious correlations). You can quickly reject these hypotheses because you bring everything you know about marriage and divorce and fishing deaths and margarine to the forefront and you apply that knowledge to the problem. You are updating your preexisting belief about causes of marriage and divorce in light of this new evidence and the new evidence isn't enough to change your belief. You realize that there must be other explanations or other hypotheses, even if the data is correct.

Yet, somehow in medicine, people abandon common sense and they return to the *correlation equals causation* myth. A study published in *JAMA* in 2016 revealed a correlation between antenatal Tylenol use and subsequent behavioral problems in children.[93] Aside from the fact that this was a poorly controlled, retrospective epidemiological study (with almost no statistical relevance and an incredibly weak pretest probability), even if it were better designed and did indeed determine that the two factors were significantly correlated, it still would be no more evidence that Tylenol causes behavioral problems in children than the above data is evidence that margarine causes divorce. There is a myriad of explanations for the correlation. Perhaps mothers who have more headaches raise children with more behavioral problems? Or children with behavioral problems cause their mothers to have more headaches in their subsequent pregnancies? Or, most likely, it's just a chance correlation like the ones we have been discussing. Correlation *almost never* equals causation.

Our medical journals are filled with these sorts of papers and they can have a huge impact on the public. Social media and the news media quickly jump to conclusions, and the assumption is almost always made that correlation equals causation. From this we learn that diet sodas make people fat or that Vitamin D cures everything, etc.

Serious and important medical journals publish this sort of junk every month. In 2013, *Fertility and Sterility*, the leading infertility journal, published an Italian paper that concluded that women with rectovaginal endometriosis were more attractive than women without it, with a leaner silhouette, larger breasts, and earlier coitarche.[94] It was recently retracted. In 2017, the *American Journal of Hypertension* published an embarrassing paper that claimed that women with higher preconceptual blood pressures (by just 2.7 mm Hg) were more likely to have a male child. That paper was picked up by the national news and was covered by dozens of news outlets. But they all make the same mistake about assuming causation from correlation, even if a true correlation exists.

Recall the Bradford Hill criteria discussed in Chapter 4; rigor of that sort is required to make any sort of conclusion that there is a causal relationship between two factors that are statistically correlated. Almost every correlation fails such rigor.

But, let's say that a legitimate finding was discovered and it appears that the proposed hypothesis and causal nature of the discovery is robust; we must next assess how large the magnitude of effect is.

4. Is the magnitude of the discovered effect clinically significant?

There are an almost endless number of discoveries that are highly statistically significant and probable, but they fail to be clinically significant because the magnitude of the effect is so small. We must distinguish between what is significant and what is important. In other words, we may accept the hypothesis after these first steps of our analysis but still decide not to make use of the intervention in our practice because it has no clinical significance.

The degree of statistical significance does not necessarily indicate that there is a large effect. In other words, a P-value of 0.00005 doesn't necessarily mean that the difference in the two groups is greater than if the P-value had been 0.05. If we use a large enough data set, we can establish that even the most minuscule differences are statistically significant differences. P-values are a separate consideration from **magnitude of effect**. We will discuss in Chapter 17 how to better communicate risks and benefits to patients and our colleagues; but, briefly here, we need to think in terms of magnitude of effect rather than statistical significance. Terms like *number needed to treat* and *number needed to harm* are far more useful. We can also think in terms of quality-life years added or quality-years per amount of money.

I saw a patient once who was taking a statin drug for elevated cholesterol. She was an otherwise healthy woman with no cardiovascular risk factors. She had complained to her primary care physician about side effects but he insisted that she keep taking it. He gave her the impression she would die of a heart attack if she did not. Her myalgias from the drug were preventing her from playing tennis and exercising as much as she once did. We used a risk and benefits calculation tool from the Mayo Clinic website[95] to determine the real value of her continuing to take this drug for the next 10 years.

It turns out that if she continued to take the drug every day for the next 10 years, given her other risk factors and health indicators, she would reduce her risk of having a nonfatal myocardial infarction by one percentage point in the next decade. Nothing that her primary care physician had told her communicated to her how little risk reduction the drug afforded her, and it's likely that the physician himself didn't understand this. In fact, her total cardiac risk might likely be higher since she was now exercising less because of the drug. So, it would have been more effective to say to her that if 100 women like her took the drug every day for ten years, the medicine might help prevent one nonfatal myocardial infarction. Viewing the outcomes in this way can help give us perspective and help us balance total risks and benefits. In her case, the lost exercise was almost certainly a bigger risk factor for MI than not taking the statin drug. She stopped the medication.

This drug's impact in reducing her risk of MI, though statistically significant, was not clinically significant, particularly when weighed against other factors. But the drug company never educated physicians about the magnitude of effect, just that it reduced the risk of heart attack for patients who took it.

The term *statistically significant* has done much harm in science. This was another phrase coined by R.A. Fisher. It is highly perverted today. Fisher just meant to imply that when data was different enough that there was less than 1 chance in 20 that those differences were due to chance, we should look there for meaningful observations. In other words, don't waste your time with findings that are so common as to occur more often than 1 in 20 times; but when you observe something less common, that should pique your interest. It doesn't mean that it is a true difference or that it is important, but it's worth investigating. It was simply a way of cutting through the noise in order to find signal; it didn't mean there was a signal. But today, a P-value < 0.05 has become the prize itself. We will talk more about this in the next chapter. We might call this the naïve interpretation of P-values. Importantly, we will also discuss in the beginning of the next chapter the American Statistical Association's 2016 position statement on P-values.

For our purposes here, it should suffice to say that something that might be "statistically significant" is not necessarily clinically significant.

5. What cost is there for adopting the intervention into routine practice?

The patient I just mentioned stopped her statin drug because the magnitude of effect didn't justify taking it for her specific situation. But what if she hadn't had the myalgias? Should she have continued taking the medicine? The drug at the time cost $270 per month. Invested at a 6% interest rate, this represents lost savings of $44,504 over the ten-year period. Since there was only a one percentage point reduction in her risk of a nonfatal MI, then a total of $4,450,400 would be spent to prevent one nonfatal MI (since one hundred women like her would have to use the medicine for ten years). That cost doesn't make sense.

If all you care about is reducing the risk of heart attacks, then you will advocate for using this drug. While the statement, "Taking this medicine will reduce your risk of a heart attack" is true, it doesn't tell the whole story. We must appreciate how much risk reduction occurs and at what cost. Most patients would rather have the money spent on that drug to help their children or grandchildren get an education than to reduce their risk of a nonfatal MI by one percentage point. We might have better spent that money elsewhere. For the cost of preventing one nonfatal MI, we could've paid for 150,000 flu shots (at $30 per shot), preventing about 4,500 cases of the flu. Some of those flu cases would have been fatal. Even more to the point, that $4.5 million dollars could have been used to financially incentivize those 100 women to stop smoking, lose weight, or exercise regularly and that would have a far greater impact on reducing heart disease than using the statin drug.

Cost comes in two forms: economic and noneconomic. The $4.5M cost of preventing one nonfatal MI in ten years is likely not to be considered cost-effective by any honest person. But noneconomic costs come in the form of unintended consequences, whether physical or psychological. Would, for example, an intervention that decreased the size of newborns by 3 ounces in mothers with gestational diabetes be worth implementation? Well, maybe. That 3 ounces might, over many thousands of women, save a few shoulder dystocias or Cesarean deliveries. So, if the intervention were free and had no significant unintended consequences, then yes it probably would be worth it. But few interventions are free and virtually none exist which don't expose the patients to unintended harms. So, the cost must be worth the benefit.

Studies do not usually account for all the costs, intended and unintended. More often, the burden of considering the costs of implementation fall upon the reader of the study. Deciding whether to use any intervention is a shared decision between physician and patient and one which respects the values of the patient. The patient must be fully informed, understanding precisely both the risks and the benefits in terms she can understand.

Finally, choosing whether to adopt a new treatment or test into medicine must comport with our ethical standards or moral framework, like value-based care or the Quadruple Aim. Cost is an ethical consideration which accounts for the principle of Justice.

The five questions we have answered are complex but they are necessary questions that must be satisfied, particularly before a new intervention is applied to patients. Let's briefly apply these five questions to the paper I mentioned earlier that claimed that higher antenatal blood pressure resulted in more male children.

A sample paper

Let's apply what we have learned by answering the questions we have proposed. This is not an interventional trial; it is Level 3 evidence. But it is an example of a paper that has been published in a highly-respected journal, widely-publicized, and accepted by casual readers as quality science. You can see the paper for free at the link.

1. How good is the study?

This Canadian study was done on women in China (where there is a great interest in selecting male fetuses). Out of 3,375 women, they analyzed data regarding 1,411 women who had preconceptual data collected and went on to deliver a singleton pregnancy. This cohort of women went on to deliver 672 females and 739 males, which skews the data towards an unexpected number of male infants. The investigators measured several independent factors preconceptually that were analyzed to see if they differed between the group having boys and the group having girls: age, years of education, smoking, passive smoking exposure, preexisting hypertension, preexisting diabetes, BMI, waist circumference, systolic blood pressure, diastolic blood pressure, mean arterial blood pressure, total cholesterol, LDL, HDL, triglycerides, and glucose. The blood pressure readings were further divided into quintiles, each of which were analyzed.

After adjusting for factors that the authors thought might otherwise confound hypertension, they found that systolic blood pressure was an average of 2.7 mm Hg higher in the group of women who had male babies and they claimed that this was statistically significant with a P-value of 0.0016. No other comparisons were statistically significant. The rest of the article draws misleading and unsupported conclusions from this data and provides unwarranted speculation that I won't repeat here. They are even so bold as to make the claim that a woman with a systolic blood pressure greater than 123 is 1.5 times as likely to have a male as compared to a woman with a lower pressure. If this seems to smack common sense in the face, there's a reason.

So, is this a good study? No. The main problem is the claim that the observed difference in systolic blood pressures (a difference of 2.7 mm Hg) is meaningful just because there is a P-value < 0.05. So why don't I care about this (the P-value, after all, was 0.0016)?

I could start by pointing out that a difference of even 1.3 mm Hg, given the number of patients in this trial, would have produced a P-value of less than 0.05 (0.0414 to be precise). Does that seem absurd? It is. The expected deviation of automated blood pressure cuffs is about ± 6 mm Hg, so the study was powered to find a difference in blood pressure smaller than the margin of error of the machine used to measure the blood pressures.

But this study has no power analysis. P-values can only be interpreted in the context of a power analysis. If a study is over-powered, then even a tiny difference will appear statistically significant, even if it is not actually significant. Let's say the authors wanted the study to be powered to find a difference of 5 mm Hg between the two groups, if such a difference existed; in this case an $N = 90$ (rather than 1,411) would have been appropriate. The easiest way to fake statistical significance when there isn't any is to over-enroll; the higher the *N*, the lower the P-value. This is what is called P-hacking and we will discuss it in detail in the next chapter. Given sufficient *N*, even a difference of 0.1 mm Hg would be considered significant. These decisions must be made *a priori* or at least considered *a posteriori*. They did neither. If you want to learn more about the basics of how to perform a power analysis, check out the link at the QR code.

Next, the adjustments used for the adjusted odds ratios were selective and artificially limited. They adjusted for age, for example, but they chose not to adjust for exposure to second-hand smoke (which could reasonably affect blood pressure). They also did not adjust for average time from enrollment in the study to pregnancy (which was the biggest difference between the two groups, and nearly five weeks different; this difference could have easily reflected a difference in staff or equipment or calibration). Since we don't have the data, we cannot perform the adjustments ourselves, but it is very reasonable to assume that a complete adjustment may have easily resulted in a loss of the purported statistical significance.

Finally, the authors have a serious multiple comparisons problem which we will also discuss in the next chapter. This study tested multiple hypotheses, and at least 22 hypotheses were put forth as preconceptual factors that might have influenced gender determination, ranging from maternal age to triglyceride levels (including the quintiling of blood pressures). We must make an adjustment to the nominal P-value to adjust for over-enrollment, and we can be *very conservative* and use a new nominal value of 0.025 as a threshold for significance (the actual threshold would vary on effect size expected but would by much lower; remember, only 90 patients would be necessary to find a difference of 5 mm HG). If we apply the Bonferroni correction to this value, accounting for 22 comparisons, then an adjusted P-value of 0.0011363 becomes the new threshold for significance. Using Šidák's correction, we get a P-value of 0.0011501 as the threshold for significance. By either adjustment, we learn that the claimed P-value of 0.0016 is *not* statistically significant, and that's assuming that their P-value is adequately adjusted (it almost certainly is not).

More importantly, if no adjustments at all are made, and the paper is taken as is, then it should be remembered that the chance of finding one or more statistically significant differences when 22 hypotheses are tested is 67.65%. Thus, it was more likely than not that at least one of the 22 probed variables would turn up with a *false positive*. Therefore, the multiple comparisons adjustments must be made; when they are, the claim of significance goes away. We will learn about all of this shortly.

The study has numerous other embarrassing problems, but suffice it to say that an unadjusted blood pressure difference in the two groups of 2.7 mm Hg is not statistically or clinically significant.

2. What is the probability that the discovered association is true?

Usually, I stop with the first question if no significant data is found, but for the sake of playing along, let's answer this question as well. Recall that this question is based upon first determining what the pre-test probability of the hypothesis being tested is. In this case, given all that we know with a high degree of certainty about how sex is determined, what would be a fair estimate that a difference of 5 mm Hg (let alone the observed 2.7 mm Hg) in blood pressure measurements many weeks before conception would affect the eventual sex of the fetus?

Again, don't worry about the specific number. But such a hypothesis would fundamentally challenge everything we know about gender selection. It's fine just to think in terms of low, intermediate, and high probability. I'm fine picking a 1% prestudy probability of this hypothesis being true (though it probably is *far, far, far* less; I am being generous). Given this low prestudy probability, even if a high quality, unbiased trial were to find a statistically significant result, it would only change the probability of the hypothesis being true to about 14%; if we introduce even a smidgen of bias to that good trial then the probability drops to only 5%. But, alas, this trial is not good and did not produce a significant result.

3. What are other hypotheses that explain the data?

Of course, even if the data is real, we must consider what other hypotheses explain the data. So, let's pretend.

Recall that just because we might accept that the data is valid doesn't mean that any one particular alternate hypothesis is true. Why else might the authors of this paper have found the observed difference (assuming it is a real difference) besides random chance?

Well,

- Different blood pressure machines may have been used (especially since the two arms were enrolled about 5 weeks apart on average).
- Different operators (maybe the person who worked with the group in which the male fetuses were more likely to be enrolled placed the cuff differently or liked to chat with the patients more causing an inappropriately high reading).
- Perhaps someone committed fraud in collecting the data.
- Perhaps someone committed fraud in analyzing the data.
- Perhaps that 5 weeks' temporal difference in the two groups reflected a different atmospheric pressure, temperature, environmental toxins or chemicals, or other unknown factors which either independently affected the patients' blood pressures, the fetuses' gender differentiation, or both.

We could go on for a while, but you get the point. Knowing nothing else, and raising no question as to the quality of the paper's data or its methods, we know at the outset that the most likely cause of the observed difference was chance alone, making *that* the most likely hypothesis. Again, this assumes that there even was a valid finding, and there was not.

4. Is the magnitude of the discovered effect clinically significant?

No, for three reasons. First, there was no discovered effect. Second, 2.7 mm Hg is well within the margin of error of the measurement device. Third, unless a prospective, randomized, triple blinded, placebo-controlled trial produces a result that says that raising a woman's blood pressure by about 3 points causes her to be more likely to have a male fetus, then I have no clinical use for this data. Even if such a trial occurred and a result corroborated the magnitude of effect suggested by this paper, I would have to raise the blood pressure of 1,000 women to produce 17 more males, and I doubt that this is something that anyone wants to do (maybe that increased blood pressure will lead to more miscarriages and the net effect is less males overall; there are so many potential unintended consequences that only an appropriately designed, prospective randomized trial can answer).

5. What cost is there for adopting the intervention into routine practice?

I'm not going to deal with the fifth question relating to the cost of adoption of the intervention since it obviously doesn't apply, but you will have an opportunity to do this on your own with every interventional study you read in the future.

Such "science" is dangerous. This was one of the most read papers of the year, mainly by lay-people. It was published in an open access journal so it is free to read and it was heavily promoted by the news media. Some patients might even be tempted to take medicine to raise their blood pressures while trying to conceive because they badly want a boy. You would be right to wonder why the editors or peer reviewers didn't point out the multiple problems I have mentioned (and many, many more). Why was this paper promoted by the journal with press releases to major media outlets? Where is the editorializing that urges caution in drawing any meaningful conclusions from such preliminary work?

We will discuss some of these issues as well in the next chapter which will provide you with the tools you need to critically analyze studies like this one in detail.

Other equations

Now let's look at the details of the equations we used for the calculations needed to answer the second question of the chapter.

Recall the basic parameters we introduced earlier:

- π = prestudy probability that alternative hypothesis is true
- a = Type 1 error rate (probability of rejecting null hypothesis when it's actually true or false positive rate)
- β = Type 2 error rate (probability of failing to reject null hypothesis when it's actually false or false negative rate)

Figure 30 is a tree diagram showing how to find the probabilities in the table that follows:

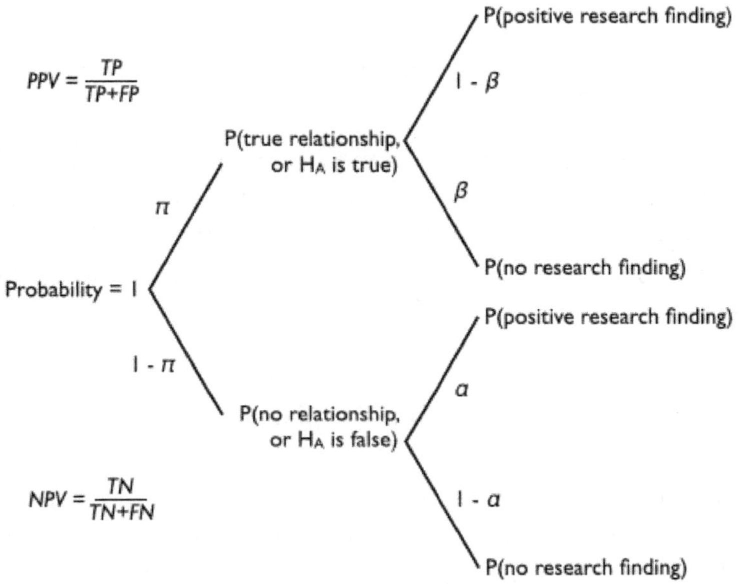

Figure 30. Probability tree for true and false research findings.

	True Relationship	No Relationship	Total
Pos. Result	$\pi(1-\beta)$	$(1-\pi)\alpha$	$\pi(1-\beta) + (1-\pi)\alpha$
Neg. Result	$\pi\beta$	$(1-\pi)(1-\alpha)$	$\pi\beta+(1-\pi)(1-\alpha)$
Total	π	$1-\pi$	1

Table 14. True and no relationship versus positive and negative experimental result.

Based on the tree and chart above, we see that the positive predictive value (PPV) is given by,

$$PPV = \frac{true\ positives}{total\ positives} = \frac{\pi(1-\beta)}{\pi(1-\beta)+(1-\pi)\alpha}$$

Note that this is also called the *positive post-test probability;* this is the probability that a positive research finding is actually true, given the pretest probability of the hypothesis.

The negative predictive value (NPV), or probability that a negative research finding is true, is given by:

$$NPV = \frac{true\ negatives}{total\ negatives} = \frac{(1-\pi)(1-\alpha)}{\pi\beta + (1-\pi)(1-\alpha)}$$

As we've discussed, it can be difficult to precisely assign a value for π; but that doesn't negate the importance of using a Bayesian approach to assess post-test probabilities. One approach is to consider R, which is the ratio of "true relationships" to "no relationships" in the field of study:

$$R = \frac{true\ relationships}{false\ relationships}$$

This ratio can be based upon generalizations of a given field of research (such as genomics, epidemiology, etc.) or specific information already known about the hypothesis from previous evidence.

Note that if we think of π, our pretest probability, as the ratio of true relationships to all possible relationships in the field of study, we obtain:

$$\pi = \frac{true\ relationships}{true\ relationships + no\ relationships} = \frac{R}{R+1}$$

Or,

$$R = \frac{\pi}{1-\pi}$$

Our equations for PPV and NPV can be expressed in terms of R:

$$PPV = \frac{R(1-\beta)}{R(1-\beta)+\alpha}$$

And,

$$NPV = \frac{(1-\alpha)}{R\beta + (1-\alpha)}$$

Now let's look at what happens when we introduce a bias parameter into the equations. Determining the degree of bias is usually subjective, unless methods are used for estimating the bias prevalent in particular fields or for particular subjects (what is called the *prevailing bias*).

As stated previously, the bias parameter u can be thought of as the proportion of studies that would be reported as negative results in an unbiased system that are instead reported as positive research findings, due to the bias of the researchers. For example, if there were 50 positive findings and 100 negative findings in an unbiased system, with a bias coefficient of $u = 0.10$, the reported positive findings would be 60 and the reported negative findings would be 90. Determining the amount of bias present in a field or for a given study is complicated. For now, just focus on appreciating how bias can affect the results of a study.

Unbiased system:

	True Relationship	No Relationship	Total
Pos. Results	40	10	50
Neg. Results	30	70	100
Total	70	80	150

Table 15. Sample results of an unbiased system.

Biased system ($u = 0.10$):

	True Relationship	No Relationship	Total
Pos. Results	43	17	60
Neg. Results	27 (↓ 10%)	63 (↑ 10%)	90 (↓ 10%)
Total	70	80	150

Table 16. Sample results of a biased system with $u = 0.10$.

Our previous probability table, with a coefficient of bias incorporated, looks like this:

	True Relationship	No Relationship	Total
Pos. Result	$(1\text{-}\beta)\pi + u\beta\pi$	$(1\text{-}\pi)\alpha + u(1\text{-}\alpha)(1\text{-}\pi)$	$(1\text{-}\beta + u\beta)\pi +$ $(1\text{-}\pi)(\alpha + u\text{-}u\alpha)$
Neg. Result	$B\pi(1\text{-}u)$	$(1\text{-}\pi)(1\text{-}\alpha)(1\text{-}u)$	$\beta\pi(1\text{-}u) +$ $(1\text{-}\pi)(1\text{-}\alpha)(1\text{-}u)$
Total	π	$1\text{-}\pi$	1

Table 17. True and no relationship vs positive and negative experimental result with bias.

The formula for PPV, with bias incorporated, is given by:

$$PPV = \frac{(1-\beta)\pi + u\beta\pi}{(1-\beta+u\beta)\pi + (1-\pi)(\alpha+u-u\alpha)} = \frac{(1-\beta)R + u\beta R}{(1-\beta+u\beta)R + (\alpha+u-u\alpha)}$$

Notice that NPV is unaffected by bias, because both the number of false negatives and total negatives have decreased by a factor of u.

Now let's consider the PPV of a single positive study out of n total independent studies. As we discussed previously, negative research findings are often not reported, which distorts the impact of a single positive study. Assuming that n independent studies are conducted (with no bias), if there IS a true relationship, then the probability of n false negatives is β^n; if there is no relationship, the probability of n true negatives would be $(1\text{-}a)^n$. We can use these values to fill in our table:

	True Relationship	No Relationship	Total
(At least one) Pos. Result	$\pi(1\text{-}\beta^n)$	$(1\text{-}\pi)[1\text{-}(1\text{-}\alpha)^n]$	$\pi(1\text{-}\beta^n) +$ $(1\text{-}\pi)\,[1\text{-}(1\text{-}\alpha)^n]$
(All) Neg. Results	$\pi\beta^n$	$(1\text{-}\pi)(1\text{-}\alpha)^n$	$\pi\beta^n +$ $(1\text{-}\pi)(1\text{-}\alpha)^n$
Total	π	$1\text{-}\pi$	1

Table 18. Equations for at least one positive result out of n studies.

Therefore, an upper bound for the PPV of a single positive research study (out of n independent studies) is given by:

$$PPV = \frac{\pi(1-\beta^n)}{\pi(1-\beta^n)+(1-\pi)[1-(1-\alpha)^n]} = \frac{R(1-\beta^n)}{\pi(1-\beta^n)+[1-(1-\alpha)^n]}$$

In practice, this means that the PPV of a positive study can be dramatically reduced by other independent negative studies; however, such negative findings are often not mentioned in a paper reporting a positive result.

The False Positive Report Probability (FPRP) can be calculated simply as 1–PPV.

Calculators for these equations can be found at www.howardisms.com/calculators/ or by scanning the QR code.

Exercises

15.1 Pick a recently published study. Go through the five questions and assess the value of the study.

15.2 Go to the calculators at the QR code on this page. Select the "Positive Predictive Value of a Study" calculator. Use this calculator to determine the PPV of the study you analyzed in 15.1. See how changing the bias of the study (u) and the prestudy probability (π) alters the PPV.

16

Pitfalls in Interpreting the Literature

I recall a discussion once with a Ph.D. statistician about the best ways to teach medical students how to scrutinize and analyze published scientific literature. He interjected that this wasn't a skill that needed to be taught to medical students because by the time a paper was published, it had already gone through a rigorous review process that prevented shoddy research from ever being published in the first place. He was serious.

Oy vey.

If we have learned anything so far, hopefully it's that most published research is wrong and will fail to be replicated in time. The great trick is trying to understand which studies hold promise and which are more on the junk side of things. Bayes' theorem helps with this tremendously. But so too does replication, understanding the magnitude of effect of the finding, the strength of the P-value, study methodology, etc. In this chapter, we will focus on a few pitfalls that affect experimental studies which should alter the confidence you have in any given paper. We will discuss several biases present in the literature that we need to be aware of as well, but we will start with a further discussion of P-values and P-hacking. This subject alone should keep you out of trouble most of the time.

P-hacking

Whether deliberate or unintended, manipulation and inappropriate interpretation of P-values is perhaps the single greatest pitfall in interpreting literature. A 2015 analysis concluded that P-hacking was widespread throughout scientific literature.[96] The naïve interpretation of P-values seems to roughly follow two axioms:

1. P-values less than (or near) 0.05 indicate true associations;
2. Statistically significant P-values reject the null hypothesis and confirm (or at least tend to support) the alternative hypothesis.

We have already discussed these points in some detail in the last chapter. It should suffice to mention again that P-values cannot be used to determine the probability of a hypothesis, i.e., the probability that a hypothesis is true given the observed data. Only Bayes' theorem can do this (refer to Chapter 15, Section 2). Also, remember that the 0.05 cutoff is purely arbitrary. It roughly represents data contained within ± 2 standard deviations ($\pm 2\sigma$). While it's true that about 5% of the data lies more than 2 standard deviations from the mean, there's no specific reason to think that two standard deviations is the magic boundary point between truth and falsehood and that was never the intent of the threshold. It might be acceptable to be 95% confident about a probable hypothesis (such as smoking causes lung cancer) but it is not acceptable to be 95% confident about a low probability hypothesis (Elvis causes brain cancer). Fortunately, this issue falls away when we apply Bayes' theorem to determine the probability of the hypothesis. The truth is, the P-value becomes less important when Bayes' theorem is appropriately utilized because we can consider any P-value, whether significant or not, when updating our confidence about a given hypothesis.

One of the most important developments in recent years regarding P-values was the 2016 publication of the American Statistical Association (ASA) *Statement on Statistical Significance and P-Values,*[97] edited by Ronald Wasserstein. The ASA took the unprecedented step of drafting a directive document about a statistical topic due to the rampant misuse of and confusion about P-values. The document emphasizes the following six points:

1. P-values can indicate how incompatible the data are with a specified statistical model (or perhaps hypothesis).
2. P-values do not measure the probability that the studied hypothesis is true, or the probability that the data were produced by random chance alone.
3. Scientific conclusions and business or policy decisions should not be based only on whether a P-value passes a specific threshold.
4. Proper inference requires full reporting and transparency.
5. A P-value, or statistical significance, does not measure the size of an effect or the importance of a result.
6. By itself, a P-value does not provide a good measure of evidence regarding a model or hypothesis.

The authors also suggest that Bayesian methods are better suited for testing hypotheses than the tests that produce P-values. In the years following 2016, nothing much changed in our journals or in our research centers. These misconceptions and misuses of the P-value seem as prevalent as ever. In response to this, in 2019, Wasserstein and colleagues wrote a follow-up editorial entitled *Moving to a World Beyond "$p < 0.05$"* in which they address the intellectual inertia surrounding P-values.[98]

This new paper is essential reading for anyone who practices or consumes science. It is accessible freely at the link provided with the QR code. It is the best current statement of evidence-based practice for not only using and interpreting P-values but for conducting hypothesis testing science in general.

Here, I would like to just highlight a few things from this article.

The paper introduces 43 other papers in the same issue which were selected to move us beyond the current state of misuse of P-values that is rampant in science. There are several important points made by Wasserstein et al.

First, and foremost, "don't" is not enough. Telling scientists not to base their conclusions on P-values or not to assume that P-values alone tell us about the chance that a true association exists has not worked. Papers are still published and interpreted with this dichotomous view of the world, where the P-value is the sole arbiter of truth. Instead, we need to focus on what to give research scientists as an alternative way forward.

Second, the authors tell us to not use the phrase "statistically significant" anymore: "Don't say it and don't use it." In their words, the phrase has become meaningless and a tyrant that dichotomizes results artificially. This phrase is misunderstood and abused in almost every paper and almost every meeting almost every day. It just has no meaning – but its misuse leads to dangerous and harmful conclusions.

Third, the authors list some things that we should do: accept uncertainty, be thoughtful (including thoughtfulness about the big picture, context, and prior knowledge – a Bayesian perspective), be thoughtful about how we communicate confidence in a research finding, be open and transparent (though sharing data, reporting negative findings, using pre-publication registries of research projects, etc.), and be modest (by understanding that multiple models may explain given data, by encouraging replication of findings by others, by accepting uncertainty, etc.).

Fourth, the authors point out that deep institutional reform is necessary to move beyond the misuse of P-values. We need to change incentives that exist structurally in our journals and publishers as well as our academic and educational systems. We need to reward reproducible and replicated science, not P-values < 0.05. They argue that massive changes are needed to our editorial and publication practices. They suggest this instruction to authors (taken from Hurlbert, Levine, and Utts[99]):

> There is now wide agreement among many statisticians who have studied the issue that for reporting of statistical tests yielding *p*-values, it is illogical and inappropriate to dichotomize the *p*-scale and

> describe results as 'significant' and 'nonsignificant.' Authors are strongly discouraged from continuing this never justified practice that originated from confusions in the early history of modern statistics.

The authors also review many of the reasons why research scientists and others have failed to reform. They point out that there is nothing new about any of this, but most practitioners of statistical sciences continue to misuse and abuse the P-value and null hypothesis significance testing in general. They include this sage remark from Steven Goodman from another wonderful 2019 article that is well-worth reading (provided at the QR code):[100]

> This brings us to the question of why eliminating P-value is so hard. The basic explanation is neither philosophical nor scientific, but sociologic; everyone uses them. It is the same reason we can use money. When everyone believes in something's value, we can use it for real things; money for food, and P-values for knowledge claims, publication, funding, and promotion. It does not matter if the P-value does not mean what people think it means; it becomes valuable because of what it buys.

Both articles are essential reading for anyone who uses null hypothesis testing and P-values. Unfortunately, they also outline the long history of resistance to use of appropriate methods. The first article highlights, as I said, an additional 43 papers which provide a way forward. I personally believe that it is unacceptable to continue, in any form, such abuse and misuse of P-values and null hypothesis significance testing. Nevertheless, I do believe that the P-value will remain in some manner and, perhaps more importantly, there are millions of research papers already published which rely on inappropriate utilization of P-values. So, this book has attempted to provide a way forward – to contextualize and make sense of published P-values and also to become expert at recognizing their misuse.

That brings us back to P-hacking. I would like to now discuss how authors and study designs can manipulate P-values. Authors of scientific studies are highly incentivized to find a "significant" P-value. It is interesting to note that much of this incentivizing would go away if we used Bayes' theorem appropriately to discuss the probability of hypotheses, but in a world where most people take a naïve view of P-values, the lust for a "statistically significant" P-value is overwhelming. Journals prefer to publish novel, positive findings. Academic careers are predicated upon such findings and the publications and grants that accompany them. Drug companies stand to make billions of dollars from a "significant" P-value. Purveyors of quackery and alternative medicine love to claim scientific standing based on an often contrived or fortuitous P-value.

It is unfortunate that the P-value has been turned into a finish line which, once crossed, signals victory – with naïve interpreters of P-values hearkening to the gods of science to support whatever pet theory they might have. But, alas, this is the current state of affairs, and therefore authors will do almost anything to find a P-value < 0.05. This has been popularly called **P-hacking.**

By playing with various **researcher degrees of freedom** (things like how much data is collected, which variables are controlled for and which are not, which outcomes are combined or excluded, etc.), it has been shown in computer models that with totally random data a P-value < 0.05 can be obtained 61% of the time with a few simple manipulations.[101] As data near significance, it becomes even easier to get the desired P-value.

In general, P-values will become smaller with:

- A larger difference between the two groups of data
- A smaller standard deviation
- A larger sample size

Sample size is very important. The higher the enrollment, the more likely it is that a significant P-value will be discovered. Many people assume that a study with a very large sample size is preferable. However, the focus should be on a sample size that is the *appropriate* size. A **power analysis** serves the purpose of determining how much data should be collected, and therefore a power analysis should normally be performed before the study is started. Part of a power analysis is an *a priori* decision about how much of a difference between study groups is considered meaningful.

Let's assume that we are interested in a new intervention that might shorten the length of stay for patients with pneumonia. Before undertaking the study, in order to construct a power analysis, we must consider the **magnitude of effect** that is desirable. For our patients with pneumonia, we might decide that a half a day difference in length of stay (as a minimum) would be desirable and then pick an appropriate number of patients to enroll based upon our intended rates of Type I and Type II errors (*alpha* and *beta*, respectively). If we select a smaller magnitude of effect, such as a 10-minute difference in length of stay, then we are much more likely to find a statistically significant result. If we don't do a power analysis at all and if we enroll enough patients, then we are very likely to find some difference in the two groups even though the magnitude of effect might be very small – if even real. No two data sets are ever identical, so if you are willing to consider any difference – even very small differences – as significant, then you will find statistically significant data.

The key to P-hacking is either narrowing the data analyzed to some small subset that shows significance (even though, on the whole, the data are no different) or expanding the data enough so that even some marginally different data may appear statistically significant. If you want to play with these ideas yourself and read a great article, visit the link at the QR code.

Scientists who perform studies truly believe in their theories. We call this **naïve realism**, which is the idea that we believe things because they are true (else we wouldn't believe them!). This naïve realism fuels the confirmation bias. Thus, we tend to design studies to find what we are looking for and we take any sliver of evidence as verification of the truth. We don't feel bad if we change the variables or manipulate things a bit in order to "prove" what we already know is true. Of course, this is completely contrary to the scientific method which requires us to attempt as vigorously as possible to disprove our hypotheses.

Humans have a strong tendency toward **pareidolia**. This is the phenomenon of seeing patterns and explanations for things that may not be there. It is a type of **apophenia**, which is our tendency to perceive connections between two unrelated things. An example of pareidolia might be seeing shapes in clouds or perhaps Elvis on a piece of toast.

But we do it with data and we are very often wrong, particularly when pareidolia is fueled by the confirmation bias or by naïve realism.[102]

Figure 31 is a collection of truly random data points. Do you see any patterns? This is the sort of random data that when sliced the right way can yield a significant P-value at least 61% of the time. What if I focus on a subset of the data? Or I stop collecting complete data when the incomplete data reveals something that looks like a pattern? I can do this by playing with exclusion and inclusion criteria, under-enrolling, or performing various subset analyses.

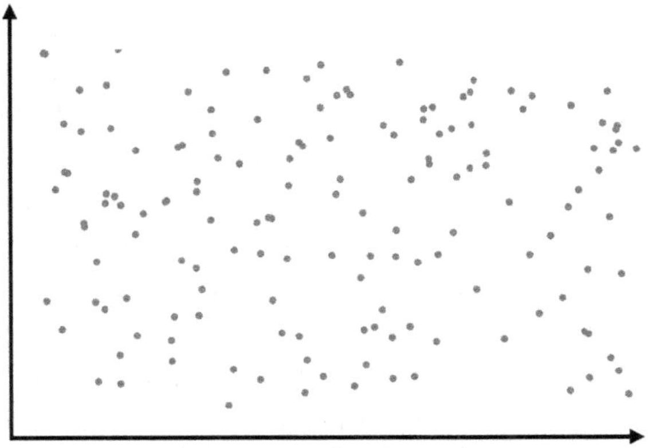

Figure 31. Random data points.

Look at the next two figures. Do you see a pattern now?

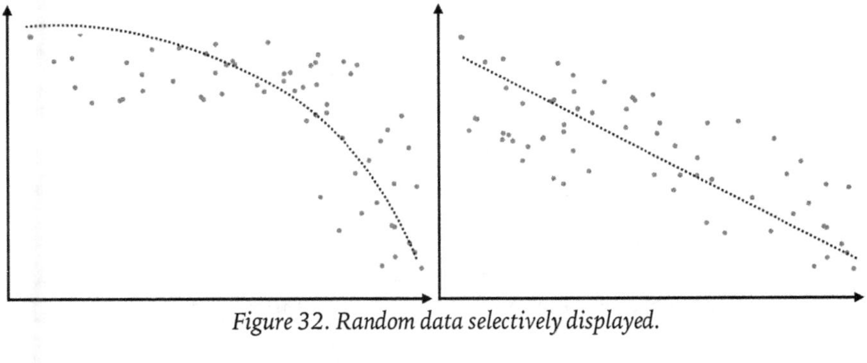

Figure 32. Random data selectively displayed.

These are the same data, just sliced and diced in different ways. I create these slices by playing with researcher degrees of freedom: inclusion and exclusion criteria, how much data I collect, when I stop, etc. When I do these things after I have collected data, I can almost always find *something* statistically significant.

Thus, it should be clear that a power analysis with a predefined desired magnitude of effect is very important, as well as *a priori* decisions about inclusion and exclusion criteria and pre-stipulated statistical methods.

There are several other issues that affect P-values besides these ideas that have been called researcher degrees of freedom. Some of these problems include decisions about the type of statistical tests and corrections that are used, one- versus two-tailed P-values, and, perhaps most importantly, the multiple comparators problem. Let's look at each of these briefly.

Statistical tests and corrections. Researchers need to make the right decision, given the type of data collected, about which tests to use to analyze data. Commonly used tests include the z-test, t-test, and f-test for data that follow a normal distribution and Pearson's chi-squared test for data that do not follow a normal distribution. Most researchers get this part correct, though it does happen that more novice researchers will select the wrong test accidentally or purposefully to find a significant P-value. Researchers can make mistakes or incorrect assumptions about the type of distribution that describes the data (e.g., normal distribution, t-distribution, or f-distribution). With sampled data, we sometimes make incorrect assumptions about the distribution and therefore use incorrect analysis tools (see Figure 33 below).

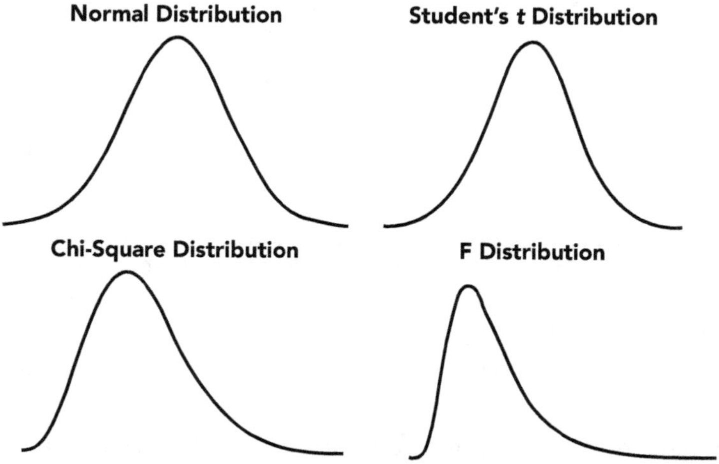

Figure 33. Normal, Student's t, Chi-Square, and F Distributions.

But even when the appropriate distribution and test is selected (before data collection begins), there are sometimes corrections that need to be applied and we need to consider whether the sampled data is representative of the total data set.

P-values are **stochastic** – not deterministic; that is, they are randomly distributed and cannot be determined precisely.[103] Given a set of experimental data, the observed P-value comes from a distribution of possible P-values. P-hacking can occur even when strict processes are followed in the context of small sample sizes when the random discovery of a P-value is made near the limit of the distribution.

As replication occurs or more data are collected, the observed P-values will tend toward some mean. This mean is typically higher than the true value with smaller data sets or fewer observations because smaller data sets may be more representative of the upper bounds of the data and not the true data set. This is analogous to regression to the mean. If there is no difference between two data sets (in reality) but a difference is observed (in an experiment), then if the experiment is repeated, that difference is likely to not be observed again or be as significant. Thus, P-values tend to increase (favoring the null hypothesis) as experiments are repeated. The smaller the data set, the less likely the data is to be representative of the true distribution of data; therefore, the smaller the data set, the more likely any observed differences are due to chance (and therefore the more likely that the P-value is artificially lower). P-value thresholds of significance need to be adjusted downward (often by an order of magnitude) in order to correct for this (or the sample size needs to be greatly increased).

Replication, then, is crucial to providing the context needed to interpret the meaning of a given P-value. It should not be surprising when initial studies or underpowered studies fail replication; this is actually expected. The surprise should be when significant P-values are supported with replication.[104]

One method of correction for small number of subjects is the **Yates correction** for the chi-squared test; this should be used to correct for potential errors in the assumption that a continuous distribution can be accurately modeled by a small data set (this is beyond our scope here and involves the Central Limit Theorem and how the chi-squared test assumes that the data collected belongs to some larger, normal distribution; when the data collected is small, then significant error can occur). Don't worry about the math; this is something that even statisticians argue about. Statisticians also argue about the threshold number of subjects beneath which the correction is appropriately used, with some arguing for 5 and others for 10. Other such corrections exist as well. My purpose here is to highlight the difficulty and controversy in applying null hypothesis significance testing to comparisons of very small data sets where our assumption that the collected data are representative of the larger data set might fail.

A paper published in October 2019 claimed that maternal stress during pregnancy was associated with fewer male children being born.[105] This paper was poor on many fronts and is a great paper to use for a journal club looking at many of the issues described in this book. The authors provide the following data comparing healthy women to "stressed" women:

Sex	Total (187)	Healthy (125)	Psychologic Stress (32)	Physical Stress (30)
Male	89	69	12	8
Female	90	54	18	18

The authors reported that the data represented in the table shows that stressed women have fewer male children, which supports their underlying hypothesis called *male fragility*.

The authors had to make some decisions about which data to compare. The P-value for healthy women compared to psychologically stressed women was 0.113, so they didn't use that since it was above P = 0.05. Instead they chose to expand the data to include women who reported physical stress as well. While this might make sense to them on some level, it's still a decision without a firm methodological foundation that affects the results (an example of a researcher degree of freedom). Nevertheless, when the stressed groups are combined, they found P-value gold: 0.011. Was this decision to combine two different pathologies made *a priori* or during analysis of the data? One wonders.

But notice also that 8 patients are lost to follow-up: 2 from the healthy group, 2 from the psychologically stressed group, and 4 from the physically stressed group. How hard was it to find these women who enrolled in the study and ask them what their child's gender was?

It is easy to see that it can be a very bad assumption that this experimental data is representative of a larger data set, which the chi-square test presumes both because of the small total number of enrolled subjects and because of the number who were lost to follow-up. The spirit of the Yates correction or tests like it is to correct for the error that can occur with these assumptions, though that correction doesn't apply to this study. However, my point is to illustrate that the smaller the number of patients in the study, the greater the likelihood that the data is not representative of the larger data set that the chi-square test assumes and potential disproportionate loss to follow-up only makes matters worse.

Let's suppose that the authors found these missing subjects and determined that both the healthy-group babies were female, and 5 of the 6 stressed-group babies were male – which is not an implausible scenario (and might be likely given the normal distribution of genders):

Sex	Total (187)	Healthy (125)	Psychologic Stress (32)	Physical Stress (30)
Male	94	69	14	11
Female	93	56	18	19

In this scenario, there are no significant P-values to be found, in comparing the psychologically stressed group to healthy (P-value = 0.24), the physically stressed group to healthy (P-value = 0.068), or both stress groups combined compared to healthy (P-value = 0.055). Each P-value is above the magic threshold of 5%. The paper's main purported result vanishes into thin air, just by supplying probabilistically reasonable missing data values.

Dozens of different statistical analyses were used in the paper to torture the data in various ways until some "significant" result appeared that supported the theory of male fragility. Such analyses should be decided upon *a priori* and conducted by a blinded statistician. The null hypothesis and alternative hypothesis need to be stated clearly and in advance, with the prestudy probabilities considered. There is no well-evidenced biologic theory of how physical or psychological "stress" would affect gender selection of offspring in a small study where nearly 10% of subjects were lost to follow-up. Nevertheless, ignoring such scientific rigor, it was widely reported in the national news and on social media at the time of the paper's release that the study had shown that stressed women had fewer male children.

Another sort of correction that often needs to be made is to correct for sample size over-enrollment. Remember, one of the easiest ways to make a P-value lower is to enroll more patients or collect more data. We can consider the BEAM trial as an example.[106] This study claimed to show that magnesium sulfate administered to women who delivered before 32-weeks' gestation was associated with fewer surviving children with cerebral palsy. The P-value for this association in non-anomalous children was 0.04397. The authors, however, only reported one significant digit, instead truncating to 0.04. Why did they truncate? One reason might be related to the correction for over-enrollment.

The power analysis for the BEAM trial found that they needed to enroll 2000 women in order to find a difference in outcomes when the P-value chosen for significance was below 0.05. However, like a lot of clinical trials, they over-enrolled. This can be a good thing because the authors don't want to miss desired enrollment due to loss of follow-up or patients dropping out. In the BEAM trial, they enrolled more than 2200 patients. That's okay. When this happens, however, it is necessary to perform an *a posteriori* power analysis correction, which will result in a new P-value that should be used as the threshold for statistical significance.

The authors did this using an **O'Brien-Fleming spending function** analysis and they concluded that a P-value of less than 0.043 should indicate statistical significance. Yet, they stated that "since the adjustment is minimal" they chose not to use this value. Since they chose to report only one significant digit for their P-values, then 0.04 appears to be below 0.043 and so they concluded that magnesium sulfate reduced the risk of cerebral palsy among survivors born before 32 weeks.

But, what we have discovered is another form of P-hacking: not applying appropriate corrections to the data and selective reporting of results. By their own criteria, they did not find a statistically significant result but that hasn't stopped widespread adoption of this non-scientific intervention, particularly by those already biased to believe that magnesium sulfate is somehow a valuable intervention for treatment for preterm labor despite overwhelming scientific evidence that it is not. There are two things I need to point out before going on: technically, the corrected P-value threshold of 0.043 doesn't apply to subset analyses (the threshold would be even lower). It really only applies to the primary outcome upon which the power analysis is based. I am fully aware of this. But authors who misuse P-values are not worried about that detail. In this paper, if the authors had declared that the P-value of significance was 0.043, then the reader would have not considered a P-value above 0.043 as significant. The abuse and misuse of P-values has created a complex tapestry of confusion, and that confusion aids the aims of P-hacking.

The other point I will make for completeness involves misallocated twin pregnancies. In this study, there were four sets of twin pregnancies where one twin either was stillborn or died shortly after birth while the surviving twin developed cerebral palsy. In the event where one twin dies or suffers a significant hypovolemic or hypoxic insult *in utero*, the risk of the surviving twin developing cerebral palsy is very high. These types of twins were misallocated in the two groups: one set was in the magnesium group and three were in the placebo group, which increased the risk of discovered cerebral palsy in the placebo group disproportionately as compared to the magnesium group. If these four pregnancies were excluded, then the P-value for non-anomalous children who developed cerebral palsy in the magnesium group compared to the placebo group is 0.067. This is another example of a researcher degree of freedom. It is truly striking that an entire specialty would introduce a new practice based on such thin data, where moving one or two patients to the other arm of the study could have such profound implications, but this is what happens when we misuse and abuse P-values. We will see another trick the BEAM authors used in a few moments, but first let's discuss one-tailed versus two-tailed P-values.

One-tailed versus two-tailed P-values. A two-tailed or two-sided P-value assumes that either intervention in a comparison might prove superior for a given outcome. For example, a blood pressure medication studied against placebo may or may not prove superior in lowering total mortality. The drug might lower the risk of death from cardiovascular events but it might raise the risk of death from accidents and head trauma from falls, so it is an open question whether the drug or placebo might prove superior for total mortality. It could go either way and therefore a two-tailed P-value is utilized and this is the P-value normally seen in studies where we assume statistical significance with a $P < 0.05$.

But what if we know that the measured effect can only favor one arm of the study? Then it would be more appropriate to use a one-tailed or one-sided P-value. Let's say you are studying the effect on blood pressure rather than all-cause mortality for the drug and you know that the blood pressure will always be lower with the drug than with placebo; in this case, it may be more appropriate to utilize a one-tailed P-value. Importantly, though, the value for statistical significance must now be one half the nominal value or $P < 0.025$.

The threshold of 0.05 for P is based on the normal distribution:

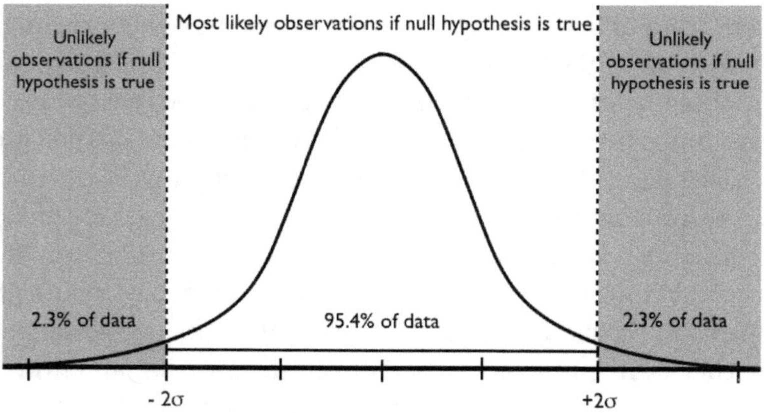

Figure 34. Tails of a normal distribution of data.

Because the two tails or sides of this graph add up to about 5%, is has been arbitrarily decided that any observation in the tails is unlikely enough that seeing such an observation would tend to invalidate the null hypothesis (that there is no difference in the two groups being compared). But this is based upon the assumption that unlikely outcomes could happen on either extreme. However, if they can only happen at one tail, then using 0.05 is inappropriate because we would then no longer be looking at data that it is more than two standard deviations from the mean. Thus, we must use 0.025 or half the nominal value.

Researchers must select a one-tailed or two-tailed P-value when performing statistical analyses. Many researchers have learned that when they discover a marginal P-value (say, 0.06) that they can rerun the analysis with a one-tailed approach and this will yield a value of half the marginal value (0.03 in this case). That's all fine, but then they may publish the smaller P-value of 0.03 while still claiming 0.05 as the cutoff for statistical significance. That's P-hacking (or just plain dishonesty).

I recently read a primer about P-values on towardsdatascience.com written by a graduate student in data sciences where he advocated, without understanding it, this very misuse of a one-tailed P-value.[107] He understood that a one-tailed P-value was appropriate, but he still used 0.05 as a cutoff for significance. It was extraordinary to me that he could make such a basic mistake. Unfortunately, his article has been well read and these sorts of errors are common even among the educated.

Multiple comparators problem. If I have a null hypothesis and test it, there is a 5% probability that I will discover a false positive finding by chance alone if we set *alpha* (or P) equal to 0.05. Each additional independent null hypothesis I test carries with it a new risk of a false positive discovery. Thus, the more hypotheses I test, the greater the chance that at least one of them will produce a positive result ($P < 0.05$) and potentially a false positive discovery. We call this the **multiple comparators problem**. It doesn't matter whether I conduct 10 different experiments in the same paper or if I conduct 10 different studies in 10 different papers; the more times I try to find a significant result, the more likely it is that I will find one.

If I want to flip a coin and get three heads in a row, and if I only flip it three times, my chance of getting three heads in a row is only 1/8; but if I do it 100 times in a row I am virtually guaranteed to get three heads in a row.

The probability of discovering a false positive result (specifically, a Type I error) in a clinical study is,

$$Probability\ of\ false\ positive = 1 - (1 - \alpha)^n$$

where α is equivalent to the P-value threshold the authors have chosen to indicate significance, so α would usually equal 0.05 and n is the number of hypotheses being tested.

Table 19 shows the chance of finding a false positive discovery for a given number of hypotheses:

Risk of false positive finding	**Number of hypotheses tested**
5%	1
10%	2
14%	3
19%	4
23%	5
26%	6
30%	7
34%	8
37%	9
40%	10
54%	15
64%	20
72%	25
92%	50
99%	100

Table 19. Probability of a Type I error for a given number of tested hypotheses.

None of this is controversial; yet, most modern published studies include numerous (in some cases even dozens) of comparisons. Think of a study that compares two surgical interventions. The authors collect several data points and list these in a table. Perhaps for a comparison of surgeries this might include things like length of stay, length of surgery, estimated blood loss, postoperative hemoglobin, postoperative pulse, pain scores, amount of pain medication required, amount of pain medicine prescribed at discharge, presence of fever, need for transfusion, nausea scores, need for antiemetic medications, incidence of ileus, rates of postoperative hemorrhage, rates of reoperation, etc. The null hypothesis is that there is no difference in the two surgeries but the authors have collected 15 different outcomes. They are, in fact, testing 15 different hypotheses (with the null hypothesis being in each case that the there is no difference in outcomes between the two interventions for each factor listed).

With 15 different hypotheses, there is a 54% risk that a P-value will fall below 0.05 for at least one of these outcomes by chance alone. It is common to see studies like this where the authors might find that the surgical approaches are similar in every regard except for one characteristic, such as postoperative ileus. The authors then conclude that one intervention is superior to another because of that one characteristic. It's like betting that you are going to roll 10 on a pair of dice but then taking 15 turns trying.

Authors must adjust for these multiple comparators in order to make the chance of a false positive finding (a Type I error) still be 5% or less. The multiple comparators problem was not widely discussed until the late 1990s so, unfortunately, it is rare to see authors perform these adjustments. Several techniques have been developed to make these corrections. The **Holm-Bonferroni method** is probably the most validated method while the **Bonferroni correction** is the easiest and likely most widely used method. If you are reading a paper, you can do the basic Bonferroni correction in your head or on the back of a cocktail napkin: simply take the nominal P-value (usually 0.05) and divide it by the number of hypotheses being tested. In our imaginary paper then we simply say 0.05/15 and find a new P-value threshold of < 0.0033. More sophisticated methods account for the relatedness of the hypotheses. Postoperative hemoglobin and need for transfusion have a definable relationship so they might be handled differently than pain scores and need for reoperation which are more likely to be truly independent.

Number of hypotheses tested	***Bonferroni corrected P-values***
1	0.05000
2	0.02500
3	0.01667
4	0.01250
5	0.01000
6	0.00833
7	0.00714
8	0.00625
9	0.00556
10	0.00500
15	0.00333
20	0.00250
25	0.00200
50	0.00100
100	0.00050

Table 20. P-values with Bonferroni correction applied for numbers of hypotheses tested.

If you pay attention to the multiple comparators problem, you will find that it is the source of a large portion of published novel and significant discoveries – mostly false discoveries. Table 20 shows corrected P-values to use for a number of independent hypotheses.

Data-mining type papers are the worst offenders. If an author writes a program to discover any data points that show statistical significance out of a large mass of data, she will find several "associations" almost all of which are false discoveries. The technique can be useful but the results must be corrected and then the findings should be replicated in a well-designed and appropriately-powered study.

Still, even in well-designed studies, this issue occurs rampantly. Think again about the study which claimed that stressed pregnant women have lower rates of male children. The study is full of different hypotheses and tested associations. Even in the narrow aspect of whether stress relates to gender of offspring, there were three tested hypotheses: the difference in gender outcomes between healthy women and women with psychological stress, the difference between healthy women and those with physical stress, and the difference between healthy women and women with stress of any sort (physical or psychological). But the authors also looked for differences between these groups regarding term or preterm birth, mode of delivery, birth weight, aggregate birth complications, NICU stays, neonatal length of stay, maternal length of stay, fetal heart rate movement coupling, and reactivity of fetal heart rate. In other words, they tested 12 different hypotheses giving them a 46% chance of at least one false positive discovery related to pregnancy stress.

A Bonferroni correction would give a new significant P-value threshold of 0.0042. The only finding that met this threshold was the association between physically stressed pregnant women and fetal heart rate-fetal movement coupling and that may very well be a true finding; it may have no clinical significance but it at least passes the initial threshold of statistical significance. The next step would be to design and power a study to replicate that finding (ideally in a prospective manner but in this case that would be unethical).

How about the BEAM Trial? The original outcome studied was a combined outcome of death plus cerebral palsy in fetuses before 32-weeks' gestation exposed to magnesium versus those not exposed. The authors found no difference in the two groups. Then they tested several other hypotheses: death alone, cerebral palsy alone, as well as four separate measures of infant development. In other words, they tested seven total hypotheses giving them a 30% chance of discovering a P-value < 0.05 (and a 26% chance of discovering an association below the adjusted P-value of 0.043). Recall that they did not find an association with the lower P-value, but disingenuously reported an association with a P-value that was between 0.043-0.05. Yet, once again, an appropriate Bonferroni correction would have been 0.043/7 or 0.006. Were the authors just unaware of the multiple comparators problem? No. In fact, they state very matter-of-factly in the methods section, "no adjustments were made for multiple comparisons." The BEAM Trial is a *tour de force* in P-hacking and there are many, many other problems with the study. Long term follow-up data from several trials now have shown no differences in developmental outcomes among the children exposed to magnesium versus placebo groups.[108,109] Still, despite such evidence, this paper has led to a dangerous change in the standard of care in obstetrics. Yet even a cursory analysis of the methods the authors used would have revealed the numerous problems with the paper.

Subset analyses. Subset analyses often create situations where P-values can't be trusted for the reasons we have been describing. Almost always there are multiple comparators involved and usually the data set is subdivided in different ways that may be unfair. Also, importantly, the power analysis of the study would have been constructed to avoid Type I and Type II errors for the primary outcome being studied; when the data is subdivided into smaller subsets, the power analysis doesn't usually apply or even make sense. One might think that if a significant association were discovered with a smaller data set, then that's great; the risk is not so much a Type I error but a Type II error and that is often the case (not that Type II errors aren't equally important). Still, this isn't always true.

Recall again the BEAM Trial. The main outcome studied and the outcome for which the power analysis was constructed was the combined rate of death and cerebral palsy. This outcome was selected because the prevailing thought at the time was that magnesium might lower the rate of cerebral palsy among surviving preterm neonates by killing off the weakest ones. Recall that the authors found no difference in this combined outcome between infants exposed to magnesium versus placebo. But in the subset analysis, they looked at death alone and cerebral palsy alone and the purported finding was that cerebral palsy rates were different while death rates were not. Of course, we have already discussed why this assertion was false. However, let's pretend for the moment that the rate of cerebral palsy *was* lower and that this finding *was* truly statistically significant. Even then, the issue remains that the subset analysis was underpowered to detect a difference in death. There were more deaths in the magnesium group (a finding that had been present in other studies) but it didn't rise to the level of statistical significance. Is this a Type II error? Perhaps. I should note that Type 1 errors can happen with underpowered studies. A researcher may stop collecting data once a P-value < 0.05 is observed; but, with more data collection, the significant P-value might vanish. This could happen because the smaller amount of data simply isn't representative of the larger data set. Power is important. Still, Type II errors would be more expected with insufficient power.

Type II errors are often overlooked since the focus is always on P-values. Table 19 showed the probability of a Type I error (a false positive discovery) with multiple hypotheses. The rate of Type II errors (a false negative discovery) is much higher. The rate of Type II errors depends on the β, which is commonly set to 0.10, 0.15, or 0.20, meaning a 10%, 15%, or 20% chance of not discovering a true association. Table 21 shows the risk of Type II errors with multiple comparators:

Risk of false negative finding			***Number of***
β=0.*10*	β=0.*15*	β=0.*20*	***hypotheses tested***
10%	15%	20%	1
19%	28%	36%	2
27%	39%	49%	3
34%	48%	59%	4
41%	56%	67%	5
47%	62%	74%	6
52%	68%	79%	7
57%	73%	83%	8
61%	77%	87%	9
65%	80%	89%	10
79%	91%	96%	15
88%	96%	99%	20

93%	98%	>99%	25
99%	>99%	>99%	50
>99%	>99%	>99%	100

Table 21. Probability of Type II error for a given number of tested hypotheses.

In the BEAM Trial, the authors elected for a power of only 80%. This means that since they looked at seven different comparisons, they had a 79% chance of a false negative discovery (like magnesium doesn't increase the risk of death). The actual rate is higher than this because the smaller numbers of the subset analysis didn't provide 80% power.

In any event, subset analyses are usually underpowered, magnifying both the Type I and Type II error rates. They also suffer from problems related to multiple comparators which in turn affect these error rates. Subset analyses rarely have any sort of power analysis. For these reasons, any findings from a subset analysis should be considered preliminary and further, appropriately powered studies should be performed to validate the findings.

What are some solutions to P-hacking? Many advocate for a movement away from P-values altogether, but this seems impractical.[110] Rather, we need to educate researchers and consumers of scientific literature about the limitations of P-values. For those analyzing data with tests that produce P-values, here are some things that can help reduce the risk of purposeful or inadvertent P-hacking:

- Decide which statistical tests will be performed on which data beforehand, and report any changes to methodology and the reasons for the change.
- For interventional studies, always include a power analysis and perform an *a posteriori* analysis if necessary after data collection. Decide beforehand when to stop collecting data and what degree of over-/under-collection warrants an *a posteriori* analysis. Some researchers stop collecting data once a desirable P-value is obtained; had they collected more data they might have discovered that the alleged statistical significance would have disappeared.
- Correct for multiple comparators and if a novel observation is made about other comparisons, or in a subset analysis, replicate the experiment with an appropriate prospective design to validate the finding.

- Clearly distinguish between pre-specified data analyses and those performed after collection of data has started or with "fishing" methodologies.
- Routinely use "**triple blinding.**" This occurs when the statistical analysis is performed by someone blinded to the source and identification of the data. In other words, the analyst doesn't know which data belong to which arm of the experiment.
- Interpret all P-values in a larger context using Bayes' theorem as previously discussed.

Other important biases

In Chapter 3 we discussed several cognitive biases that affect how we gather and interpret data. Many of those apply to performing and consuming scientific studies as well. Here let's discuss some important biases that are specifically problematic for the design and analysis of scientific studies.

Optimism bias. In a broad sense, the optimism bias causes someone to believe that a negative event is less likely to happen to him compared to others (it's also called **comparative optimism**). In a more specific sense related to this book, it means that people tend to believe that they are less likely to make critical errors of thinking or suffer from biases compared to others. This is true of individuals and of generations of people. For example, people alive today believe that they are less likely to make critical scientific mistakes compared to generations that have lived before, or that they are less likely to suffer from bias than previous generations. People also may assume that they have more understanding of complex issues compared to previous generations.

A result of this bias is that physicians can easily scoff at the silliness of prior generations of physicians (*"Can you believe they used to bleed people with leeches to treat preeclampsia"*) while being almost completely blind to their own misperceptions, biases, and blind spots (like using magnesium to prevent cerebral palsy). Only time will tell which current beliefs will be tossed into the trash bin of medical/scientific rubbish, but appropriate use of statistics and critical analysis of hypotheses can help us avoid serious blunders.

Recall the concept of naïve realism: no one walks around believing that they are wrong about what they believe. The optimism bias convinces us that we are, at the very least, more correct than others and especially more correct than our predecessors. But the published scientific papers of our predecessors were, on the whole, more correct than those published this year. In other words, we are more likely to be wrong about science today than we were fifty years ago. Does this sound absurd? Let me explain.

Most of the low-hanging fruit in science has been investigated. Fifty years ago, there were far fewer journals and scientific publications with millions fewer articles published each year. Scientists were more likely to study hypotheses with higher pre-experimental probabilities (things like whether cigarettes and asbestos are oncogenic). Today, with millions more papers published each year and a lower threshold than ever required to get published, and with most of the low-hanging fruit already plucked, scientists are studying hypotheses with a much lower probability of being true (things like talc as a cause of ovarian cancer or glyphosate as a cause of lymphoma, both of which are almost certainly not true but are popularly believed at the time of this writing). So, if you look at the conclusion of a paper published in 1970 and compare it to one published in 2020, with the same statistical power, the 1970 paper is more likely to be "true."

Solutions: Assume that most of what we know today to be true will in time be invalidated. Consider Bayesian updating to understand the probability of a given hypothesis and vigorously challenge our beliefs, even – and especially – those that are widely accepted.

Prevailing bias. Biases aren't necessarily bad; many can be protective. When I see a bear in the mountains, I tend to walk the other way because my implicit bias tells me that I might become lunch; however, some bears are quite docile so my bias about a specific bear may be incorrect. But, overall, my biases serve me well. Bias, however, needs to be accounted for when determining the veracity of a belief generally or a scientific study specifically.

We are good at defending what we believe in and we are good at defending what is commonplace, *status quo*, or widely accepted. We are reluctant to deviate from that to which we are accustomed. I call this **cognitive inertia** or **normalcy inertia**. We see the world not as it is, but as we *need* it to be.

We are slow to change our views about anything that we are deeply invested in. We are comfortable with what we know, what is familiar, what is near, and what is common. We are biased towards these things. This is cognitive inertia. We are attracted to what we are told is normal or what seems normal. This leads to normalcy inertia. It is comfortable to think that everyone around us, our parents or our friends, have a good grasp of reality. If our parents or our elders or those whom we view as smarter than us or our leaders are wrong about the world, what chance do we stand? We are taught culturally what is normal, and we can hardly see the world otherwise.

Normalcy inertia and cognitive inertia exist in science and we call this **prevailing bias**. If the prevailing scientific bias is that the sun revolves around the earth, then all evidence (even evidence that refutes the idea) will be seen as supporting that prevailing belief. It was Copernicus who made the first serious, modern suggestion that the earth revolved around the sun and not the other way around in 1543. In doing so, he challenged the prevailing view (the prevailing bias) of Ptolemy that had reigned relatively unchallenged for nearly 1400 years. Copernicus did not provide "proof" of this theory, but he did do something that I think is very interesting: he showed that the same celestial observations (the same data) which were accounted for by the Ptolemaic system could also be accounted for by his system. In other words, he showed that there was more than one hypothesis that could fit the data. Tycho Brahe quickly added yet a third hypothesis that was compatible with the data as well.

When Copernicus first showed that all known data could be explained by his system as well as by the Ptolemaic system, "science" should have immediately stopped and admitted that, barring any new evidence, both explanations were equally likely. This is a foundational principle of Bayesian updating. But, of course, that is not what happened. His contemporaries couldn't objectively view the issue due to the prevailing bias and the cognitive inertia that went with it, just as many modern physicians quickly discount literature which challenges dogmas like the cholesterol hypothesis. In fact, wide acceptance of the Copernican theory didn't occur until after 1700.

Prevailing bias is neither good nor bad, but it must be accounted for when analyzing data. It is our current bias that cigarette smoking is bad for you, and this bias affects the way studies are designed and how we interpret the data generated by them. In other words, we tend to over-estimate just how bad cigarettes are due to this bias (please don't accuse of me supporting cigarette smoking).

The impact of prevailing bias on the scientific literature has been studied. John Ioannidis has found that in some fields of research, like nutrition, "the extent that observed findings deviate from what is expected by chance alone would be simply a pure measure of prevailing bias."[111] This means that prevailing bias alone could account for the appearance of statistical significance in many publications, particularly in low probability hypothesis fields like nutrition and psychology. So, we must attempt to account for this bias (which we did mathematically in the equations at the end of Chapter 15).

None of this is that important if the prevailing bias is correct, but when it is incorrect, it can have a huge impact on the interpretation of literature. Take, as an example, Vitamin D. There has been a prevailing bias over the last decade that Vitamin D deficiency is a cause of or a contributing factor to dozens of different medical problems. This belief never had a solid foundation in science but was popularized by a handful of charismatic authors and supported by the vitamin industry. A decade later, the scientific literature is clear that the enthusiasm that has been afforded Vitamin D screening and supplementation was woefully misplaced. Yet, individual physicians, fueled by the optimism bias and this prevailing bias, have been stuck in a morass of cognitive inertia, blind to the fact that prevailing bias alone is the source of their beliefs on the subject. Indeed, it tends to take almost 20 years before new evidence is integrated into the majority of clinical practices.

Solutions: We can attempt to measure and account for the prevailing bias in meta-analyses and introduce a correction factor for the result itself or for the magnitude of effect as described in the previous chapter.

Publication bias. This occurs anytime the choice to publish results of an experiment depends upon its outcome. Journals tend to favor what is new and "statistically significant," meaning that they steer away from replication studies or negative studies. This can skew the interpretation of literature in a given field because mostly positive studies are published. An analysis of those positive studies (removed from the context of the many unpublished negative studies) can lead to a false belief that the science supports something that it does not.

Researchers, too, may choose to not submit a paper for publication if it didn't result in a positive finding or support their hypotheses. This is called the **nonresponse bias**. They may also submit papers for publications that include only favorable interpretations of the dataset and exclude others interpretations that were performed that weren't favorable.

John Antonakis has cleverly described five "diseases" which affect and threaten scientific publications:[112]

1. *Significosis:* the inappropriate focus on only statistically significant results.
2. *Neophilia:* a love of only novel findings.
3. *Theorrhea:* a mania for new theories or novel explanations for data.
4. *Arigorium:* this describes the pervasive lack of rigor in hypothesis development and testing and the widespread misuse and misinterpretation of statistical methods.
5. *Disjunctivitis:* this describes the incredible volume of low impact, poor quality publications produced because of perverse incentivizing for academics to produce papers.

The resultant literature is skewed and can make secondary analysis of the literature in the form of systematic reviews and meta-analyses fundamentally flawed. Imagine if I am interested in finding evidence that Vitamin D helps treat depression due to the prevailing bias that low Vitamin D is associated with depression. So, I design a small study of 100 patients and randomize them to take Vitamin D versus placebo, but ultimately find that it doesn't help. At the same time, 9 other researchers do the same thing, all influenced by the same prevailing bias. Let's pretend that 9 of us find that it doesn't work and one researcher finds that it does. Recall from Table 19 that we would expect this to happen 40% of the time by chance alone. The problem is, though, that only the false positive finding gets published, either because the researchers didn't submit the data for publication or because journals simply rejected papers which didn't present a novel finding or promote a new theory. Subsequently, a meta-analysis or systematic review of the literature would tend to support Vitamin D use for depression because such studies aren't aware of the unpublished results.

This sort of publication bias has been found to be widespread among meta-analyses.[113]

Publication bias is sometimes called the **file-drawer bias** with the idea that the negative studies just get crammed into a file-drawer somewhere.

Solutions: We can help control for publication bias by requiring studies to be registered in a public database before any data collection occurs and for all statistical analyses to be pre-specified, and outcomes should then be made available, whether published or not. Hypotheses should be clearly specified and authors should avoid theorizing after data has been collected. All collected data should be made available, and academic models should de-emphasize the need for publications when considering promotion and tenure.

Channeling bias. This bias tends to occur in studies that lack true randomization, and is in reference to how researchers may "channel" certain patients to one arm of a study versus another. For example, in a non-randomized study of patients undergoing one of two surgical options, the surgeons might implicitly choose certain types of patients for one arm, e.g., obese women for a vaginal hysterectomy instead of a laparoscopic one, or "healthier" patients for a novel drug while less healthy ones get the standard, proven treatment. Of course, randomization should correct for this, but most published literature is not randomized. Thus, we expect to see this type of bias in case cohort and retrospective type studies; while authors may attempt to correct for this bias, they are often not successful. In some cases, researchers may not even realize the "lurking variable" that requires adjustment.

Solution: randomization.

Priming. In studies that involve any intervention, double-blinding helps prevent priming from occurring, wherein a researcher may "prime" the subject to report more positive or negative results based on knowledge by either the subject or the researcher about whether the subject received the intervention or placebo. Accordingly, any time blinding does not occur, a significant priming effect in the results can be anticipated. But priming also occurs with questionnaires and patient interviews. Wording on questionnaires can easily favor a more negative or a more positive response. Non-blinded, retrospective studies and those involving questions with subjective answers or scores are most vulnerable.

Solutions: Emphasize objective results over subjective ones and strive for triple blinding in study design.

Interviewer bias. This is a subtype of priming wherein the personality, social style, intonations, reactions, and emotional tone of the interviewer influences the answers of the respondent.

Solutions: Use a tone-neutral script for interview questions and avoid personal interviews (use a computer or written queues). Don't allow the primary researcher to also conduct interviews with subjects.

Forced choice format bias. This is one example of dozens of problems that can arise with questionnaires. A simple example of this bias might be asking a respondent "Do you agree" and then providing only "yes" or "no" choices. If they must pick an answer, they will be biased to select "no" even though they may agree with most of the statement or may not understand the statement. An additional answer might be "I don't know" or "Mostly agree" or "Mostly disagree," etc. There are dozens of ways that survey questions can be designed to yield the answers the researchers want, and this is one of the chief problems with surveys, questionnaires, political polls, etc. Choi and Pak have provided a nice list with examples of these issues in their 2005 paper, "A Catalog of Biases in Questionnaires"[114] which is available online by scanning the QR code above.

Solutions: Use only carefully designed and standardized tools for questionnaires and emphasize collection of objective data provided without constraint by the tool.

Recall bias. Case-control and retrospective studies are particularly vulnerable to the recall bias, where the faulty or selective memories of patients influence what data is collected. A parent of a child with autism might incorrectly recall the temporal association between the onset of the child's symptoms and the date of the MMR vaccine. The mother of a child with Type 1 diabetes is more likely to recall a viral illness the child might have had in the preceding months of the diagnosis than a mother of a child who does not have diabetes. Recall bias is rampant in studies with retrospective data collection.

Solutions: Conduct studies prospectively and when retrospective data collection is necessary, use objective sources like medical records rather than patient memories.

Sampling bias. The sampling bias is also called the **ascertainment**

bias. The sampling bias involves how data are collected, as opposed to the **selection bias** which involves how collected data are processed. When authors include or exclude some data based on criteria chosen after data collection, this could form a selection bias. Sampling biases happen before the data is even collected.

There are many forms of sampling biases. Many studies suffer from **self-selection** sampling biases. For example, a study might recruit patients through commercials or advertisements; those patients who respond to the advertisement may be very different from patients who do not respond to the advertisement. We will talk about this soon when discussing the placebo effect. Many online polls on websites suffer from this problem. If a conservative news website polls its readers, such a poll has first excluded people who would never go to that website for news in the first place, but even among its own readers it has excluded individuals who just don't like interacting with online polls or who aren't passionate about the topic at hand, for example.

Some studies may include **prescreening** either in the form of a predefined exclusion criteria or just by recruiting at certain places. For example, if the study is only recruiting patients with a history of depression who have failed SSRI use, this might be a rather different population compared to patients who have had depression that responded to SSRIs. Or a study measuring the effects of a vegan diet on cardiac disease that advertises for patients at Whole Foods in an affluent neighborhood will find fundamentally different patients than if the recruitment happened at different locations and at different stores. Perhaps you study Vitamin D levels in roofers and compare them to Vitamin D levels in age-matched men who work in accounting. The roofers will have higher Vitamin D levels because they are sun-exposed but they are also likely more physically fit than the accountants, not because they have high Vitamin D but because they do manual labor. These sorts of variables can form a **healthy user bias**.

The **spectrum bias** occurs when the group of patients studied doesn't well represent the spectrum of diseased patients of interest. For example, studying the outcomes of COVID-19 infections only in healthy non-smokers provides little value to understanding the impact of the disease on the general population and might lead to biased epidemiological implications and projections.

Another type of sampling bias is **Berkson's paradox**, where a negative correlation of two related positive traits occurs. This happens, for example, when studies are conducted in hospitals or clinics instead of among a random population at large. Some call it the **hospital admission bias.** Diabetes is a risk factor for acalculous cholecystitis; if we study people who get acalculous cholecystitis and people who have diabetes in a random sample in the general population, we will find there is a significant correlation between the two variables. However, if we study those populations in a hospital, we may not find a correlation because diabetes is over-represented in hospitalized patients. There just aren't as many people without diabetes in a hospital so the observed correlation washes out. Finding diabetes among hospitalized patients (like those with cholecystitis) just isn't special.

Overmatching occurs in some types of studies that require matching control and intervention groups statistically; it occurs whenever something that is a result of the exposure is unintentionally matched in the control group even though it is actually a mediator of the result. For example, if a researcher chooses patients for a case-control study to determine the incidence of postpartum suicide among women who suffer from depression, she might choose to match the two groups for substance abuse disorders since she views that as an independent cause for suicide. However, such a design would not account for the fact that depressed women often turn to illicit substances as a means of dealing with untreated depression. In other words, depression can lead to substance abuse and, in turn, lead to suicide related to substance abuse. Overmatching for substance abuse disorder would tend to minimize the relationship between depression and suicide in this example.

The **survivor bias** is yet another type of ascertainment bias wherein enrollment occurs only among survivors. For example, a researcher designs a study to determine risk factors for myocardial infarction; he finds a group of patients who have suffered a myocardial infarction and compares them to a group of healthy, matched peers. He determines that listening to jazz saxophone is protective against myocardial infarction. The problem is that his study doesn't include people who died of myocardial infarction before being enrolled. Had those patients been included, he might have discovered that listening to jazz saxophone is actually associated with fatal heart attacks compared to matched peers. Survivor bias is very pervasive and affects the way we think about a lot of things. I live in a 115-year-old house that is well-built and really quite nice. It is built far better than new homes. One might assume that *they just don't build homes the way they used to,* but my home survived the last 100 years and other crappy shanties did not. In reality, the average new home is far better than the average home built 115 years ago.

Remember the flawed BEAM Trial we discussed earlier that examined the effect of magnesium sulfate exposure on the risk of developing cerebral palsy? The theory that magnesium sulfate might prevent cerebral palsy was based upon a paper from 1995 in which the authors performed a retrospective, case-control study in California of 155,636 children followed to age 3.[115] They found that 7% of very low birth weight (VLBW) babies who developed cerebral palsy (CP) had been exposed to magnesium in utero while 36% of VLBW babies who were not exposed to magnesium went on to develop CP. In other words, VLBW babies who did not develop CP were more likely to have been exposed to magnesium *in utero*. Note that only 117 babies out of 155,636 children were candidates for analysis. This is a subtle sleight of hand authors often use to make studies sound more impressive: they looked at over 150,000 children but they only found 117 that mattered and who were subsequently included in the statistical analysis. In any event, the real problem here is the survivor bias: since only children who lived to age 3 were included in the analysis, the authors failed to account for children who may have died *because* of magnesium exposure. Later studies provided evidence that antenatal magnesium exposure is associated with more neonatal deaths and therefore fewer survivors with cerebral palsy (since the infants likely to die were those most vulnerable to developing cerebral palsy anyway).

The **nonresponse bias** is another form of ascertainment bias which is related to the reasons why people may choose not to respond to a survey or answer certain questions. For example, imagine you send a questionnaire to patients in your practice asking about techniques they have used to help deal with anxiety; not everyone will respond to the survey, even if you have chosen a good representative sample. Those who choose to not respond may be very different than responders. Perhaps people who haven't done well with anxiety choose to not answer more often than those who are doing well. It may look as if an intervention works well for treating anxiety simply because most of the people who failed a trial of the intervention didn't answer the survey.

Even among those who do choose to answer, there may be a **response bias**, wherein respondents give inaccurate answers for some reason. Few people are entirely comfortable giving truthful answers about very sensitive subjects such as sex, drug use, embarrassing habits, or even religious or political issues – especially if they feel judgment from the interviewer or fear exposure. Similar to this is the **unacceptability bias**, which relates to any form of measurement which causes unacceptable pain to the subject – physical or psychological. Subjects may underreport things that are painful to admit, such as things they feel guilty about, and they may also refuse tests which are invasive and painful or which they fear will bring bad news.

Lead-time bias. The lead-time is the difference in the time from detection of a disease with some new test and the usual detection of the disease based on clinical symptoms or some old test. Lead-time bias occurs when the lead-time in detection makes life expectancy or five-year survival rates appear higher. For example, the average five-year survival rate for women with invasive breast cancer is 91% based upon current diagnostic, detection, and treatment practices. If a new proteomic test were developed that could detect breast cancer three years sooner than normal, then five-year survival rates might increase to something like 98%, which is roughly the current two-year survival rate, based upon the lead time difference alone. However, an author might claim that the new detection method should become universal because the test is associated with significantly better five-year survival. This conclusion might be due entirely to lead-time bias.

Funding bias. The funding bias or **sponsorship bias** relates to the tendency for studies to support the agenda or interests of the organization or group paying for the research. In many cases, this may be overt and due to actual malfeasance and dishonesty. However, more commonly, it is unintentional. Researchers may not even be conscious that they tend to design their studies or interpret data in a way that is more likely to make conclusions supportive of their sponsors. Obviously, the interests of the sponsor and the researcher are often aligned (that's why the researcher applied for the grant from that sponsor). The researcher starts with a predetermined conclusion and looks for data to support it; she may choose to not publish data or even perform analyses that are unfavorable, as we've discussed previously. One 2010 review found that research sponsored by a pharmaceutical company was four times as likely to find favorable results for a drug than research sponsored by an independent party.[116] A 2017 Cochrane review confirmed that industry-sponsored studies were not only more likely than non-industry sponsored studies to find that a drug or product is effective, but also the magnitude of effect of the drug or device is uniformly greater in industry-sponsored studies.[117]

Oversampling and undersampling. Data science is already very complicated; indeed, you would be accurate if you assumed that most people who do research for a living make serious mistakes and have significant misconceptions. However, we are now seeing more and more studies that utilize "big data" and **machine learning** to aid in scientific research. For example, several studies in the last five years have employed machine learning techniques to develop predictive models for preterm birth by studying a variety of physical and chemical predictors of birth. These studies have reported near 100% success in accurately predicting who will deliver early. These sorts of studies involve complex computer models and programming – skills well beyond the understanding and bailiwick of even advanced health science researchers. A 2020 paper showed what we all suspected: the results of eleven such papers were too good to be true.[118] They all suffer from the same important methodological error: **oversampling**.

Machine learning techniques rely on large datasets called **training sets** and **testing sets**. The training set is what the machine learns from, and the testing set is used to validate and confirm what has been learned. These sets must be different, so before analysis they are partitioned, but they also need to be matched to ensure that the outputs of the sets are similar. To match the records, some of the data may be oversampled to make up for a lack of one type of record or, less commonly, undersampled to reduce the number of another type of record. The terminology and math are borrowed from signal processing science. These adjustments must be done with the correct proportions and only after the data is partitioned. If the resampling is not performed after partitioning, then some of the same data can be located in both the training and testing set and this will make the performance of the machine learning algorithm look better than it is (since it just analyzes the same data twice, essentially). This is what occurred in the 11 papers analyzed. All the conclusions were false based on this simple misunderstanding, but all passed multiple peer review processes and all involved doctoral level data scientists.

Streetlight effect. This describes our tendency to look for answers where it is easiest to see them (if you're looking for something you lost on a dark street, you will probably look under the streetlight first and foremost). Health science is littered with streetlight effect problems. If we are interested in studying preterm birth, for example, the data that is easiest to collect and readily available include things like maternal demographics and BMI, smoking and substance abuse status, family history, parity, and other items that are easily found in the chart. This means studies are likely to focus on comparisons of those sorts of data with risk of preterm birth. Yet, the real solutions may lie in the dark, down the street: genetics, connective tissue disorders, the inflammatory cascade, etc. Related to this is the **McNamara fallacy**, named after US Secretary of Defense Robert McNamara who lost the Vietnam War by making decisions based only on what he could measure (things like number of casualties) and ignoring things he couldn't measure (like the resiliency of the Vietnamese people). In clinical medicine, for example, we might promote a drug because it leads to prolongation of life (something we can measure) while ignoring quality of life or other more holistic outcomes which are difficult to measure.

Surrogate outcomes and **lurking variables.** We have discussed these issues already. Both are related to the concepts we have been considering here. In Chapter 12, we discussed surrogate markers at length. Recall that a surrogate marker or outcome is some easy-to-measure outcome that is felt to correspond to the outcome that actually matters, such as cholesterol level (the surrogate) instead of all-cause mortality; we are assuming that lowering cholesterol is a good thing, but that's not what we actually care about. In the same way, lurking variables are the unknown variables that are either difficult to measure or just unknown. We study what is easy, which means we study the obvious and the short-term, not the obscure and long-term.

The **Hawthorne effect** occurs when subjects change their behaviors or responses because they know that they are being observed. This effect can also impact clinical trials when others involved in the experiment, say nurses or doctors or other clinicians, act differently than they might otherwise because they know that a trial is being conducted. In this way, trial results might not apply to the real world.

Placebo and nocebo

Most interventional studies contain a comparison to a placebo which is meant to be an intervention (such as saline, a sugar pill, or even sham surgery) that has no therapeutic benefit and therefore should not influence patient outcomes. When the intervention and the placebo are compared to a third group of no intervention at all, the difference between the placebo and no intervention group is called the **placebo effect**.

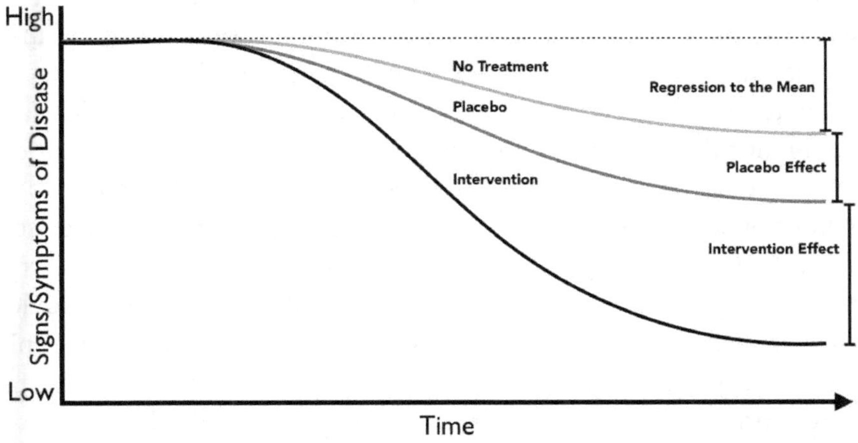

Figure 35. The placebo effect.

The difference between outcomes with placebo and outcomes with intervention is the **intervention effect**. These terms are commonly misused. Both the intervention effect and the placebo effect look larger if they are compared to the initial state of the disease, marked by the dotted line in Figure 35. However, disease states have their own natural history, and the symptoms and signs of most diseases will improve over time due to normal convalescence or regression to the mean, as we discussed in Chapter 4. In other words, the magnitude of the placebo effect should be determined by comparison to the disease not at time of enrollment but at the time of the conclusion of the study, while the intervention effect shouldn't be compared to the baseline at all. It is difficult, however, to observe a true "no treatment" group so these comparisons are not often made.

We should never compare an intervention to no treatment because such a comparison doesn't account for the placebo effect at all. If data for no treatment, treatment, and placebo are collected, these should usually be compared against each other using **ANOVA** or *analysis of variance* testing. However, it has been estimated that this testing does not occur in more than half of published research.[119] This leads to significant overestimation of the magnitude of effect of the intervention arm. We should never compare an intervention to nothing, but to placebo. A drug might be no better than placebo, but if the drug is compared to no intervention or to the disease baseline it might look great. If it is compared to nothing, it shouldn't be the initial state of the disease but the disease over time, since the natural history of the disease has likely led to changes (sometimes improvement, sometimes worsening), though this still won't account for the placebo effect.

Therefore, a placebo arm is very important because it essentially accounts for the changes of the disease over time without intervention and hopefully accounts for other aspects of treatment (but remember, half of published literature fails to make the appropriate comparison with ANOVA testing). Those other aspects of treatment are important: a patient may receive counseling, education, or other advice that impacts care. It is rare that the studied intervention is the only thing the patient benefits from. This is especially true for mental health problems, where just seeing the physician and talking is itself a form of treatment. These elements must be perfectly matched in both the treatment and the placebo arms. What's more, a person who seeks treatment and enters a study is a different person than someone who doesn't, so even being enrolled in a study will show differences compared to those with totally untreated disease.

Let's imagine that we develop a study for a new intervention to aid in quitting smoking. We advertise for people to join the study and then we give either a placebo or our new drug to the participants. We note that 36% of people quit at 6 months with the drug and 25% quit with the placebo. We also find that 10% of patients who decided not to do the study after presenting quit smoking in the next few months anyway. We find another study that says that only 2% of people who smoke typically quit smoking in any given 6-month period. The drug company that we work for encourages us to say that people who use our drug are 18 times more likely to quit smoking over the next 6 months compared to people who don't. This sort of thing is commonly done in published literature but it is a lie.

The baseline risk of 2% is an unfair comparison because we don't know if it's relevant to the panel of patients who chose to enroll in the study. Even if it is a demographic match, there is something different about people who answer an advertisement to join a study to quit smoking compared to those who do not. They are motived, and this is a form of selection bias.

But if we compare it to the group who entered the study but didn't complete it, where 10% quit smoking, this is also unfair. They differ from the placebo group in many ways: they are motivated but perhaps not as motivated as the other participants; they may not have had follow-up visits where they received encouragement and other advice; etc. The only fair comparison is to the placebo group.

The **placebo effect** is large and is seen in virtually every study that measures it. But what is it? Is it even real? These questions are important because many drugs are prescribed *only* for the placebo effect. This includes many alternative medicine treatments. In medical school, I was taught that even if alternative medicines had no benefit whatsoever in a scientific assessment, we should not discourage patients from using them because they still might at least benefit from the placebo effect. It might surprise you to know that the placebo effect may not even exist. Refer to Figure 35. In theory, the placebo effect is the difference seen between the *no treatment group* at the conclusion of the study and the *placebo group*. But "no intervention" is almost never measured in this way. Instead, no intervention is usually represented by the state of the disease at enrollment into the study. By using this zero-state assessment, we fail to account for regression to the mean or even the normal convalescence of the disease.

An early study in 1955 defined what we mostly think of as the placebo effect: that an inert treatment can influence the course of disease through the power of the mind.[120] But a large systemic review of 130 studies comparing placebo to no treatment found almost no evidence for a placebo effect.[121] In other words, almost all the placebo effect is likely due to regression to the mean and other unaccounted for treatments, like counseling, office visits, patient motivation, etc. that occur as part of either the placebo or intervention group.

Let's consider the placebo effect for different sorts of diseases. If I conduct a study of insulin versus placebo for treatment of Type 1 diabetes, I would expect all the participants in the placebo arm to be dead in 6 months and almost all the participants in the insulin group to be alive. There is no placebo effect for Type 1 diabetes: the natural course of the disease and the placebo arm both result in death.

On the other hand, if I conduct a study in Type 2 diabetes, I likely will see some placebo effect resulting in better blood sugar control; but this is almost certainly due to factors like an improved diet and daily schedule, more intense blood sugar monitoring (and greater awareness of blood sugar excursions), more regular office visits with more feedback and counseling, etc. There is no scientific reason to think that it is due to the power of the mind over blood sugars.

This doesn't mean that the power of the mind has no effect over the course of disease. It almost certainly does. It just means that lying to patients about the effect of a drug or an alternative medicine isn't the right way to achieve that response. Let's say we do a study for treatment of depression. These sorts of studies show a very large placebo effect. But, again, when treating diseases like depression, a patient who has sought therapy, decided to receive therapy, received hope and encouragement from the visit with the physician, etc., will feel better whether they take a drug or not. In this case, the patient decided to take a medicine that she was told will help her and there is no doubt that this was associated with altered brain chemistry, new found hope and optimism, etc. that makes the mood better. But the effect size of altered brain chemistry alone associated with the belief that she is receiving a treatment is almost negligible. In fact, even when this does happen, it's not the course of the disease that is being affected but rather the brain's perception of symptoms.

Remember, most of the magnitude of the placebo effect is exaggerated because placebo is so often compared to the wrong baseline. We can summarize then that the placebo effect is almost entirely due to:

- Other therapeutic factors that occur during delivery of care;
- Regression to the mean;

- Patient motivation;
- Operant conditioning (building-up expectations to the patient that something will change);
- Unintended treatment from inert ingredients; and,
- Study bias (response and reporting biases).

Nocebo effect. Just as important as the placebo effect is the *nocebo effect*. We noted above that operant conditioning of the patient can cause a placebo effect; that is, telling them to expect to feel improved can lead to them feeling improved. This would be an example of **positive conditioning** and it should be utilized by clinicians to maximize patient outcomes. It's the one part of the placebo effect that can be weaponized to help treat patients. At the same time, though, we often cause **negative conditioning**, too, when we warn patients about negatives experiences they might encounter with an intervention or just with the course of their disease process. This negative conditioning leads to patients being more likely to report (and experience) the negative effect than had they not been conditioned.[122] This is called the **nocebo effect.**

If we tell patients that something is going to hurt badly, this may lead to anxiety and then up-regulation of cholecystokinin; this messenger actually causes a worse perception of pain.[123] This doesn't mean that we shouldn't tell people that what we are about to do is going to hurt, but how we say it is very important. I think about these principles when I do procedures like cervical biopsies or placement of intrauterine devices. I maintain a jovial demeanor with patients and tell them to expect some discomfort but I don't over-emphasize it; in turn, they hurt less.

In the same way, we must be careful about how we present the risks of surgeries or warning signs about medical conditions to patients; over-emphasis of negative outcomes can *cause* those outcomes. When we prescribe a drug, careful delineation of every possible side-effect, even unimportant ones, can cause those side effects. We will talk more about communicating risks to patients in the next chapter.

The same effect occurs in clinical studies. When researchers carefully detail to patients all the possible negative effects of the intervention, they will be more likely to experiences those negative effects. In a well-controlled RCT, the nocebo and placebo effects will be found in the placebo arm as well as the intervention arm. While we understand that we must subtract the placebo effect to determine the true positive impact of the intervention, we must also subtract the nocebo effect to determine the true negative impact of the intervention.

For example, let's say that we warn patients about sexual side effects of a new drug that treats depression. In the treatment group, 15% of patients reported sexual side effects while in the placebo group 13% of patients reported the same side effects. In the "no treatment" group, 5% of patients reported similar sexual issues. In this situation, it may be that the nocebo effect regarding negative sexual symptoms is around 8% percentage points and this may not be statistically different than what was observed in the treatment group (15% versus 13%). In other words, the drug likely has no sexual side effects even though 1 in 6 patients reported sexual side effects. Rest assured that companies marketing against the drug will be quick to point out that 1 in 6 patients reported sexual side effects even though those reports were secondary to negative operant conditioning. A 2014 study of trials evaluating depression medications found that 1 in 20 patients in placebo arms dropped out of studies due to side effects from the placebo – in other words, because of nocebo.[124]

The nocebo effect is magnified in clinical medicine due to reporting of drug and intervention effects on social media, websites, etc. In my practice, a patient may have a symptom she doesn't understand such as hair loss or weight gain; the only medication she is taking is her birth control. She can quickly find other women on the Internet who have falsely attributed their hair loss or weight gain to the same birth control and then it is almost impossible to convince her that her birth control is not the source of her problem. What starts as false attribution for one patient leads to the nocebo effect for the next patient, as social media and websites falsely list symptoms that a patient might expect with a new medication. I don't mean to imply that she is imagining that her hair is falling out; rather, the problem has another cause (such as normal postpartum changes), but its impact is greatly magnified due to the nocebo effect. Other symptoms like nausea might be purely related to the nocebo effect without any underlying pathology. The problem is also compounded by pharmacies that give patients long lists of potential medication side effects when the prescription is filled; most of these side effects are not present when the medication is compared to placebo.

Anecdotal and alternative medicine are particularly vulnerable to placebo and nocebo. A doctor may have used a drug for years and then is asked to switch to another drug because of a change in hospital or insurance formulary. When he talks to his patients about changing the prescription, he may display his displeasure with the change and warn the patient that the new medicine is not what he would have normally chosen for them and it may not work as well or it might have more side effects. In doing so, he is almost guaranteeing that his patients will report decreased efficacy or more unpleasant side effects. When they report this to him, he will increase the vigor with which he warns patients and his anecdotal confidence that the one medicine is better than the other; in turn even more patients will be harmed by his negative operant conditioning of them.

Alternative medicine thrives off pitting the placebo effect against the nocebo effect. An alternative medical provider will emphasize to patients the real or imagined negative effects of an evidence-based medical intervention (thus increasing the nocebo effect) while simultaneously over-selling any real or imagined benefits of the alternative treatment, thus increasing the magnitude of the placebo effect. Through skillful use of positive and negative conditioning, the alternative medicine practitioner makes money selling woo-woo like acupuncture, homeopathy, naturopathy, vaccine denial, etc.

Peer Review

Remember the conversation I related at the beginning of the chapter: the statistician who thought the peer review process ensured that we only get good papers published. We should end this chapter by talking about that problem specifically. Many scientists have far too much confidence in peer review, which, unfortunately, is next to worthless in most cases.

Remember at the beginning of the last chapter we talked about a study that was published in the *British Medical Journal* (BMJ) about remote intercessory prayer? Don't hold that against the BMJ; it was published with a purpose. It's a good journal, as far as journals go, and has done a lot to advance the cause of evidence-based medicine. In fact, its editor, Fiona Godlee, along with a former editor and a trio of reviewers for the journal, published one of the most important studies about peer review in 2008.[125]

The group wanted to study the effect of training on the quality of peer review. They had a control group and two intervention groups with two types of education (in person and self-taught). They created three test papers and each paper had nine major and five minor methodologic flaws. The subjects were reviewers for the BMJ, not a shoddy bunch – mostly distinguished, academic physicians. The majority were researchers themselves. Half had post-graduate training in statistics and sat on editorial boards of journals. Without training, at baseline, the 607 peer reviewers found only an average 2.58 of the nine major errors and only 0.91 of the five minor errors. What's worse, their abilities to detect errors didn't improve with either type of educational intervention.

Were these difficult, obfuscated errors designed to trick researchers with advanced degrees in statistics? Not at all. They were easy to spot errors – not complicated efforts at statistical malfeasance or P-hacking – just basic, simple mistakes.

A similar study was performed with 262 reviewers of the *Annals of Emergency Medicine* where 10 major and 13 minor errors were included in a manuscript and two-thirds of these errors were missed.[126] This state of affairs has not improved much, and it is disheartening to know that training doesn't result in improved peer review quality. A review in *JAMA* found that peer review was of dubious value at most.[127] Richard Smith, a former editor of the BMJ, concluded:

> So peer review is a flawed process, full of easily identified defects with little evidence that it works. Nevertheless, it is likely to remain central to science and journals because there is no obvious alternative, and scientists and editors have a continuing belief in peer review. How odd that science should be rooted in belief.[128]

If you've read this far, you realize it isn't odd at all. Douglas Altman has reported that among papers he has analyzed from high quality journals, 63% use incorrect analysis of multiple observations, 61% give inadequate information on harmful consequences of interventions, 58% use an incorrect method of comparison for subgroups, 89% fail to report the mechanism used to allocate interventions, 51% fail to state whether blinding was utilized, and 25% fail to mention the eligibility criteria.[129]

Many journals, including the BMJ, have moved to an open and transparent peer review process and other efforts to raise the stakes of peer review; but, so far, there is no evidence that much has improved. Many of us feel that a truly crowd-sourced peer review process, facilitated by open and free publication on crowd-reviewed websites, may be the best solution.[130]

In the current free-for-all world of PubMed, old articles are never removed, corrections rarely issued or linked to the original publications, there is no annotation of which articles have failed replication, etc. It truly is the wild, wild west. False conclusions stand unchallenged and unchecked, except in the relatively rare case that a paper is retracted. Meta-analyses and other flawed attempts at systematic review continue to incorporate faulty research years after the science has moved on. This creates inertia in scientific progress. Today, scientific papers are archived dead documents. An open-source, crowd-reviewed platform could help make these problems better.

Worse, though, we have only been speaking here of top tier journals with top tier reviewers; most biomedical publications don't have those constraints. There are now over 10,000 open access or pay-to-publish journals. The quality varies widely, but many will publish any paper submitted for a small fee within a couple of hours of receipt. Luckily, half of all published scientific papers are read only by their authors and the editors.[131]

Still, this doesn't stop someone from finding bad articles in a database like PubMed and assuming that it must be valuable since it was peer reviewed. The static publishing and reading cycle doesn't even work that well for articles that have been retracted since retraction gets little if any publicity and other derivative papers, textbooks, and opinions are usually unaffected. Almost 700 retractions and 12,000 errata are published each year.[132] When was the last time you recall seeing one?

The bottom line is that peer review shouldn't change your opinion of an article, nor should you be overly impressed that something has been published in an important journal. You must analyze the paper for yourself or rely on trusted third party reviewers.

Exercises

16.1 Select a recent interventional study. Did the authors make multiple comparisons? Do they address this issue with a P-value correction? If not, calculate an appropriate corrected P-value. Does this change any of the conclusions the authors made?

16.2 Are there any biases that you've learned about in this chapter that you suspect might affect the paper you analyzed for the exercises at the end of the last chapter? Did the authors make any errors in interpretation or manipulation of the P-values discussed in the study?

16.3 What are some examples of the placebo effect you are aware of in your field of practice? How about the nocebo effect?

17

Communicating Outcomes to Patients

Ultimately, our job as physicians is to offer the most benefit with the least harm as possible for our patients. This involves mutual decision making, during which we share with patients the risks and benefits of different treatment options (including not treating) and then make a decision together in light of the patient's values and desired goals.

What we say matters; it alters our perceptions of risks and benefits and heavily influences the decisions that patients make. Our words can actually change patient outcomes through positive and negative conditioning (placebo and nocebo), as we learned in the last chapter. Many times, we focus on the cons (the negative risks) without focusing also on the pros (the positive risks or benefits) and this leads to cognitive **risk distortion**. Everything in life has risk; in medicine, we seek to minimize risk while providing maximum benefit. Without understanding the benefits of what we do (and which benefits are important to each patient), we cannot balance this equation correctly.

How we understand the literature and how we communicate it to our colleagues and to our patients matters, too. At the heart of clinical medicine is uncertainty: uncertainty about the diagnosis, uncertainty about the best treatment, uncertainty about the magnitude of the benefits and risks of treatments, uncertainty about the validity of the scientific literature, etc. Humans innately don't like uncertainty. Physicians don't like appearing uncertain in front of patients.

To compensate for the uneasiness that uncertainty affords them, physicians act too confidently. They seek to make things black and white: they tell patients they have a certain diagnosis even though other diagnoses may be nearly as likely; they choose a best treatment even though others may work nearly as well; they decide that some scientific paper is better than another even though both may have serious flaws.

We need to involve the patient in our decision-making calculus. She has a right to be fully informed about the diagnostic and treatment options that we are considering and how certain we are about the risks and benefits of each option. It is our ethical duty to deliver this information in a way that is as accurate as possible, as free of bias as possible, and respectful of her personal beliefs and wishes. We do this poorly. In this chapter, we will explore a few areas that can help us better communicate risks and benefits to patients, namely,

- Communicating uncertainty;
- Choosing noncoercive words;
- Describing statistics effectively; and,
- Using words therapeutically.

Before diving into these subjects, we need to address two basic principles important to understanding how patients make decisions and how they perceive what we say: Prospect Theory and the Cooperative Principle.

In a theoretical rational model, we would assume that humans would make decisions according to the **expected utility hypothesis**; this hypothesis generally assumes that humans behave as rational agents who dispassionately weigh risks and benefits and make decisions rationally based on expected gain or loss. Historically, expected utility theory arose from **expected value theory**, which presumed that individuals would derive the greatest benefit from choosing options with the highest expected net payoff (*expectation* or *expected* *value*). A rational individual, given a choice between a 95% chance of winning $10,000 or a 100% chance of winning $9000, would choose the first option given its higher average payoff ($9500 vs. $9000). Over time, expected utility theory evolved to become much more nuanced, acknowledging that *utility* is what a rational person will try to optimize, not just expectation.

Utility can be thought of as the whole value of an option for an individual, considering their risk tolerance and personal values, preferences, and goals. An individual might stand to benefit the most economically from certain risky investment strategies, but lose sleep at night with concerns about the risk; for this individual, the risky investments do not optimize personal utility. Another person may leave a lucrative profession to dedicate time and resources to aiding a marginalized group; beneficence factors into this person's utility. Utility is a complex concept, incorporating emotional, psychosocial, socioeconomic, and other factors, and can vary widely for different individuals faced with the same proposition.

As we have discussed previously, decision-making in clinical medicine is full of examples of utilitarian considerations. A given intervention might bring a slightly increased life expectancy, but at the cost of unacceptably diminished quality of life; a rational patient might well forego the intervention. A new drug or treatment might provide marginally better outcomes than an old one, but not enough to justify a ten-fold increase in cost or burden to the health system; a rational doctor might choose to stick with the old treatment.

One important aspect of decision-making that is incorporated into expected utility theory is the idea that potential gains and losses are considered in proportion to an individual's current assets (where "assets" are wealth, health status, etc.) For example, someone with a net worth of $100 views potentially losing or gaining $50 differently than someone with a net worth of $100,000. In medicine, we generally realize that an intervention offering six more months of life means more to a terminally ill patient than a healthy 30-year-old. Note that "current health status" is not just objectively measured status, but perceived status: two individuals with clinically comparable rheumatoid arthritis symptoms may perceive their health conditions very differently, due to differences in pain tolerance, mental health, personal beliefs and values, the effects of social support or lack thereof, lifestyle goals, etc. Therefore, two individuals with the same severity of the same condition may rationally make different choices based on their individual perceptions of their current health status.

Of course, anyone who has observed actual humans for any length of time realizes that people often *don't* make decisions rationally or dispassionately; we see examples of this in the stock market, politics, love, and just about every other area of life. While expected utility theory presumes that people will act rationally based on their individual utility parameters, **Prospect Theory**, developed in 1978 by behavioral economist Daniel Kahneman and cognitive psychologist Amos Tversky, seeks to model human decision-making based on empirical observation as opposed to assumptions about rationality.[133] You may recall Kahneman's name from our discussion in Chapter 3 about System 1 and System 2 thinking; he won a Nobel Prize for his work on Prospect Theory in 2002.

	Potential Gain	Potential Loss
High Probability	**High probability to gain a lot** • *Risk averse behavior* • Fear of disappointment Examples: • Choose 100% chance to win $9000 over a 95% chance to win $10,000 • Quitting smoking • Losing weight • Covid-19 vaccine	**High probability to lose a lot** • *Risk seeking behavior* • Hope to avoid large loss Examples: • Choose 95% chance to lose $10,000 over a 100% to lose $9000 • Surgery for terminal cancer • Higher-risk alternative medical treatments
Low Probability	**Low probability to gain a lot** • *Risk seeking behavior* • Hope of a large gain Examples: • Choose 5% chance to win $10,000 over a 100% chance to win $550 • Play lottery, gambling • Alternative medical treatment • Vitamins, supplements	**Low probability to lose a lot** • *Risk averse behavior* • Fear a large loss, "better safe than sorry" Examples: • Choose 100% chance to lose $550 over a 5% chance to lose $10,000 • Insurance • Unnecessary medical tests

Table 22. Prospect Theory.

There are two very important insights of Prospect Theory, in relation to decision-making in clinical medicine. The first is that we view losses and gains *asymmetrically*, fearing loss more than desiring gain. Facing two propositions with objectively equal expected payoffs, an individual will be more likely to make the choice that avoids loss; this preference persists even when the payoffs are not equal, so that an individual may go to an irrational extreme to avoid loss. The second observation is that we tend to underestimate the magnitude of high probability prospects, and overestimate the magnitude of low probability prospects. Putting these two ideas together leads to the conclusion that we are irrationally reluctant to take a risk when there is a high probability of a large payoff, or a small probability of a large loss. Conversely, we are irrationally over-eager to take a risk when there is a low probability of a large payoff, or a high probability of a large loss.

Now let's consider how each of these cognitive biases might play a role in the decisions our patients make regarding their health.

Our patients may be *too reluctant* to take a risk on a *high-probability large gain* ("risk averse"). One of the most obvious and important examples we see of this is reluctance to try to stop smoking, to try to quit alcohol or drugs, or to try to lose weight. Smoking, substance abuse, and obesity are some of the greatest health issues threatening our patients, and altering the associated behaviors carries a very high probability of very significant gain. So, why are patients often so resistant to act? Obviously, the reasons are numerous and complex, but a big part of the issue is that patients underestimate the magnitude of the payoff. They may seem content with the status quo of their health, not yet suffering the direst effects, and unwilling to undertake the hard work of behavior modification and risk of failure. A lifetime smoker may not realize how great they could feel with healthier lungs; an obese person may underestimate their improvement in energy and mood, and decrease in joint pain, made possible by weight loss.

Another good example of this phenomenon is the rising resistance to vaccines in our culture. Vaccination is by far one of the safest and highest yield medical interventions of modern times, yet we see patients averse to taking this high-probability, high-payoff risk. Again, many underestimate the magnitude of the true benefit available from vaccination, instead focusing on avoiding the relatively minuscule risks of loss presented by vaccines (e.g., reaction, injection site soreness, etc.) Patients choose to maintain their status quo, in which they feel unreasonably safe from diseases like measles and meningitis, which have become very rare precisely because of vaccination.

Our patients may be *too eager* to take a risk on a *low-probability large gain* ("risk seeking"). This is the same tendency that leads people to buy lottery tickets. A classic medical example of this would be the popularity of alternative medicine treatments that have a very low probability of potentially large benefits. Chronic migraine sufferers are too eager to get a tragus piercing for migraine: even when they realize there's a very low chance that it will work, the slight potential for a significant gain (ending chronic pain) leads them to take the risk. The same could be said of an arthritis sufferer using acupuncture, a fibromyalgia sufferer using essential oils, or any number of other fad treatments. Often these treatments are perceived as carrying low risk of loss or harm – sometimes accurately and sometimes inaccurately. Sometimes the risk of harm is not in the treatment itself, but in the patient's willingness to pursue the alternative treatment in the place of traditional medications or therapies. Of course, the placebo effect, as discussed previously, as well as regression to the mean, may lead patients to believe these fruitless "treatments" are efficacious, further distorting their perceptions of real risk and benefit in the context of evidence-based medicine.

Our patients may be *too eager* to take a risk on a *high-probability large loss* ("risk seeking"). A terminally ill patient may be desperate to avoid the natural progression of disease, instead choosing risky surgery with a high chance of complications which may end up increasing morbidity or shortening total lifespan. The risk of surgery seems preferable to the "certainty" of natural disease progression and death. As discussed earlier, the patient is considering the choice in proportion to his (poor) current health status, with a sense of "what do I have to lose?" Unfortunately, the patient may not recognize the high probability of losing what little he does have, in the form of increased pain and suffering and decreased quality of life in the time he has left. These conversations can be the most difficult to navigate with patients, but honesty tempered with compassion and patience is the best approach. It is our job to help our patients realistically face the choices they have, even when all choices offer a poor prognosis.

Our patients may be *too reluctant* to take a risk on a *low-probability large loss* ("risk averse"). We see this when patients prefer a high-premium, low deductible health insurance because they are overly afraid of large out-of-pocket medical bills. They might be certain to come out ahead financially by choosing a cheaper, high deductible plan and putting the money saved on premiums into an HSA; but the irrational fear of unexpected large bills prevents them from making the more reasonable choice. Another common example is patients' interest in getting unwarranted medical tests for very rare but serious conditions. Patients often do not understand the concept of "net harm" in medical screening for low-risk conditions, instead taking the stance "better safe than sorry." They are averse to taking the risk of a low-probability large loss (serious disease) by foregoing the testing.

So, in communicating with our patients, how can we account for and overcome these cognitive distortions surrounding risk and potential benefits and loss? (And remember – we, as physicians, are also humans subject to the same cognitive distortions. Be aware!)

The answer is that we must be very mindful of how we frame information and how we contextualize choices that we present to a patient, bearing in mind the proclivity to avoid loss above seeking gain. In presenting information about a health factor, focusing on avoiding potential loss may be more motivating than focusing on a hope for some potential gain. I'll give an example related to smoking cessation. Here are two true statements you might make to a patient:

> Currently, your smoking hasn't had much impact on your health. You are still active and able to do the things you want to in life. But as you age and continue smoking, your health will almost certainly continue to deteriorate. You'll be unable to do the things you do now; you may need an oxygen tank; you might develop cancer and die an early death.

> Though your smoking hasn't had much impact on your life right now, you could be doing more. If you quit smoking now, you will breathe better, have more energy, and be even happier. Your clothes will smell better and you'll have more free money to spend on fun things.

The first statement focuses on loss of status quo, the second on improvement of status quo. A patient is more likely to quit smoking with the first statement than the second. **We fear loss more than we covet gain.** Framing is important.

The second basic principle we need to understand relates to some unspoken rules of communication. When we communicate with another person, we both assume some basic rules that allow efficient and meaningful conversation. This is called the **Cooperative Principle**. Paul Grice described four ideas, now commonly called **Grice's Maxims**, that define some boundaries for avoiding misunderstandings during communication that might occur from what our manner of conversation might imply intentionally or unintentionally (called conversational implicature):[134]

1. Maxim of quantity
   - Make your contribution as informative as is required (for the current purpose of the exchange)
   - Do not make your contribution more informative than is required
2. Maxim of quality: try to make your contribution one that is true
   - Do not say what you believe is false
   - Do not say that for which you lack adequate evidence
3. Maxim of relation (or relevance):
   - Be relevant
4. Maxim of manner: Be perspicuous
   - Avoid obscurity of expression
   - Avoid ambiguity
   - Be brief (avoid unnecessary prolixity)
   - Be orderly

A fuller discussion of these principles is best left to Grice himself, but a few examples here that relate to medical applications will suffice to illustrate the importance of these concepts for our purposes. Here I just want you to be aware of these concepts and think about them in this chapter and in your practice. When you see poor or misleading communication, think about which maxim was violated.

Fully informing a patient of all the relevant risks, benefits, and other information necessary to make a decision is one of the keys to not violating the first Maxim of Quantity. When we present risks out of proportion to benefits or vice versa, the cooperative principle is violated and instead we are participating in coercion. We also must provide enough information to ensure that a patient's understanding is complete. If a 25-year-old woman has a papillary thyroid cancer, she cannot make an informed decision about treatment approaches without being informed that this type of cancer is usually indolent and that surveillance is often a better choice than surgery. Yet, too often, the patient is not adequately informed.

Most people understand the Maxim of Quality and most doctors would not intentionally lie to a patient. Yet, doctors every day say things for which they lack adequate evidence. If you don't have high-quality, replicated scientific evidence that an intervention works – say, Vitamin D supplementation for fatigue – don't tell a patient that it does. Neither white lies nor pseudoscience have a place in medicine. In fact, even more than in regular social circles, physicians must abide by this maxim carefully. Patients do not come to see you for your opinions; they come for you to provide evidence-based, scientific answers to their problems. They assume that what you are saying is true ***and*** based on good evidence. We will discuss this concept more generally in Chapter 19. Obviously, much of what we do is in the realm of uncertainty, and this present chapter will help us communicate uncertainty more effectively.

The Maxim of Relation or relevance is critically important. Patients assume that everything we tell them is relevant and that we tell them every relevant thing. Advertisers often flout this maxim intentionally to sell products. A medical example of this is the Paragard intrauterine device (IUD). The Paragard is a copper IUD. For years, Paragard has been marketed as the only "prescription birth control that's 100% hormone-free!" Its current website (as of July, 2020) features a splash video of young, attractive women dancing happily; they seem to be celebrating one gleefully happy woman's choice of getting a Paragard. Over the video is this caption: "Birth control that's absolutely, positively 100% hormone freeeee." The marketing team behind these ads are deliberately flouting the Maxim of Relation. They haven't said anything that is technically incorrect; they have not lied. But what they have said is irrelevant. They know that women who view the advertisement will assume that the message is relevant and then assume some things to explain that relevance.

The obvious implicature is that hormones are bad, that they have negative side effects, and that the attractive dancing woman has finally been liberated from their injurious and harmful effects. Had the company said any of those things, they would have been lying; so, instead, they just let the viewer assume those things. The science says that the hormonal IUDs are superior in every way: they have a six-fold lower risk of pregnancy, a lower risk of uterine, ovarian, and cervical cancer, a lower risk of pelvic inflammatory disease, higher rates of patient satisfaction, dramatically higher rates of normal or favorable bleeding patterns (eumenorrhea), higher continuation rates, easier insertion, fewer problems with pain and cramping, etc. This dishonest advertising has had far-reaching effects on how patients perceive hormones.

The key is to understand the difference between what is said and what is implied. Patients will search for meaning and take inferences from everything you say. How much you talk about a given issue will infer to patients how relevant it is. When I was a medical student on my Pediatrics rotation, I spent some time with a Pediatrician who specialized in treating children with developmental disorders like cerebral palsy. When we saw a new patient with CP, a careful history was taken with a great deal of time spent on the delivery of the child. There were many questions of this form: *How was your delivery? Did the baby have decelerations? Did you have a lot of bleeding during your labor? Was there a time when several nurses ran into the room and changed your position or put oxygen on your face?*

The implicature to the patient is that CP is often caused by obstetric accidents (or malpractice); because 70%+ of the interview focused on the labor and delivery process, a patient might reasonably infer (perhaps informally or subconsciously) that around 70%+ of CP is due to these events. For reference, only around 10% of cases of CP are due to events that happen around the time of delivery. About 80% of causes are related to issues earlier in the pregnancy and 10% are related to events after delivery in the neonatal period.[135]

Our actions also imply relevance (or importance). When doctors check the cervices of asymptomatic pregnant women routinely at prenatal visits late in the pregnancy, the patients search for relevance and they assume (incorrectly) that cervical dilation is important to know or that it is a predictor of when labor will occur (it is not). When we order a Vitamin D level after a patient complains of fatigue, the implicature is that Vitamin D replacement is an effective treatment strategy for fatigue (it is not) or that Vitamin D deficiency is a cause of fatigue (it is not).

When a Pediatrician or nurse rounds on a newborn baby and reports to the mother that the baby has lost 6% of his body weight since birth and then immediately asks the mother in the next breath how her breastfeeding is going, the implicature is that the mother is inadequately breastfeeding even though a 7-10% weight loss is considered normal for breastfed babies. I once worked at a hospital where women in labor were routinely asked on admission if they performed self-breast exams; the implicature was that this was something they should be doing even though we haven't recommended routine self-breast exams in many years. I'm sure you can think of other examples.

Finally, the maxim of manner demands that we avoid obscurity and ambiguity. In medical care, this requires that we gauge the medical literacy of our patients and ensure that they adequately understand what they have been told. For example, if a newborn has viral encephalitis due to human herpesvirus 6 (HHV6 or roseola), it would be a violation of this maxim to simply tell the parents that the baby has an encephalitis due to herpes. The implicature, due to ambiguity, is that HSV1/2 is the cause of the encephalitis since this is the extent of most patient's knowledge of herpes. In this case, the clinician was technically correct but failed to assess the patients' knowledge of virology.

This maxim is flouted commonly when physicians use terminology that they believe is understandable to the patient but which is actually misleading; the irony is that physicians believe that they are practicing good communication skills by trying to speak at the patient's level of understanding. However, the job of the physician is not to dumb things down so that patients understand them but rather to identify knowledge gaps and provide education. For example, many patients are told that they have "low iron" instead of being told that they are anemic. The implicature is that their anemia is due to low iron stores – and it may be in many cases – but "low iron" and anemia are not the same thing. One can be anemic with normal iron levels and one can have low iron stores and not be anemic. In cases like this, there is no reason not to use the specific and accurate term, even if it means educating the patient about what that term means.

The same thing is true for perpetuating incorrect pathophysiologic concepts. For example, a physician and patient might discuss mood disorders using the theoretical framework of "hormones." Hormone levels and mood disorders are not related – at least not directly. But because we often culturally discuss mood disorders in terms of hormones, patients (and now many physicians) seek to treat mood disorders with sex hormone replacement despite a lack of scientific evidence for this approach. Patients may be more comfortable talking about estrogen deficiency rather than depression, but our job is to educate (and, in this case, destigmatize), not perpetuate faulty or false understandings of diseases. It is our ethical duty to speak accurately and unambiguously because patients assume that we will do so.

The consequences of poor communication are broad and destructive. Poor communication leads to noncompliance, dissatisfaction, lawsuits, poor care, poor decisions, anxiety, fear, and perpetuation of false beliefs. Let's dive into some specifics.

Communicating uncertainty

If we do not deal with uncertainty in a fair way, then we will distort risk and benefit. For example, if we say that *statin drugs prevent heart attacks*, that sounds too absolute; it implies that taking the medicine is more important and more powerful than it is. If we say that *they may help prevent heart attacks*, that still tells us nothing about the magnitude of effect the drug has in reducing heart disease. We need to use statistics to do this effectively, and we will discuss the best ways to do that momentarily.

As mentioned already, we need to not oversell our certainty about diagnoses or what test results mean. Being too certain about a diagnosis or treatment plan can lead to many of the cognitive biases we discussed in Chapter 3, such as premature closure, confirmation bias, the sunk costs fallacy, etc.

When we talk to patients or colleagues, it is important to consider how our phrasing and words impact their perception of our certainty. Pathology and radiology reports are famous for "hedging," not committing to a diagnosis, or recommending "clinical correlation." Physicians are often frustrated by this, but these tactics are far preferable to overselling certainty about a specific diagnosis. Worse is the electronic health record, which contains a list of other people's diagnoses (maybe *lupus,* for example) but does nothing to tell us how certain that diagnosis is. Remember, virtually no diagnosis is 100% certain.

We must strive to not imply certainty or confidence where it does not exist. Think about a pathology report with non-committal wording. A 2011 study asked clinicians to rate what degree of certainty was communicated by the following phrases:[136]

- *The features are diagnostic of ...*
- *The features are indicative of ...*
- *The features are in keeping with ...*
- *The features are probably those of ...*
- *The features are compatible with ...*
- *The features are suggestive of ...*
- *The features raise the possibility of ...*

I have listed those phrases from the most certain to the least certain according to clinicians' perceptions. How confident are you when you read each one? What percent probability would you assign to each phrase for the diagnosis the pathologist might suggest? Among physicians who participated in the study, the confidence for the first phrase ranged from 97-100% while for the last phrase the confidence ranged from 10-90% (with other phrases in the middle collectively ranging from 20-100%). Obviously, phrases like these don't communicate well and are understood very differently by different clinicians. Having a 20% confidence about a result versus a 100% confidence has huge implications for treatment and obviously affects Bayesian updating.

Another option would be to list, whenever possible, a specific numerical confidence, e.g., *these features are consistent with the diagnosis X% of the time.* However, such precision is not always known and using a number may also imply a degree of confidence that doesn't exist since we are often uncertain about how uncertain we are. Exasperated? Don't be. These sorts of problems are why people avoid the statistical methods described in this book in the first place. But acting certain when we are not doesn't make these problems go away. Ignorance might be bliss, but it isn't good patient care.

As I have suggested throughout this book, one approach to dealing with this issue is simply to divide our level of confidence into perhaps five categories. Here is a suggestion for five categories, with the addition of extremes of confidence (certain and impossible):

- Certain (100% confidence)
- Very likely (85–<100% confidence)
- Likely (55–85% confidence)
- Roughly even chance (45–55% confidence)
- Unlikely (5–45% confidence)
- Very unlikely (>0–5% confidence)
- Impossible (0% confidence)

I'm not suggesting this is perfect in the least. But something like this conveys probabilities more precisely than the verbiage shown before, while also allowing clinicians to not settle on a specific number which might imply spurious confidence. The clinician then is forced to decide how to act, given a certain level of confidence (and understanding that uncertainty exists unless the terms *certain* or *impossible* are used, which should rarely be the case).

Let's discuss how the results of scientific studies are reported. I get a daily email from my professional organization (American College of Obstetricians and Gynecologists) which summarizes new literature. On the day of writing this (April 14, 2020), here are some headlines:

- *Participation of immigrant women in cervical cancer screening* ***may be increased*** *by interventions ...*
- *... Aromatase inhibitor* ***may be associated*** *with decreased risk of development of certain age-related neurodegenerative diseases ...*
- *Lack of regular visits to physicians during pandemic* ***reportedly could have*** *negative impact on some patients ...*

I have bolded the language which has been used to imply uncertainty. In the first two headlines, it would be just as truthful to say *may not be increased* or *may not be associated*. The author is careful not to imply any degree of certainty because none exists. The studies being reported in these articles were poorly controlled, retrospective studies with tenuous data. They report associations without evidence of causation. If I were discussing cigarette smoking, it would be accurate for me to say *Cigarette smoking **is** associated with an increased risk of cancer* but when discussing Elvis, if I'm being honest, I'd could only say *Listening to Elvis music **may be** associated with an increased risk of brain cancer*. When reading conditional language, it always works to put in the opposite condition as well: *may not* rather than *may*.

Nevertheless, most people who read headlines like these tuck away in their brains that new data has proven something to be true when that is definitely not the case. We should avoid, whenever possible, using this sort of language. In almost every statement where *may* is appropriate, we could substitute *may or may not* to better convey uncertainty to the listener. For example, if I were using a new experimental treatment on a patient, I could say something like *it may or may not help you* and then proceed to explain the evidence for risk and benefit. It will change the way you think about what you are communicating, as well, if you do this routinely.

Science writers and physicians also misuse the word *prove*. Remember, the purpose of scientific experiments is not to prove but to disprove. No study has ever *proven* anything. Rather, studies lend weight to certain hypotheses or support certain conclusions. Better statements might be *studies support* or *studies have demonstrated*, etc. Be careful to not use words that imply certainty that does not exist.

Recently, a patient asked me if taking a baby aspirin during pregnancy *prevented* preeclampsia. I told her that it does not prevent preeclampsia, but rather *reduced the risk* of preeclampsia by 24% in women with select risk factors. Is that 24% figure accurate? Does it apply to her situation perfectly? Probably not, but it is a lot better than just saying *it prevents preeclampsia,* which she had been told matter-of-factly by another doctor. We should strive to be as precise as possible while avoiding words that imply an overinflated sense of certainty.

Choosing noncoercive words

Whether we know it or not, when we speak to patients we are *persuading* them. For example, we believe that a patient should get a colonoscopy, so we present the idea in a way that convinces her that it's the right thing to do. But we must be careful and we must recognize the power we have over patients. Our job is not to talk someone into something, but to support her in making the best decision for herself by providing accurate information that honestly presents the risks and benefits.

At least once a week, I meet a patient who tells me that she had cervical cancer when she was younger. It might even be recorded in the EHR that way. When I ask, she tells me that she had a simple destructive procedure like cryotherapy or a loop electrical excision procedure. I then explain to her that she did not have cervical cancer but instead that she had cervical dysplasia which, even if untreated, returns to normal over 80% of the time. With treatment, dysplasia almost never progresses to cancer. Yet, some doctor used the word *cancer* and probably thought he was doing a good thing by sufficiently scaring her into compliance with his recommended treatment. But this is unethical. Worse, many times the treatments performed were incorrect and unethical as well. If you want to talk someone into an unnecessary hysterectomy, the word *cancer* will get a lot more mileage than *dysplasia.*

Coercion like this happens both intentionally and unintentionally. Using the wrong words or using scary words like *cancer* distorts the patient's perception of risk, and may change the decision she makes about a treatment. For some types of cancers of the breast, thyroid, and prostate, surveillance is a better choice for many patients than active treatment. Yet patients, faced with the scary c-word, rarely choose surveillance (if it's even offered). For this reason, for example, there is a movement to rename ductal carcinoma *in situ* of the breast, indolent thyroid cancers like papillary thyroid tumors (particularly in young women), and prostate cancers with a low Gleason score.[137] Avoiding the word *carcinoma* will not only change the way patients perceive the diagnosis but will also alter the way physicians perceive the diagnosis.

Other examples of coercive words might be *tumor* or *mass* when describing benign processes like fibroids or lipomas. To avoid coercion, we should use words that accurately reflect the diagnosis and its risks without causing risk distortion. We don't need to make things sound better or worse than they actually are.

Describing statistics effectively

When communicating risk and benefit to ourselves, our colleagues, and our patients, it's not enough to be technically correct; we must also be mindful of how the message is received and perceived. Humans are not good at understanding things like large numbers, probabilities, and rates. It's why we are so commonly taken for rides by loan agents, lotteries, insurance salesmen, and politicians.

Deception, unintended or otherwise, may result from the use of terms or comparisons that tend to shroud meaning, or from using misleading scales or other counterintuitive or misleading statistical terms. Sometimes we do this graphically with charts that are inappropriately scaled or misleading (e.g., not showing 0 on the Y-axis). This obfuscation is often on purpose, particularly when a drug company, a device maker, or a research scientist wants to sell us something.

I attended a conference a few years ago in Hollywood where I was bombarded by presenters and advertisers with the word 'accuracy':

- "Our device improved diagnostic accuracy by more than 50%!"
- Another tool "improved diagnostic accuracy more than 120%!"
- "Our test is 99.4% accurate!"

It all sounded fantastic – and many eager attendees excitedly gathered as much information as they could so that they could return to their practices and continue being amazing doctors who utilized the latest and greatest and best of everything. However, recall that,

$$Accuracy = \frac{True\ Positives + True\ Negatives}{Total\ number\ of\ patients}$$

Put another way, accuracy is the total number of true results divided by the total number of results. Now imagine that 1 person among 500 has some disease. Some new test fails to detect the one person with the disease (FN=1) and at the same time falsely identifies 2 people as having the disease (FP=2). The other 497 patients are true negatives (TN=497). There are no true positives. What is the accuracy of this test?

$$Accuracy = \frac{0 + 497}{0 + 2 + 1 + 497} = 99.4\%!$$

So, a test with 99.4% accuracy failed to identify the one person with the disease and falsely identified two people as having the disease who did not. If it had correctly identified the one person with the disease, it would have been 99.6% accurate, even though 2 out of 3 positive results were false positives. In both cases, we see that the term *accuracy* does not mean what we might think it does intuitively.

Advertisers and authors of studies love the word *accuracy* because it is technically true but misleading. It's not enough to be technically true if the perception of the audience is distorted.

What about the other examples at the conference? Well, a test whose accuracy increases from 96% to 98% could be advertised as having improved accuracy by more than 50% (since it reduced inaccuracy by 50%). The wording of such a claim is, in fact, as ambiguous as the word *accuracy* itself. And what does it mean to improve diagnostic accuracy by 120%? Let's say that a test was already 80% accurate; is it now 176% accurate? Clearly not. Statements like these are purposely unclear and ambiguous. My guess is that they meant that accuracy went from 80% to 96% (since 96 is 120% of 80), but, clearly, this is misleading and deceptive. A more normal statement for this would be that it increased accuracy by 20%.

Another often abused statistic is the *false positive rate*. This term confuses doctors and patients alike. As an example, the reported false positive rate for the 'Quad screen' is 5%. The Quad screen is an antenatal screening test for fetal Down syndrome. Its 5% false positive rate is repeated over and over again in nearly every description of the test. Conscientious doctors will counsel patients that the test has a 5% false positive rate. Patients incorrectly assume that this means that if the test comes back positive, there is a 5% chance that they don't have the disease (that it was a false positive result). This is patently wrong.

I recently watched an industry-sponsored video on ReachMD featuring Dr. Lorraine Dugoff which was intended to demonstrate effective patient counseling for patients considering options for antenatal screening. She tells the patient:

The false positive rate, which is the chance, if the test comes back positive, it's the chance that your baby really doesn't have Down Syndrome, it's 5% with that test whereas it's 0.2% with cell-free DNA screening.[138]

Dr. Dugoff has misstated the meaning of the false positive rate and demonstrates this common misunderstanding. She has confused the false positive rate with the false discovery rate (which, recall, is equal to 1 minus the positive predictive value). Her sentence would be correct if she instead said, "the false discovery rate is the chance, if the test comes back positive, that your baby really doesn't have Down Syndrome."

Let me explain. The Quad screen is considered positive for any result which gives a probability greater than 1/270 for fetal Down Syndrome. The 5% false positive rate means that 5% of tested women who are not carrying a Down Syndrome fetus will falsely test positive. Before you take the test, you should know that if you are not pregnant with a Down Syndrome fetus, there is a 5% chance you will falsely test positive.

However, if you are looking at patients who have already taken the test and received a positive test result, about 97.3% of those positive results are *false positives*. They are false discoveries; the false discovery rate is 97.3%. This concept is very confusing and misleading, so let's look at it graphically in Figure 36:

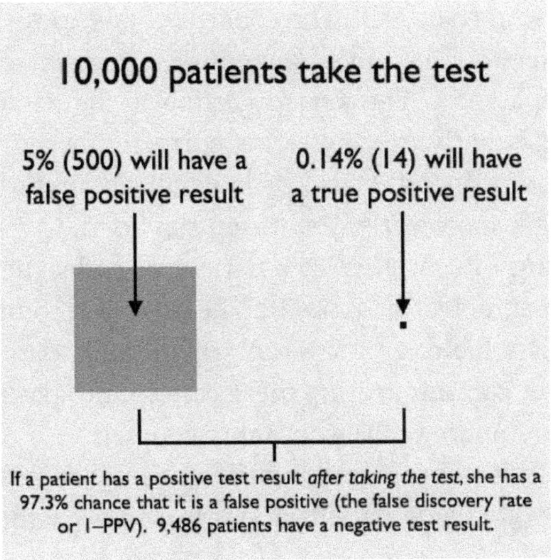

Figure 36. "Quad Screen" test results.

This is the heart and soul of conditional probability. Knowing about the 5% false positive rate isn't helpful at all in meaningfully interpreting the positive test result for the patient sitting in front of you asking about her positive result, nor does it help you counsel a patient who might be interested in the test. Recall the concepts discussed in Chapter 5: for a patient with a positive test result, the most meaningful values are the positive predictive value (the chance the result is right) and the false discovery rate (the chance the result is wrong.) Those are the primary numbers you want to share in conversation with your patient, because those values are actually relevant to her. (For a patient with a negative test result, you want to share the negative predictive value and its complement, the ***false omission rate.***)

A patient may ask questions about other values she may have learned about, like accuracy, false positive rate, etc., and you should answer those questions as clearly, simply, and accurately as possible while emphasizing that PPV (or NPV, in the case of a negative test result) are the relevant values for her now considering her test results. Make sure you're answering the question that is relevant to the patient, because the patient will assume you are answering that question, even if you are not.

You might be surprised to know how many physicians don't understand or remember the differences between these test values and how to interpret them properly. Of course, a physician who does not understand these values, and how to explain them, will have patients who are misinformed as well. The concept of false positive rate in general is misunderstood by physicians, nurses, research scientists, statisticians, and patients. The key to getting it right is to study many examples, consistently use the terms correctly in your own practice, and even to offer respectful correction when encountering a colleague using the terms incorrectly. Pay attention to the wording you use when explaining PPV or NPV to a patient or colleague; is it accurate and as simple and clear as possible? Learning to communicate with patients well is a lifelong process of growth and adaptation, and the starting point of communicating these concepts to patients effectively is a solid understanding of the concepts yourself.

For practice, let's compare accuracy to the false positive rate and false discovery rate using data from our previous example:

- **Accuracy:** 99.6% (the test was correct in 498/500 tests performed)
- **False positive rate:** 0.4% (two false positives per 500 tests performed)
- **False discovery rate:** 66.67% (two of three positives tests were false)

Don't confuse the false positive rate or test accuracy with the false discovery rate.

It's also important, when communicating with patients, to express absolute risk and relative risk in appropriate ways. Usually, when studies or doctors or drug companies describe the benefit of a treatment, they give a statement comparing the risks with and without treatment: "You have a 10% chance of the disease without the treatment and only a 5% chance with the treatment." Sometimes the statement is given by expressing one rate as a percent increase or decrease of the other: "You are 50% less likely to get the disease if you take the treatment." Both are technically correct ways of expressing the data but the former is more meaningful than the latter to the patient. That's because it is expressed in absolute rather than relative terms. Recall the previous discussion in Chapter 5 about absolute versus relative risk. We should always attempt to express risk and benefit in absolute terms to patients, not relative terms.

For example, which of the following paints an accurate portrait of risk and benefit: *You are 240% as likely to get pancreatic cancer if you smoke* or *If you smoke, your risk of getting pancreatic cancer is 1 in 27 but only 1 in 64 if you don't smoke*? Even though 240% sounds scarier, the truth is a patient is more likely to quit smoking when she sees actual, absolute numbers.

The terms 'number needed to treat' (NNT) and 'number needed to harm' (NNH) are useful for describing absolute risks to both patients and ourselves. Remember the patient who stopped her statin drug because of myalgias? Her individual relative risk reduction for heart attack might have been something like 33%; but her individual absolute risk for a heart attack was very low. Recall that she would have had to take the medicine for ten years to have a 1% lower risk of heart attack. This 1% value is called the absolute risk difference; by calculating 1/0.01, we can interpret this to mean that 100 women like her would have to take a statin for 10 years to prevent one heart attack (or alternately, 1000 women would have to take a statin for one year). We call 1000 the number needed to treat (NNT); in this case, the NNT does a phenomenally better job of explaining her risk and benefit than telling her that it will reduce her risk of a heart attack by 33%. At the same time, the NNT from a couple of hours of cardiovascular exercise each week to prevent one fatal heart attack is something less than 100. Remember, she quit playing tennis due to myalgias associated with using the statin drug. When we consider both issues and put them in terms relevant to her, we see that taking the statin may have *increased* her risk of death by 10-fold or higher (since it led to her exercising less). If you want to look at summaries of evidence described in terms of NNT and NNH, visit thennt.com.

One famous example of abuse of these concepts involves the drug Lipitor or atorvastatin. In advertisements for this drug, it was claimed that "Lipitor reduces risk of heart attack by 36%" with the footnote "if you have risk factors such as family history, high blood pressure, age, low HDL or smoking." This "fact" about Lipitor was a large reason that the drug had $125 billion in sales. This metric comes from a study called ASCOT-LLA, a trial involving 10,305 patients who had hypertension plus at least three other risk factors for heart disease.[139] Half of the patients received a placebo and half received Lipitor. The authors chose to end the study after just 3.3 years (it had been planned for 5) because of "the impressive 36% reduction of fatal CHD and nonfatal MI." In the placebo group, 3% of patients suffered a heart attack while 1.9% of patients in the treatment group suffered a heart attack. This means that the absolute risk reduction (ARR) was 1.1%. However, (3%-1.9%)/3% = 36%; this is the relative risk reduction (RRR). Obviously, 36% sounds much more like $125 billion than does 1.1%.

Still, 1.1% is something, right? But there is more deliberate ambiguity and misleading implications. The authors said that it reduced the risk of fatal CHD and nonfatal MI by 36%, but in fact, neither the rates of all-cause mortality nor cardiovascular deaths were significantly reduced. In other words, only the risk of nonfatal MI was reduced; but by making an aggregate endpoint that included fatal CHD, the authors implied that fatal CHD was also reduced. The authors could have combined any nonsignificant finding with nonfatal MI and made the same conclusion. Also of note, there were no differences in reduction of nonfatal MI in patients with diabetes, left-ventricular hypertrophy, previous vascular disease, women, patients under age 60, or those with renal dysfunction or metabolic syndrome – in other words, virtually every patient to whom the drug was commonly prescribed (including me).[140]

If you feel like you need a statistician with you every time you counsel a patient, don't worry. The key is to understand these statistics yourself, and then focus on telling the patient what you would want to know if you were the patient. If you understand these terms correctly, your counseling will be great.

When communicating with patients, our goal is to avoid **risk distortion.** Risk distortion occurs anytime the perception of risk and benefit doesn't comport with reality. Depending on a person's personality, she may over-emphasize negative effects compared to positive effects or vice versa. Often, drug companies market a drug in terms of its relative risk reduction (*our drug will lower your chance of a scary disease by 50%!*) while discussing the disease itself in terms of absolute risk (*you have a 1 in 60 chance of developing this scary disease*). The effect is risk distortion: it makes the drug seem more powerful than it is. That's certainly the case with statin drugs where the last thing the patient remembers is that it lowers her risk of a heart attack by 33%.

Risk distortion also occurs when we or the patient assume that we have more control over outcomes than we do. The illusion of control is pervasive among physicians. Do people who don't smoke, regularly exercise, have normal cholesterol, aren't obese, and have no cardiac risk factors ever die of heart attacks? You bet they do. Our goal is to reduce the risk of heart disease to as low as possible – not to eliminate it – while adding to the number of healthful years a person might have. An example from my specialty might be miscarriage. How much control do we or does a pregnant woman have over miscarriage? Almost none. Virtually all miscarriages are due to chromosomal, genetic, or immune issues for which we currently have no treatment. A tiny fraction of miscarriages is due to things likely poorly controlled diabetes, but for the average woman without diabetes or thyroid disease and who has an anatomically normal uterus, we have next to no control over whether she miscarries. Yet, we tend to focus on things we do have some control over (like smoking or level of activity) rather than things we have no control over (like how old someone's ovaries are). Thus, when we talk to patients about miscarriage prevention or causes of miscarriage, we may spend a lot of time talking about cigarette smoking. This causes a distortion in the patient's mind where she thinks that her miscarriage was due to her smoking or that smoking plays a large role in causing miscarriage. Smoking may play *a* role in the risk of miscarriage, but the magnitude of effect is unclear. Whatever role it plays is small. But we must avoid distorting the perception of risk, particularly when we might cause **guilt sanctioning** whereby the woman may believe that she has caused the miscarriage or that she could have prevented it.

None of this means we shouldn't work on reducing modifiable risk factors. It means, though, that we must provide context to patients about both modifiable and non-modifiable risk factors to avoid risk distortion. It also means we should use real, absolute numbers as much as possible.

Back to statistics.

We have already learned how non-intuitive odds are to most people, so using probabilities to describe risk is preferable. But just like the previous examples, how we express probabilities matters. We can express the same probability in three different ways and they are all exactly the same mathematically. For example, ¼ = 0.25 = a 1 in 4 chance. People intuitively understand *1 in 4* more easily than they do fractions or decimals. However, we could also say 25 in 100 or 250 in 1000 and these relate precisely the same ratio but they are perceived in different ways by our brains. This is called **ratio bias**.[141]

The use of larger numbers (like 250 and 1000 rather than 1 and 4) is called **scale expansion** and it distorts our perception. If someone were told that he has 250 chances in 1000 to win a lottery, he will perceive his chance of winning as greater than had he been told that he had a 1 in 4 chance. In most cases, using the lowest ratio that explains individual risk makes the most sense. For example, if a patient has a 1 in 10 chance of developing a condition, use 1 in 10 rather than 0.5 in 5 or 10 in 100.

In epidemiological studies, we often use scales on the order of *per 1000* or *per 100,000* to communicate risks and benefits of certain treatments or screening tests since that is the minimum scale needed for rarer diseases. It is also important, then, that we express risks and benefits using the same scale. For example, the pregnancy rate for a patient with an intrauterine device is 1/1000 per year while if she uses birth control pills it is 84/1000 per year. I could express this pregnancy rate with birth control pills as 1 per 119 but doing so distorts the comparison. I need to use the same scale to avoid risk distortion when patients are making a comparative choice. I don't need to use a larger scale (like *per 100,000*) in this case and doing so would only lead to ratio bias.

We also make a choice about whether we focus on negative or positive aspects of a treatment or intervention. We should attempt to balance these elements. A physician can manipulate a patient into doing almost anything by counseling patients in a biased and skewed way; a doctor may even tell the truth, but if the information is presented in a certain way, the patient almost always will make a certain choice – and this choice may not be the best choice for the patient.

Let's consider the example of external cephalic version (ECV) for non-cephalic pregnancies at term. ECV is not offered to everyone who is a candidate, but even when it is offered, women often don't choose it. Why don't they choose it? It's all in the counseling. If the obstetric provider doesn't want to do it or has a negative opinion of the practice, it will be obvious in the counseling, which will focus on potential negatives and risk factors for failure rather than potential positives and risk factors for success. We manifest what we focus on, and what we focus on is not necessarily based on scientific evidence. A lot of doctors just don't like doing ECV or don't feel that the reimbursement is worth the time, so their counseling reflects their disdain for the procedure. Performing ECV on all eligible patients is the right and evidence-based thing to do, but most providers don't. Imagine this conversation between an obstetric provider and her patient, who is a 37-weeks' pregnant gravida 1 with a breech baby:

> Version is an option, but I'm hesitant because you have an anterior placenta. I don't like to be too aggressive with anterior placentas. The small chance that it works and that you would go on to have a

vaginal delivery isn't worth the risk of an emergency Cesarean.

This type of counseling is just daring the patient to say no to ECV. It's full of coercive words like "aggressive" and "emergency" and uses qualitative and vague words like "small chance" rather than concrete and quantitative words. This is persuasion, not informed consent.

Briefly consider some actual data regarding ECV. While some studies have suggested that there is a higher failure rate for ECV with nulliparity and an anterior placenta, these are not contraindications to the procedure (and not all studies show a reduced success rate for patients with these factors). There is no evidence that performing an ECV on patients with an anterior placenta is any riskier than performing it on patients with any other placental location. Success rates for ECV are around 60% for all-comers, not just those who are good candidates. Those who are good candidates (lots of fluid, multiparous) have much higher success rates. Abruption, cord prolapse, ruptured membranes, and even fetal death have been reported as consequences of ECV, but these are very rare outcomes (less than 1% added together) and the negative consequences are largely mitigated by performing the procedure in a labor and delivery unit where Cesarean delivery can be performed if needed. More than 100 ECVs are performed before a single emergency Cesarean is required. Since ECV reduces the risk of Cesarean delivery, it reduces the risk of maternal mortality and other negative outcomes, as well as lowers the cost of healthcare. So here is a fairer statement to make to this patient:

> If we do nothing, you are almost guaranteed to have a Cesarean delivery; this might be planned at 39 weeks' or might be emergent if you present in labor before then. Either way, Cesarean delivery carries substantially higher risk for both the mother and baby, including increasing the risk of maternal death from 3.6/100,000 to 13.3/100,000. If you want more children in the future, each pregnancy will be riskier if you have a Cesarean compared to having a vaginal delivery in this pregnancy.
>
> Even though you have some factors which make an ECV harder to do, you still have about a 60% chance of success. There is about a 1 in 100 risk that something will happen at the time of the ECV that

will require an emergency Cesarean delivery, but we will be in the hospital and ready for this. In fact, think of it this way: there is maybe a 1% chance that you will need an unplanned Cesarean at the time of ECV, but, if you do not attempt the ECV, there is almost a 100% chance you will have a Cesarean delivery. Since you might go into labor before 39 weeks' gestation, there is about a 10% chance that you will have to have an emergency Cesarean delivery if you don't do the ECV.

The goal is to present positives and negatives in a balanced and comparable way, not to talk the patient into what we want them to do.

Risk distortion happens due to the cognitive biases we have previously discussed, particularly base-rate neglect. When we focus on negatives at the expense of positives, we cause base-rate neglect (believing that negatives are common while positives are rare). Imagine if the only thing you knew about planes was that they crash; you'd likely never fly. The news media distorts our perceptions by focusing mostly on negative stories, even if they are rare. Thus, we think that crime and tragedy and bad things in general are more common than they are. Social media platforms like Facebook and Instagram largely show only the exaggerated happy moments of people's lives while rarely showing the blemishes; this distortion leaves a record number of us feeling depressed because the perception is that our lives aren't quite as awesome as everyone else's. Social media also impacts our patients' perception of the risks and benefits of medical interventions.

When I discuss intrauterine devices (IUDs) with patients, many of them are immediately put off by the idea of getting one because of what they have read on Facebook. They have the perception that IUDs are absolutely horrific, maiming patients and causing infertility day in and day out. The reality, though, is that IUDs are the best-tolerated and safest method of birth control in existence, with the highest success rate on an intent-to-treat basis, highest continuation rate, and best side-effect profile. But the difference is that the only stories the patients read on Facebook are the bad ones, and those bad ones are usually exaggerated, presented out of context, and then magnified with shares and reposts. In fact, through the nocebo effect, these stories lead to some patients having a *perception of harm* even if no harm exists, or lead some patients to misattribute problems they might have to their IUDs even though the IUD is not the cause of those problems. It takes a lot of work to appropriately contextualize risks and benefits and avoid base-rate neglect.

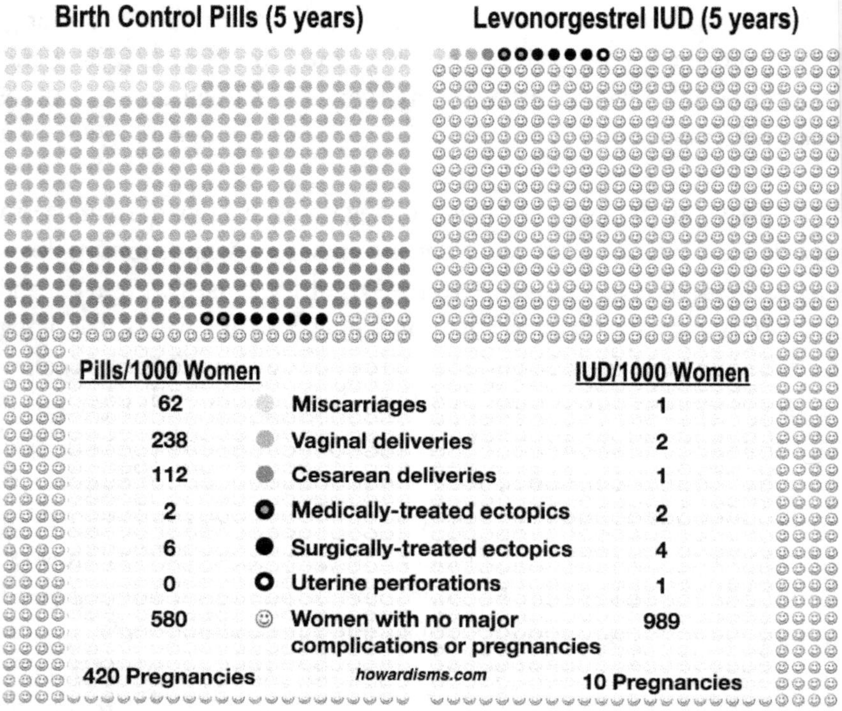

Figure 37. Frequency chart comparing birth control pills vs levonorgestrel IUD.

One effective way of presenting statistical concepts is through the use of graphics. Charts that have adjustable scales, like column or bar graphs, can be very misleading; they can make differences look larger than they are or exaggerate relative differences by not showing the absolute scale. Pie charts are fairer in the sense that these scale issues don't typically matter. The most effective graphics use individual symbols to represent people. Different symbols or colors can represent different outcomes.

Figure 37 is a black and white version of a color graphic I use in my office to reorient patients whose views of IUDs have been biased by social media. It is highly effective at contextualizing risks and benefits. I also use similar graphics to discuss the risks and benefits of mammography, use of statin drugs, etc.

Using words therapeutically

Our words have the power to harm and the power to heal. They can bring both comfort and pain. They can provide both hope and despair. The scientific explanation of this was provided in the last chapter in the discussion of the placebo and nocebo effects. We use operant conditioning every day whether we are aware of it or not. Great doctors are mindful of this and use their words to encourage, provide hope, and ease the discomforts of their patients.

Great doctors are great communicators. They listen to the patient and let her express her concerns. They probe deeper into issues that the patient is hesitant about. They allow the time needed to communicate clearly and address each issue fully. They ensure that the patient understands everything. They enable the patient to be a true partner in shared decision making.

Many wonderful books have been written about physician-patient communication and communications styles. Here I would like to focus on some things that should not be done:

- **We should not give patients false hope.** Always be honest about information, even bad news.
- **We should not conceal important information from patients.** The patient has a moral right to know everything relevant to her case.
- **We should not sanction guilt.** Patients listen intently to every word we say; when something bad happens, even a suggestion that it was their fault can be devastating and harmful. Tell the truth, but don't imply causality when it doesn't exist. An example of this is miscarriage. It may seem innocuous to tell a patient to be on pelvic or bed rest when she is at risk of miscarriage, but neither sex nor activity causes miscarriage. Those 'innocuous' words are psychologically devastating to a patient who later miscarries.[142]
- **We should not imply a cause when it is unknown.** Physicians want to tell patients why things happened; doing so can provide closure and relief. But, too often, the explanation is unknown. There is no room for opinion or pet theories in medicine. It's

not okay to lie to patients just to give them closure. An example might be intrauterine fetal demise. About half the time a fetal demise occurs, patients are told that it was due to a 'cord accident,' but this is actually the cause only 1-2% of the time. A better answer is that we don't know why it happened. These types of white lies are the source of blatant misinformation on the internet and in the community of patients. Today, patients are worried about the 'cord wrapped around the baby's neck' when it is a rather benign, common, and innocuous condition in reality; they have this fear due to incorrect things said by doctors.

- **We should not oversimplify information.** It is critical to gauge the degree of medical literacy and numeracy of our patients so that we can effectively communicate on a level of understanding relevant to each patient. Too often, physicians either don't explain medical issues at all or dumb it down to a point of condescension. Patients want to understand and they can be educated; our job is to take the time to educate them. Great doctors are great teachers.
- **We should not give bad advice.** When we speak, we are speaking for the scientific community. We should provide only advice that clearly has high-quality evidence to support it. But the majority of advice given lacks such scientific rigor, particularly in the areas of nutrition and mental health.
- **We should not act as salespeople.** Our job isn't to convince patients to do what we want them to do and medicine is not a business. This ethical line is crossed far too often.

Think about what you say to patients, both the literal and technical meaning as well as the perception of what is said or what is implied. Make sure that the perception matches reality. Don't scare people. Don't give them false hope. Don't over-represent risks and negatives without the appropriate balance of benefits and positives.

Exercises

17.1 Visit thennt.com and look through the 'Therapy Reviews' for your field of interest. Pick an intervention that you commonly employ and read the associated review. Integrate this information in your counseling of patients.

17.2 Look through a medical journal to which you subscribe. How do the abstract summaries use conditional language? Does this language imply a certainty that does not exist? How might the conclusions be restated?

17.3 What are some examples from your field of interest of coercive terms? How about non-specific terms like "low iron"? Can you think of any drug advertising that uses deceptive implicatures?

18

Measuring Error in the Treatment Process

It's hard to know exactly, but perhaps only 20% of patients receive the right diagnosis at the right time and get the right treatment. That's a pretty poor success rate. If we had that low a success rate in something like physics, it is certain that no one would ever fly on planes or ride amusement park rides. Yet, patients trust their lives to the medical system every day.

No disrespect to physicists, but medicine is harder than physics. In physics, there are a series of well-understood laws and equations that can be used to solve virtually all problems (at least outside the realms of particle physics and cosmology). If you want to build a rocket, physicists can explain every aspect of the problem for you and give you incredibly accurate predictions about what it will take to make your rocket successful.

Most problems in physics involve **complicated systems**. In a complicated system, all relationships in the model are well-understood and accounted for, so changing the value of one variable will have predictable effects on the outcome of the whole system. Scientists can make increasingly accurate predictions by including more variables in the model (like wind resistance) and increasing the accuracy and precision of measurements. Medicine, on the other hand, is mostly made up of **complex systems**, where not all relationships in the model are well-understood, and some factors playing a role in the system may be completely unknown and unaccounted for. In a complex system, researchers may be very limited in their ability to make accurate predictions; changing the value of a single variable will generally not have predictable effects on the state of the system as a whole. In fact, the results of complex systems are often counterintuitive because our intuitions are based upon incomplete models. Relationships that appear linear in one circumstance are not in another. Sometimes opposite effects proceed from the same input given different conditions. Biological systems are mostly stable in a variety of conditions because of buffering and compensatory mechanisms. Doses and amounts matter. Linear thinking and first-order relationships break down quickly and for reasons we often don't understand completely. Prediction and intuition fail. Things often don't work the way we think they should.

First-order thinking is the natural preference for humans. We see first-order causes and associations everywhere. But even in complicated systems, first-order thinking quickly breaks down.

As the number of variables present in a complicated or complex system increases, the number of states available in the system increases exponentially. For example, if there is only one variable considered, A, there are just two states of the system: A present (A^1) or A absent (A^0). Let's look quickly at how this scales for an increasing number of variables:

No.	Possible states of the system
1	A^1, A^0
2	$A^1B^1, A^1B^0, A^0B^1, A^0B^0$
3	$A^1B^1C^1, A^1B^1C^0, A^0B^1C^1, A^1B^0C^1, A^1B^0C^0, A^0B^1C^0, A^0B^0C^1, A^0B^0C^0$

Notice that the number of states of the system = 2^N, where N is the number of variables in the system. Thus, a system with 10 variables has 1,024 possible states while 20 variables yields up to 1,048,576 possible system states. What's more, any of these variables might be dependent or independent of another variable; this creates dependency networks which further increase the degree of complexity.[143] Why does this matter?

Consider all the variables, attributes, and characteristics that might determine how a patient responds to an intervention. With first-order thinking, only one variable is often considered; in the best of studies, a few variables are analyzed. For example, let's say we do a study comparing two types of insulin therapy for treatment of diabetes. We choose as an outcome the effects of the insulin therapies on HgA1c. In a first-order model, we might assume that the insulin treatment program that resulted in the lowest HgA1c would be the best treatment.

However, in a more developed model, we might note that insulin is a complex hormone that affects the actions of many other hormones and messengers in the body. Insulin has many other actions besides lowering blood glucose levels, including increasing DNA replication and protein synthesis; regulating activity of several enzymes; increasing fat synthesis and esterification of fatty acids; decreasing lipolysis, proteolysis, and autophagy; increasing potassium and amino acid uptake; decreasing renal sodium excretion; etc. In fact, insulin affects so many other things through primary, secondary, tertiary, and *N*-ary effects that the benefit or harm of administering more or less insulin cannot be measured by looking at one outcome, such as HgA1c.

During the COVID-19 pandemic, we have seen this illustrated on a large scale. Lockdowns help limit the spread of the virus and therefore save lives, but they also increase poverty, hunger, economic failure, depression, substance abuse, anxiety, and suicide. They discourage access to medical care for many emergencies and preventative care for sometimes fatal illnesses such as breast cancer. So, lockdowns are both good and bad. Most things are both good and bad. Unintended consequences are often unforeseen consequences. How much lockdown is good cannot be predicted; it can only be observed by experimentation. This is the point of picking the outcome that matters, such as life expectancy. More lockdown is good to a point but then fails and increases the number of fatalities; so, measuring an outcome such as life expectancy helps us to see through this complexity to make decisions that make sense. Indeed, more insulin seems to be beneficial to a point (with expected lowering of HgA1c) but then becomes harmful (with increased cardiac mortality). Such an effect could not necessarily be intuited through first-order analysis. Throw in some unknown effects and unknown variables, and we have a complex system.

But even this doesn't tell the whole story. Think about the patient-specific factors which might influence an individual's response to insulin therapy. Most studies control for things like demographic factors (age, sex, ethnicity) because these are readily-available metrics – not necessarily because they are important. But beyond these demographics, we must consider how an individual's comorbid health conditions, other medications, compliance, genetic and epigenetic traits, lifestyle, diet, environmental triggers, psychosocial stresses, and a myriad of other known and unknown factors might affect that person's response to treatment.

Finally, we must individualize the treatment program considering the patient's desired outcomes. It might be our priority that longevity is desired, but for an individual patient, he might value his cognitive function above his longevity (higher levels of insulin are associated with increased memory and cognition but shorter lifespan, for example). With all these layers of complexity at work, it's clear that something as seemingly basic as determining an individual's optimal insulin dosage is far from simple.

The best way to study a complex system is to measure its initial state, change one variable, and then measure its final state. This is the point of the randomized, controlled trial.

Failure occurs when we attempt to predict the outcome of a complex system based upon how we assume variables will affect it. For example, those who question the safety of vaccines will often cite an individual ingredient of a vaccine, such as aluminum, and then cite evidence that aluminum is neurotoxic with the assumption that this means that anything with aluminum in it will necessarily be neurotoxic. While it is true that such logic ignores many important variables such as dosage, chemical structure, etc., this is not how I know that the logic is faulty. I know it is faulty because I simply look at the whole system and note that children who receive vaccines with aluminum are less likely to have neurological deficits or autism than children who do not receive vaccines. Speculating about why that is true is simply speculation. None of us understand all the variables involved and none of us have a complete model available. We simply must use empiric evidence.

One way to understand the fundamental difference between complicated and complex systems is through a math analogy. Suppose you are given this equation as a "complicated system" to work with:

$$3x + 2y = 10$$

Everything you might want to know about this "system" can be easily determined using the rules of algebra. For example, if you are told that $x = -2$, then you know that $y = 8$; you can make a table of values or graph to completely display the behavior of x and y. On the other hand, suppose you are given the equation:

$$3x + 2y + z = 10.$$

If you are told $x = -2$, but given no further information, you cannot determine values for y and z. Furthermore, in a complex system, the situation is even more opaque: we generally don't know what "extra" variables, or even how many "extra" variables, are involved. In the next chapter, we will talk more about known unknowns and unknown unknowns. In complex systems, like human biology, our abilities to make predictions based on theory alone are extremely limited; only carefully designed, single-variable empirical trials produce reliable results.

Such complexity and higher order effects explain a lot of non-intuitive observations in medicine. Here are a few things that are "non-intuitive" when only first-order thinking occurs:

- Paradoxical use of thiazide diuretics in diabetes insipidus (a polyuric condition)[144]
- Lower oxygenation status in patients with COPD who retain CO_2 leads to decreased mortality[145]
- Transfusion and higher hemoglobin targets in sepsis worsen outcomes[146]
- Medical anti-shock trousers (MAST trousers) are associated with worse outcomes in trauma patients[147]
- Early ambulation with patients with DVT does not increase risk of PE and actually improves outcomes[148]

- Deep-suctioning of infants born through meconium stained amniotic fluid does not decrease the risk of meconium aspiration syndrome and may worsen outcomes[149]
- Estrogen replacement therapy decreases the risk of breast cancer and is associated with better outcomes if breast cancer should occur[150]
- Hypoxemia does not cause dyspnea[151]
- Use of low dose dopamine to increase urine output in patients with preeclampsia and oliguria[152]
- Use of ACE-inhibitors to treat scleroderma renal crisis[153]
- Routine use of backboards in trauma patients is associated with worse outcomes[154]
- Use of Trendelenburg position is associated with worse outcomes in patients with hemorrhagic shock[155]

We could list dozens of other examples and I'm sure you can think of many from your specialty. The point is that system complexity and higher order effects need to always be considered, and we should never do anything just because it "makes sense."

First-order thinking and neglect of system complexity are the sources of most inappropriate predictions in medicine. In the same way, these ways of thinking precipitate many errors in clinical treatment, either because the treatment is based on faulty (but sometimes intuitive) reasoning, or because the diagnosis itself is rooted in bias, much of which is the result of oversimplification of complex systems. If we are to track and describe error in the diagnostic process, we must appreciate system complexity.

If a physicist makes an inaccurate prediction based upon a model of a complicated system, it is usually due to just a couple of problems. The first is simply not accounting for enough variables (an incomplete model). If you don't account for things like air resistance or unexpected gusts of wind, the rocket might fall over. The second is the margin of error in measurements. There are thousands of calculations that go into designing a rocket based upon hundreds of measurements: measurements of size, mass, forces, speeds, etc. Each of those measurements has some small degree of error associated with it and those errors can add up quickly, leading to a rocket that falls over or one that blows up, if the errors are unchecked.

In complex systems, like human biology, the model is almost always incomplete, and so predictable responses and outcomes are almost never discovered. We might mention, briefly, chaotic systems. **Chaos theory** describes the behavior of complex systems wherein very subtle differences in inputs (that is, small differences in values of variables) produce dramatically different results. In other words, the system is nonlinear. This behavior looks random but it is not; the system remains deterministic. The notion that a chaotic complex system is still deterministic was fundamental to the concept as originally developed by the brilliant Henri Poincaré in the 1880s, and was validated by the work of Edward Lorenz, who famously stated:

> Chaos: When the present determines the future, but the approximate present does not approximately determine the future.

In other words, in a chaotic system, even if we fully understood the model, and the model were accurate with all variables accounted for, we would still not be able to produce accurate predictions due to imprecision in our measurements. These small differences may even be beyond our ability to measure. Chaotic complex systems describe things like the weather and the stock market (and we don't have complete models for either), but also apply to many biological domains like irregular human heartbeat patterns.[156] These chaotic systems are complex systems with intricate self-regulation, feedback systems, high order interconnectedness, buffering and extreme sensitivity to input variables that produce nonlinear and (seemingly) not predictable results.

Error propagation

In May, 2009, a Russian military rocket launched the second Meridian satellite into orbit, but it never quite reached the orbit it was meant to; in fact, it was too low to function, and drifts in space today as a symbol of miscalculation. The fuel burn was 3 seconds short due to a miscalculation. The calculations apparently had been made incorrectly due to an error made about the mass of the payload. The physicists had failed to compensate for the margin of error in the measurement resulting in a very slight underestimation of the fuel needed.

The margin of error in a measurement reflects the degree of confidence one has about the "truth" of the result. Error begets error; that is, errors are magnified throughout calculations so that a small mistake in one place may result in a massive mistake at the end.

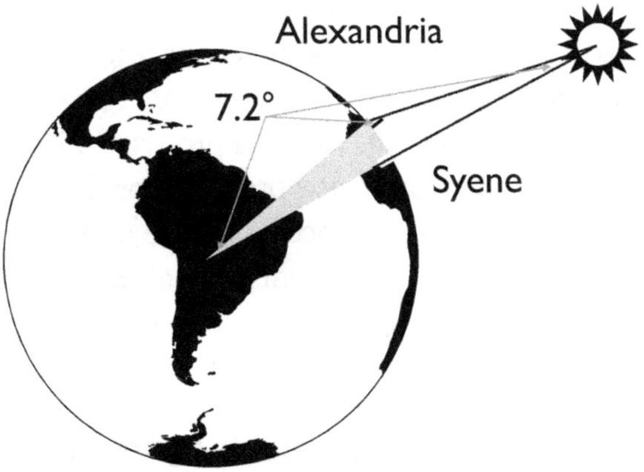

Figure 38. Syene and Alexandria, Egypt, lying 5,000 stadia apart, form a right triangle with the sun with angles 7.2° and 82.8, making the circumference of the Earth about 252,000 Egyptian stadia, or 39,375 km (compared to actual circumference of 40,041 km).

A Greek called Eratosthenes in the third century BC calculated the circumference of the Earth to be about 39,375 km (the actual value is about 40,041±34 km) using trigonometry. His method of determining this number relied on the distance between Alexandria, Egypt and a place called Syene that had a very deep well. As light illuminated the water in the bottom of the deep well in Syene, he measured the size and angle of a shadow created by a rod back in Alexandria. This was enough to solve the problem using triangles. His measurement was about 1.6% off, but that error merely reflects the magnification of very small errors in measurement (which he had no good way to account for), not an error in his logic. Small errors in measurement were magnified enormously over the larger scale. We call this **error propagation.**

What does all this have to do with medicine? Good physicists today account for the margin of error and degree of confidence in their measurements. Results are reported as some number $n = x \pm$ *margin of error*, with an associated level of confidence in the measurement. Physicians and medical research scientists usually do not report margin of error or degree of confidence. Error propagation is an enormous and often overlooked factor in medical outcomes. Let's look at how it happens and how to better account for it.

Error propagation in medical treatment

We have already discussed the idea of accounting for our uncertainty in terms of Bayesian updating; this requires us to explicitly state our confidence in certain steps of the medical process and understand how our confidence at one step (e.g., our pretest probability that a patient has a disease) affects our confidence in the next step (e.g., the post-test probability that a patient has a disease after performing a given test). We have stated our confidence in terms of a probability of belief (*I'm 70% certain that the patient has a UTI*) but we could have stated it in terms of a chance of being wrong (*I'm 30% doubtful that she has a UTI*). The truth is, we aren't even certain about the 70%; that's our average confidence. We might state that we are 60-80% sure, or 70 ± 10%.

Notice that I am using probability theory to express my confidence in a belief (*do I believe she has a UTI?*). Using probability theory as a tool of epistemology is characteristic of Bayesian approaches, but is starkly contrasted to Frequentist methods. If using probability to discuss confidence in a belief doesn't bother you, then I'm glad; you're a Bayesian. If it does, then we will discuss more about this in the Appendix. Nevertheless, let me state emphatically what I've indicated numerous times before: Frequentist methods cannot be used to express confidence in a belief. They have no epistemological value. That's why we used Bayes' theorem to determine whether a hypothesis was true rather than P-values.

I mention these issues here because we need to briefly talk about something that I have so far avoided: **confidence intervals**. Confidence intervals (CIs) are as misused and as misunderstood as P-values, and we could dedicate a whole chapter to them. You might have been reminded of confidence intervals when I discussed recording our confidence in measurements above. Could confidence intervals, for example, be used in the same way – to provide an error range for our measurements of data or for our belief in the validity of a study or test? The short answer is no.

But just to briefly correct some misconceptions about CIs, give me a couple of paragraphs. Confidence intervals were introduced by Neyman and Pearson as an alternative to the null hypothesis significance testing approaches of Fisher (P-values, etc.). They hoped and argued that scientists would be less likely to think dichotomously (i.e., either accept or reject the null hypothesis based on the P-value). What's more, a CI can help us understand the precision of a test, the magnitude or strength of an effect, and can be used for any hypothesis (not just the null hypothesis). Today, it is rare to see a quality study that doesn't report CIs and many will include graphical representations of CIs with error bars. But just as with P-values, it seems that most people misunderstand what CIs mean, even researchers and professors of statistics.

Hoekstra et al. in 2014 sent a questionnaire consisting of statements about the meaning of confidence intervals to 120 researchers and 442 students.[157] The subjects were to indicate which of 6 statements about CIs were false; all were false in reality. But, the average respondent believed that more than half of the false statements were correct and being a researcher rather than a student or indicating more experience and training with statistics didn't change the results. People just don't understand CIs, not even people who write textbooks. Their study was inspired by a similar study in 1986 that asked similar questions about P-values and found similar results.[158]

So, what does a confidence interval mean? It is a range of values that a parameter is believed to fall within, given some level of confidence (the **confidence level**, often 95% in biological sciences). But confidence intervals cannot be used to state anything about the truth or the true value of the parameter being studied. If, for example, we use a 95% confidence level, we cannot state that there is a 95% probability or confidence that the true parameter falls within the CI and we certainly cannot state that if the experiment were repeated there would be a 95% probability that the result of the next experiment would fall within the CI. We can make no such claims or derivative statements. These statements require a Bayesian view of probability theory (meaning that we use probability to determine a degree of confidence in a belief or what is called **credence**) and Bayes' theorem requires consideration of the prior probability. CIs do not account for prior probability. The CI is merely calculated based on the data in the experiment alone and therefore has no information related to prior probabilities.

Thus, it would be technically incorrect to use confidence intervals like a "margin of error." I should note that Bayesian statistics uses something called a **credible interval** that does allow for these inferences about truth to be made. In fact, the common misunderstandings people have of CIs would be mostly correct for credible intervals. In fact, that's a general theme of how people misinterpret statistics: they have an intuitive understanding that is inappropriate for Frequentist methods (CIs, P-values) but is appropriate for Bayesian methods (credible intervals, positive predictive values). This is because most people intuitively understand probability to express confidence in a belief but don't realize that this is limited to Bayesian methods.

But let's get back to error propagation or propagation of uncertainty. In the clinical procedures we have been discussing, where do error and uncertainty exist? Everywhere. Some of the errors and uncertainties are known and some are unknown. Ignoring uncertainty doesn't make it go away so we must at least acknowledge that we remain uncertain at every step of the diagnostic and treatment process.

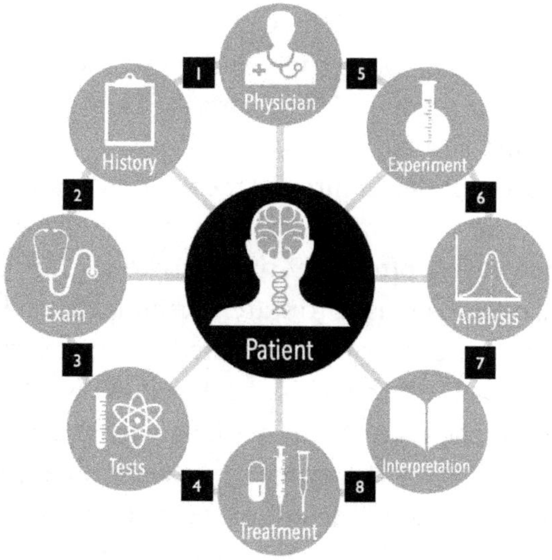

Figure 39. Points of error in the treatment process.

Figure 39 shows a summary of the diagnostic and treatment process. At the middle of everything we do is the patient who has her own genetic and psychosocial variables as well as her own wishes. Each clinical procedure that we perform, numbered 1-8 below, should be and is influenced by these patient-specific factors, many of which are unknown.

Let's consider each numbered item:

1. The physician gathers information. Does he ask the right questions? Does he read all the relevant portions of the chart? How does bias influence his process? Has he failed to consider the actual diagnosis the patient has in his differential diagnosis?
2. Does the physician have the differential diagnosis appropriately ordered? Are the probabilities assigned to each accurate? Is the physical exam selected appropriate? What are the negative and positive likelihood ratios of the selected examinations? How confident are we about those ratios based upon the original studies? Was the physical exam performed correctly? What patient and physician factors have made the physical exam suboptimal?
3. How inaccurate are the revised probability estimates for the diagnoses being considered on the differential diagnosis? Are the selected tests appropriate? What are the likelihood ratios for the tests?
4. How confident is the diagnosis selected for treatment based on this chain of confidences?
5. How probable is the hypothesis selected for experimentation? How does bias influence the study and data collection?
6. Were the methods used for statistical analysis of the collected data appropriate? How much bias is present in interpretation?
7. Is the interpretation of the analyzed results biased? How much do the results change the prestudy probability assigned to the hypothesis?
8. Was all the appropriate scientific literature relevant to the diagnostic and treatment choices for the patient assessed? What is the cumulative confidence in the decisions made for the patient based on the relevant scientific literature? What confidence is there that the individual being treated is a match to the subjects of the scientific studies applied to that individual's diagnosis and treatment?

These are the sorts of questions we have been trying to get better at answering in this book. But the answers are not easy and therefore the best way to estimate error in the treatment process is an open question. But error occurs and significantly so. It would not be wrong to estimate confidences at each step of the process and then multiply them together to determine an overall estimate of confidence but we don't, in most cases, understand the errors at each step well enough to do so.

Decision making at extremes of confidence

I'll leave this subject with a thought about how our confidence in the data should affect our decision making.

Suppose we have a new test with a positive likelihood ratio of 1.8. This is the mean value from our collected data. The *credibility interval* might be 1.8±0.3 with 95% credence. In other words, the **credible region** for the value is 1.5-2.1. How the test with this credible region would affect our decisions is different than if the credible region were 0.9-2.7 (that is, if the credibility interval were 1.8±0.9). In the latter case, the test might be absolutely useless or it might be very useful. We don't know. This range of credibility makes the test worthless.

The lesson is that we should consider our decisions at both the high and low end of the credible region. If a new cancer drug had a relative risk of mortality of 1.3±0.4, meaning that it might possibly increase your risk of death if you choose it, would you use the drug? It might reduce your risk of death by up to 70% but it also might increase the risk of death by 10%. We just don't know. At the low-end, you would refuse the drug but at the high end you would take the drug; your decision must consider the entire credible region. Hopefully, in most cases, you would make the same decision across the whole range. When you wouldn't, then usually you should reject the treatment.

Exercises

18.1 What are three counterintuitive treatments or treatment responses in your specialty or field of interest?

18.2 Consider a condition like depression. List as many variables as you can think of that might impact depression (consider environmental, genetic, physical, psychological, cultural, medical, and other factors).

19

In Defense of EBM

Evidence-based medicine (EBM) is controversial. There. I said it. Many of you might not have believed that sentence before reading the last few chapters. If you were already aware that EBM is controversial, you might not have believed that for the same reasons that I do. It might surprise you to know that the majority of practicing physicians don't believe in the basic principles of EBM. Patients certainly do not. Also, if you have read the last few chapters of this book, you might be close to wanting to abandon EBM for all its problems. But what would you turn to instead? What are the alternatives? What even is EBM?

What is EBM?

David Sackett, a pioneer in EBM, in 1996 wrote an influential letter to the *British Medical Journal* that provided his analysis of what EBM is and is not.[159] He states:

> Evidence based medicine is the conscientious, explicit, and judicious use of current best evidence in making decisions about the care of individual patients. The practice of evidence based medicine means

integrating individual clinical expertise with the best available external evidence from systematic research.

Hopefully, that definition resonates with the things we have discussed in this book. The problem, of course, is that there is significant variation in the real world regarding some of the terms in Sackett's definition. What does it mean to be conscientious, explicit, and judicious? Hopefully that evoked thoughts about medical ethics or the Quadruple Aim or maybe value-based care. How do we define individual clinical expertise? Do we have standards of competency or do we just let physicians do whatever they think they are good at doing? What is the best available external evidence? What is systematic research? When the rubber hits the road, EBM can be used and abused for almost any agenda depending on the answers to these questions.

In 2016, John Ioannidis wrote a report to David Sackett arguing, after a long career in defining and working on EBM, that EBM had been "hijacked" by various special interests, like pharmaceutical companies who have stolen the main tools of EBM (the randomized controlled trial and the systematic review) to P-hack their way to profits.[160] They profit from a misunderstanding of what evidence is and how these tools and the statistics that form the bedrock of that evidence should be appropriately utilized. It's all the stuff we have been discussing in the last few chapters. But these problems are not a reason to give up. In 2017, Ioannidis wrote a response to his own commentary called *Hijacked evidence-based medicine: stay the course and throw the pirates overboard.*[161]

The pirates are not just pharmaceutical companies. The pirates are anyone who stands to benefit from the inappropriate utilization of and misunderstanding of statistics and the basic science of EBM. Unfortunately, due to the perverse incentive structures in many areas of our society, this includes a wide variety of people who are motivated to misuse EBM for many different reasons. Academicians pursue novel research for their own self-promotion and career progression; physicians seek "evidence" to support their already existing beliefs and practices; policy-makers misuse data to slay their political opponents, who in turn misuse other data to fight back, with both groups claiming the title "evidence-based." Payers, insurance companies, and the government all misuse statistics and EBM to save money or increase profit; pharmaceutical and medical technology companies abuse P-values to sell billions of dollars' worth of new products. Of course, as we have discussed, the problem is not just faulty statistical methodology, but over-emphasis of positive outcomes without appropriate consideration of negative outcomes, along with gross misinterpretation and misrepresentation of data (even when the data are reliable).[162]

Sackett in 1996 also told us a few things that EBM is not:

- **EBM is not "old-hat."** EBM was (in the 1990s) a totally new way of practicing medicine, which we'll talk more about later.
- **EBM is not "cookbook" medicine.** Hopefully, I have made it clear how important the individual patient is to the process. The first half of this book focused on patient-specific issues on purpose. The patient comes first and then we apply the science. Patients are not data points in some RCT, and they don't always match the inclusion and exclusion criteria. Patients also have their own goals and need to fully understand the risks and benefits of a treatment and how those outcomes mesh with their own desires.
- **EBM is not limited to RCTs and meta-analyses.** Too often, we discount all the available evidence when making decisions because we believe that only the highest levels of evidence should factor into our decision calculi. Evidence comes in many forms and we've learned in this book how to better utilize any sort of available evidence to make better decisions.[163]
- **EBM is not impossible to practice.** If you're feeling overwhelmed by what we have learned together, you may feel daunted. That's okay and expected. We have turned a lot of things upside down. But don't give up. Don't be tyrannized by data pirates. Learn to practice and teach the concepts in this book and let's throw the pirates overboard.

What are the alternatives?

If physicians haven't always practiced evidence-based medicine, what did they do? What are the alternatives? What are they doing now instead? The history of medicine over many millennia has mostly centered around some central theory.[164] Indeed, I should clarify here the difference between an **empiricist** and a **theorist**. An empiricist forms beliefs based upon observation of evidence; specifically, a **skeptical empiricist** in medicine makes decisions and forms beliefs about patients or about treatments based upon evidence and experimentation, as we have discussed in every chapter of this book. On the other hand, a theorist in medicine starts with a theory of disease and then looks for evidence that confirms or comports with that theory. The theory becomes a lens through which all evidence is filtered. The theorist tends to reject evidence that doesn't comport with the already established theory. The theory forms a narrative fallacy and this biases the search for and the interpretation of new and existing evidence.

This dichotomy between empiricist and theorist is actually the central premise of this book. What we have been learning about is what we might call **science-based medicine**. The scientific process, when applied correctly, is the tool of the empiricist. It creates the evidence that is used in the decision calculus of EBM. It is empirical by design.

Everything else – ***everything*** – is just theory.

What are some of those theories? In a very broad sense, let's divide some nonscientific theories of health and healing into three categories based on how they are currently perceived: traditional medicine, alternative medicine, and religious medicine.

- **Historic traditional medicine theories:** Bloodletting (humorism), trepanning, fermentation theory, lapidary medicine, zymotic disease, Brunonian medicine, animal magnetism, etc.
- **Current traditional medicine theories:** Vitamin D therapy, Vitamin C therapy, bioidentical hormones, antioxidant therapy, dry needling, acupressure, acupuncture, therapeutic touch, nutritional medicine, adrenal fatigue, chronic Lyme disease, cupping, applied kinesiology, craniosacral therapy, Chapman's points, systemic yeast, hypnotherapy, dream explanations, etc.
- **Alternative medicine theories:** Organic diet, vaccine denial, essential oils, aromatherapy, subluxations, naturopathy, herbal-

ism, Tai Chi, vitamin infusions, homeopathy, Alexander technique, detox diets, chelation, colloidal metals, magnet therapy, colon cleansing, hydrotherapy, ear candling, reflexology, apitherapy, Ayurvedic medicine, phrenology, iridology, etc.
- **Religious medicine theories:** Reiki, faith healing, energy fields, Quigongteo, Shiatsu, Shamanism, crystal healing, astrology, meditation healing, chakras, Tarot, spirit animals, mesmerism, etc.

I've tried to include enough examples so that something you believe in or endorse is likely on the list, but of course there are many, many more examples. Did you realize that the thing you believe in on that list is nonscientific? Why did you start believing in it? What was your process for analyzing the evidence? Note that some of the things on these lists do have legitimate purposes, like Vitamin D for treatment of osteopenia; but the theory that Vitamin D deficiency has a larger influence on human health and dozens of different disease processes is just a theory. For its strongest adherents, however, it takes an almost religious fervor, and those adherents selectively interpret evidence to support the theory and are very subject to the confirmation bias and motivated reasoning.

What each thing on the list has in common is that quality, replicated studies, held to the standards espoused in this book, say that the treatments and theories of disease listed are ineffective. Thus, they are rejected completely (or in large part) by empiricists; that is, they are rejected by evidence-based medicine.

In human recorded history, there have been many *theories* of disease. In ancient Greece, people thought that human suffering and disease was related to gods being angry with a person. If you suffered from infertility, you must have upset Aphaea or Demeter. Slowly, that theory of disease gave way to a dietary theory of disease (dietetics) wherein disease was related to dietary deficiencies or unclean food. Of course, some diseases are due to dietary issues, but diet is not *the* explanation for disease in general. Later, Western medicine embraced humorism and phlegm theories. For the Palmer chiropractor, the subluxation is the theory of disease. For the acupuncturist, energy fields are the theory of disease. These overarching theories dominant all diseases for their practitioners.

You might wonder: what are the evidence-based medicine theories of disease? Perhaps the germ theory, genetics, autoimmune processes, oncogenic agents, inflammation, etc. It's true that all these are concepts which currently dominate our understanding of pathophysiology, and there is good evidence that these processes explain a lot of diseases. But they are not theories of EBM. **EBM doesn't have theories.** It is empirical. It has no dogma. We don't look for explanations based on theories, we analyze evidence and act accordingly. The theories come later.

This distinction is essential to minimize bias and to avoid dogma. We can also accept that something is true without understanding why it is true. We don't do things because they "make sense" in our theoretical framework; nor do we reject things that don't make sense. We decide what we do based on empiric evidence. I don't reject acupuncture because I don't believe in the theory of Qi, I reject it because legitimate science has found it to be ineffective in dozens of quality trials. It doesn't work.

We might also group all these non-EBM practices of medicine into what might be called **anecdotalism.** If you don't use evidence to make decisions, then your only alternative is to use anecdotes. Prior to the 1990s, this *was* the way medicine was practiced in the Western world, and unfortunately it continues to be the predominant method today. Not too long ago, we didn't have the tools to conduct high quality research; until the age of the personal computer, the computational power necessary to analyze large datasets simply wasn't widely available. If you read medical journals from the 1970s or earlier, you'll find just collections of anecdotes: experiences of a physician with a patient or a small number of patients and a guess (or theory) about what was going on. Medicine valued "experts," who were just people who had more experience with some disease process than other people did. But those experiences were not subjected to the scientific process.

The first RCT was published in 1948, one of whose authors was Austin Bradford Hill (of the Bradford Hill criteria). He is the father of the modern RCT.[165] It took another generation for RCTs to be commonplace, and they have dominated the last 25 years of medical research. Previously, we still just did what we thought worked in our own small experiences, or we did what "experts" said was best based upon their own experiences.

We now need to move into a new era of rational utilization of evidence as discussed in Chapters 15 and 16. But most physicians still do not base their practice of medicine upon empiric evidence, instead falling back on what is familiar or what they have personally experienced. Indeed, patients value physicians for their "experience" rather than their knowledge of the scientific literature. Why?

What are the pitfalls of EBM?

Why do physicians so quickly abandon evidence-based approaches and revert to anecdotalism? There are several potential reasons and these reasons are, essentially, the pitfalls of EBM. Let's think about a few of them.

Uncertainty and personality type. We have already discussed how much physicians and scientists want to appear or feel certain; this makes them reject an analytical approach (EBM) which deals with uncertainty. We will discuss this issue and the psychology of this in more detail in a few minutes.

A tendency towards first-order thinking. It is simply easier for us to imagine and understand simple solutions to our problems. Humans search for "cause and effect" paradigms to explain observable phenomena. Thus, it's easier to say and understand that "the increased rate of autism diagnosis is caused by vaccines" instead of "the increased rate of autism is related to a combination of changing diagnostic criteria, diagnostic accretion, greater awareness (that can sometimes lead to overdiagnosis and misdiagnosis), rising parental age, spatial clustering, disuse of intellectual disability diagnoses, as well as an increased prevalence of other known and unknown polygenetic and environmental causes." We tend to reject complexity for simplicity and we tend to favor first-order thinking rather than higher order thinking.

Medical education system. Our medical education system emphasizes telling us what to do but not necessarily how to determine for ourselves what we should do. In medical school, medical students learn information in the basic sciences that is often a few years out of date. They usually have too much belief in the principles they learn in the basic sciences because that information is usually not presented with any indication of the level of evidence that supports the theories. They are taught "facts" that might actually change in a few years. The "facts" assume too much permanence in their minds, especially when their futures depend on them being able to regurgitate these facts on a multiple-choice exam.

When one "fact" changes, it often has downstream effects due to the house of cards built upon it. When that fact is falsified by new data, it causes cognitive dissonance and our brains rationalize away that new data (ignoring it in most cases). You might call this being "closed-minded," that is, unable to fairly assess evidence that challenges our opinions or even being able to admit that such evidence exists. We are teaching our medical students, ironically, to be closed-minded.

How valid an opinion or idea might be is related to how good the original data was. Even long-held beliefs start somewhere; if they are based on good facts to begin with, that's better than if they are just based on what your parents or teachers or attendings told you. How well your view of the world (your internal narrative) comports with reality is directly related to how many of your "facts" are up-to-date.

In residency, physicians again learn information that is perhaps current and up-to-date or perhaps woefully out-of-date and archaic. Usually, no distinction is made between these two types of information. It seems most physicians are reticent to change learned practices after graduation from residency. The skill and quality of new physicians tends to peak just three or four years after graduation when they have gained sufficient confidence to apply what they learned in residency. After this initial period, in most cases, there begins a long and gradual drop off as those same physicians do not update their skills, attitudes, and knowledge. Knowledge and skills that may have once been relevant and reasonably up-to-date become increasingly irrelevant and incorrect. What's worse, after just three or four years into practice, most physicians have little accountability. They are no longer worried about obtaining board certification, passing exams, or impressing senior partners, and they are comfortable in their daily practice patterns. Therefore, they are also not prompted by the anxiety of independent practice to study on a regular basis.

With no incentive to study systematically anymore, physicians get their information very casually and *ad hoc*. When a physician starts independent practice, she may participate in CME events of varying quality and importance, she may browse through the journals that are relevant to her field and occasionally read the abstracts of articles that pique her interest, and she might learn new information from colleagues. Yet, an alarming number of physicians learn new information in the same way that their patients learn new information: from news stories and pharmaceutical commercials. They also may get their information, perhaps even most of their information, from sources that are carefully constructed to take advantage of the non-systematic way in which physicians learn: pharmaceutical and product representatives and industry-sponsored “throwaway journals.”

These so-called “throwaway journals” are usually free, flashy, and concise. This means that they have a large audience, they quickly garner the reader’s attention, and the message is succinctly delivered so that even those physicians with short attention spans will consume the information. Important studies and articles from other publications are abridged, giving the editors the ability to present them however they wish. Most of these journals also include effective social media marketing, and the cost of production and marketing are generously supported by advertisers. These advertisers spend money because they know how effective it is at influencing physician practice. Remember, good drugs never need marketing.

The point is that medical school does not teach students how to gather and assess data or make decisions based upon that data. What’s more, our day-to-day practice does not incentivize us to spend time doing so. When medical students or residents prefer to learn in a way that just gives them the bottom line, they are primed for a lifetime of industry pirates or special interests doing their work for them, often with disastrous results. For proof of that, think about how much of the information in the last few chapters you learned in medical school or residency, and how often you were required to use it. Or, when was the last time you had a discussion about cognitive bias on rounds or in conferences?

We must teach differently if we want different outcomes. I recall once that a friend asked me a clinical question. She was the learner and I was the teacher. I began to explain background information necessary to determine the most appropriate next management step. Eventually, she cut me off and said, “I don’t want the labor, just the baby.” She was in a hurry and just needed to take care of the patient.

That's fine sometimes. But, in general, it's a disaster. Students often just want the answer. But the answer is less important than the skills necessary to determine the answer. Search satisfying replaces learning and assessing evidence. Wrong or right answers are given too much authority. Young learners don't synthesize, they regurgitate. "Facts" are things other people tell them, not things they can determine independently. Facts then become subjective. Critical thinking goes away. People prey on this lazy thinking. Having facts is not enough. Using facts requires certain skills. A little knowledge is indeed dangerous.

Thus, we must integrate the critical thinking skills I have presented in this book into medical education in a fundamental way. We must teach students how to determine the correct answer based on current data and emphasize that the answer will change as the data does. The answer is less important than the process. Evidence-based medicine only works if we teach it and make it the system by which we deliver care.

Hierarchy. An extension of the ideas just discussed is that the medical system values position and experience more than evidence. This is a form of the *appeal to authority* logical fallacy. As students and residents, we look up to the doctors who train us. What they say, especially if they are charismatic or in positions of power, is taken as infallible truth. It carries too much weight in our minds and their words linger forever. We don't want to go against what they taught us, even twenty years later. So, what do we do when new evidence says that they were wrong? We ignore it or rationalize it away. Often, this puts physicians in a position where they are forced to attack EBM if EBM disagrees with what they believe. These attacks on EBM are often carried out by the most senior members of our departments or colleges since it is their long-held beliefs which are most likely challenged by new evidence. Because they are in a position of authority and are older, young students or residents often take too seriously the assaults made on EBM and learn the same tricks for when they need to defend something that they believe which isn't supported by science.

I once had a department chair who gave out a paper to the medical students that discussed a litany of problems with EBM and made them read it.[166] It was a good paper and thoughtful but, taken out of context, it served to teach students how to generally discount EBM. This was while I lectured specifically on how to use EBM effectively. It was a deliberate subterfuge designed to make the students not get too upset if they saw faculty members do things which were not supported by evidence, such as yearly Pap smears. It taught them that EBM was just not important and that the experience of older physicians was golden. In another department, a senior faculty member who was supposed to lecture students about childhood development instead spent his allotted hour every six weeks presenting the students ten reasons why they should not use or believe in EBM. This sort of undermining of science is rampant even in academia.

I have been to at least ten talks at meetings that I can recall where someone, right before he began to present something that he knew was unscientific, talked about the "Parachute Study." This is always done in a humorous and disarming way. This study, published in 2003 in the *British Medical Journal*, concludes:

> As with many interventions intended to prevent ill health, the effectiveness of parachutes has not been subjected to rigorous evaluation by using randomised controlled trials. Advocates of evidence based medicine have criticised the adoption of interventions evaluated by using only observational data. We think that everyone might benefit if the most radical protagonists of evidence based medicine organised and participated in a double blind, randomised, placebo controlled, crossover trial of the parachute.[167]

This paper, written by the current Head of the Department of Obstetrics and Gynecology at Cambridge University, essentially argues that the world would be better off if everyone who believes in EBM jumped out of a plane without a parachute. Yes, it's funny. The audience always chuckles and feels a sense of satisfaction that their non-evidence based practices are justified since, clearly, EBM is a failed and fraudulent approach. But the paper is foolish and it has been used for great harm. I wonder if the professor realizes what evidence is? Does he think that the same type of study is appropriate for every type of intervention? Does he think that the people who make parachutes never first tied them to a 200 lbs. dummy and threw them out of a plane before giving it a try themselves? In the same way, there was never a randomized, placebo controlled trial that showed that insulin is an effective treatment for type 1 diabetes. Instead, there was basic lab experimentation on dogs after decades of basic science research and then a dying young boy named Leonard Thompson was given a shot of insulin and it saved his life. Evidence comes in many forms and the appropriate use of EBM allows us to use any evidence to make decisions, even single anecdotes. But it also recognizes that a high quality, controlled trial is superior to and displaces even high quality anecdotes. To be clear, I realize that Dr. Gordon Smith's purpose in writing the article wasn't to cut the legs out from under EBM; he's a fine man. But it saddens me to see his article used this way in so many meetings I attend.

Conflicting guidelines. To simplify the work of EBM for the everyday clinician, practice guidelines are created that summarize the evidence and suggest an appropriate clinical pathway. As I have stated numerous times, these clinical guidelines are often not followed. Over 1,000 new guidelines are issued each year and many conflict with each other. Physicians, faced with conflicting and disagreeing guidelines, often just do what they have always done. This is indeed a problem. Clinical guidelines are often used by some of the "pirates" to support a special interest.[168]

In my field, there are currently several conflicting guidelines in the US and internationally about screening mammography. Do I screen a low-risk woman at age 40, 45, or 50? Do I screen her yearly or every other year? I can cite a "guideline" to justify almost any approach. This makes EBM seem hollow and tiring. Still, the individual responsibility for deciding which guideline is appropriate is mine, and how I treat a given patient is a decision I make with that patient. I am allowed to think for myself and understand how and why each guideline was developed. Yes, it is more work – but it's my job. That work provides the information I need to appropriately counsel patients and help them make the best decisions for themselves. Yes, we need to work to improve the process for guideline development. As many as half of all published guidelines are currently inaccurate or out-of-date.[169] But even bad guidelines are often informed by more data than the personal, subjective decision of a single clinician.

Pitfalls of RCTs. The randomized controlled trial is too often held-up as the *sine qua non* of EBM. But RCTs also have significant limitations, including the many biases and problems we have already discussed but also many we have not, such as data that fit the pattern of

Simpson's Paradox. For example, when placebo and intervention are able to cause the same benefits and harms, RCTs can cause profoundly misleading conclusions as the methods of analysis fail to distinguish the source of the harm or benefit.[170] These weaknesses in the RCT are abused by industry pirates to hide harm and magnify benefit. EBM cannot exclusively rely on RCTs if it is to be resistant to such mistakes and abuses.

The Art of Medicine. Some criticize EBM by arguing that the "standardization" that EBM affords will disincentive creativity and innovation and create stagnation. By allowing many different approaches to medicine or procedures, we can innovate and discover new things. They argue that science doesn't fully capture the "art" of medicine and some interventions may work better in some physicians' hands than others because the art varies.

Physicians are all too happy to recall the phrase "the art of medicine" because it implies that they have some special power or ability in their practice of medicine, as if they were Beethoven or Raphael. Because they are artists, they do not have to always justify their rationales or their methods. But medicine is not an art in this sense. I can train almost anyone with at least average intelligence to be a physician; but in an entire lifetime working with brilliant people, it's unlikely that I would produce even one person who had half the abilities of Beethoven or Raphael. A student can be taught to play the piano or to sculpt or to paint, and they might become technically very excellent at these skills; but that is not what makes an artist. Being a masterful technician is not the same as being an inspired artist.

Is medicine a science or an art? If it is the former, then the rigors of the scientific method and scientific analysis should govern its practice. If it is the latter, then there may be no limit to the amount of individuality and individual expression, and therefore anecdotal practice, that is allowed to enter into the discipline.

Most of us who use the term *art of medicine* believe that the art of medicine relates to the human aspects of medicine, particularly involving communication and empathy, where the pure, empiric applied science of medicine might fail the emotional and human needs of the patient. That form of the art of medicine should not spill over into things like the diagnostic process, therapeutics, and other questions of medicine that have strictly scientific answers. Sir William Osler said,

> The practice of medicine is an art, not a trade; a calling, not a business; a calling in which your heart will be exercised equally with your head. Often the best part of your work will have nothing to do with potions and powders, but with the exercise of an influence of the strong upon the weak, of the righteous upon the wicked, of the wise upon the foolish.

The lofty quotes of Sir William Osler and others about the art of medicine sound elegant and tend to enrich the physician's view of himself and of the profession, but they also reinforce the physician's belief in his special abilities or special powers that grant him the right to practice medicine in a way that may disagree with widely accepted scientific principles. Who among us would not like to think of himself as Michelangelo or da Vinci? This particular quote also emphasizes a bygone era of paternalistic medicine that most modern physicians should find abhorrent. A more interesting quote is from Paracelsus, who said,

> Medicine is not only a science; it is also an art. It does not consist of compounding pills and plasters; it deals with the very processes of life, which must be understood before they may be guided.

It appears that Sir Osler was familiar with the quote from Paracelsus and interpreted and expounded upon it to emphasize his own view of himself: that he was strong and righteous and wise and that it was his job to treat the weak and wicked and foolish. Surely we should strive for a more mature view of ourselves and of our profession than this.

So, how can we rationalize this dichotomy and understand the meaning of the "Art of Medicine"? The expression probably has its origin with Hippocrates. One of his most famous *Aphorisms*, in Latin, was "Ars longa, vita brevis," or the "Art is long, life is short," in reference to the "art of medicine." The full quote is:

> Life is short, and the art long, opportunity fleeting, experience perilous, and decision difficult.

This is an English translation from the Latin, but Hippocrates originally wrote in Greek. The Greek word for *art* or the Latin *ars* was *techné*. This Greek word means an *art* in the sense of a craft or a trade; indeed, a craft that is governed by a set of rules. It is the origin of the English word *technical* for example. It was not an art in the sense of the creative arts. This is also the meaning of the Latin word *ars*.

So, what did this aphorism of Hippocrates really mean? Well, let's expound upon what he said. The life of any given physician, and indeed the life of any given patient, is short and limited; but the discipline of medicine and the need that patients have for physicians is ongoing and it exceeds and surpasses the lifetime of any one of us, and it exceeds and surpasses the accomplishments of any one of us. Our own opportunities to learn are fleeting and therefore we need to rely upon the collective opportunities to learn from all physicians and from many different patients; indeed, from more patients than any one physician can ever individually treat. Hippocrates was arguing for EBM. Therefore, he says, our experience is perilous; our individual experience should not be trusted. It is our collective experience and our collective knowledge that has more value and authenticity. The decisions we make are difficult; they are difficult because we must learn to not rely upon our own experiences and see past our own biases when we make decisions.

It seems, in fact, that the Hippocratic origin of the expression "the art of medicine," when properly understood, is a fundamental rejection of the way many modern physicians now use the term and was rather an endorsement of evidence-based medicine. Paracelsus, whose Latin was much better than Osler's, understood this. His comment about the art of medicine, when read correctly, was an exhortation to integrate the basic sciences into the practice of medicine, which was not common in his day.

None of this is to say that some physicians do not possess more empathy and interpersonal communication skills than others. This is obvious to anyone who has ever been to a doctor. We also should not under-emphasize the importance of these skills. They are harder to teach and seemingly less subject to the scientific process than many of the other competencies and knowledge-based skills of physicians (but they can be taught and can be studied). Nevertheless, these skills are not art. They are skills which are requisite to being a good physician and skills which we should strive to teach and model for all physicians. They are no more important than the skills of critical thinking, problem solving, statistical analysis, and the scientific method.

The aphorism "art of medicine" should never be invoked by anyone to justify himself when he chooses to not practice evidence- and science-based medicine. It is not a license to be anecdotal. It should not make a physician feel like she is unique or has special powers.

As far as the field progressing and being innovative, I will say this: we do not advance the field through individual variation but through the scientific process of hypothesis creation and testing. That is at the very core of EBM.

Inertia. Finally, the biggest impediment to the implementation of EBM may be simple inertia. Physicians just do what they are comfortable doing. Deviating from their comfort zone creates uncertainty. So, let's talk more about uncertainty.

Dealing with uncertainty

Many are so overwhelmed by the seemingly endless problems with scientific papers that underlie EBM that the Mark Twain comment of "Lies, damned lies, and statistics" becomes almost an excuse to abandon science-based medicine and return to the dark ages of anecdotal medicine. Most physicians have been frustrated by some study that feels wrong but they can't explain why it's wrong. It looks like science on the surface, but it just isn't. Not understanding the principles we have outlined in this book leaves them powerless to reject the bad study as unscientific, so they reject science instead.

But Mark Twain's quote could be restated: There are lies, damned lies, and Frequentism. The scientific process is saved by Bayesian inference and probabilistic reasoning. Bayes makes no claim as to what is true or not true, only to what is more probable or less probable; Bayesian inference allows us to quantify this probability in a useful manner. Bayes allows us to consider all evidence, as it comes, and constantly refine how confident we are about what we believe to be true. Imagine if rather than reporting P-values, scientific papers concluded with an assessment of how probable the studied hypotheses are. This would leave the reader understanding that scientific studies, like lab results, are not binary. It would force them to think about what it means for a hypothesis to be only, say, 70% certain. It would force them to deal with uncertainty.

That is partly why Frequentists are so resistant to the principles of Bayesian inference. There seems to be a lot of guessing involved in using the Bayesian principles we have discussed. Remember the Elvis hypothesis? I don't precisely know how likely it is that listening to Elvis causes brain cancer. I said 1% but, of course, it is probably more like 1 in a billion. I also don't know exactly how likely it is that a patient has a UTI based upon her symptoms or exactly how likely it is that a patient has a given disease based upon her risk factors. This uncertainty bothers a lot of statisticians, so they don't want to use a system that is dependent on so much uncertainty. Many statisticians are working on methods of improving the processes for making these calculations, which will hopefully make our guesses more and more accurate. But I think what ultimately bothers not just statisticians but physicians and people at large is that the answers are uncertain. How do we make decisions with uncertain data? It feels uncomfortable. We want to organize the world into neat bundles of information that we know to be true and information that we know is not true. But that's not how epistemology works (see the Appendix for a fuller discussion).

Most folks assume that science delivers real, solid, and certain results. But it does not. The hallmark of good science is that it is never certain. The hallmark of good medicine is that it is never certain. This seems counterintuitive because this causes us to be uncomfortable making decisions. But we must learn to be comfortable with uncertainty if we are to be good scientists, good physicians, or even just good thinkers.

In 2002, Donald Rumsfeld famously made the following statement while deflecting a question about evidence for weapons of mass destruction in Iraq:

> Reports that say that something hasn't happened are always interesting to me, because as we know, there are known knowns; there are things we know we know. We also know there are known unknowns; that is to say we know there are some things we do not

know. But there are also unknown unknowns—the ones we don't know we don't know.

Rumsfeld was widely criticized for this statement as it was taken as obfuscation by the press. But the statement itself is a profound reflection on epistemological limitations stated in a cleverly simple way. Rumsfeld didn't make it up; this was a summary of a principle widely taught in the intelligence community based upon something called the **Johari window**.

The Johari window was developed by two psychologists in the 1960s called Joseph Luft and Harrington Ingham as a tool for introspection and self-awareness (the name was formed from the first syllables of their first names).

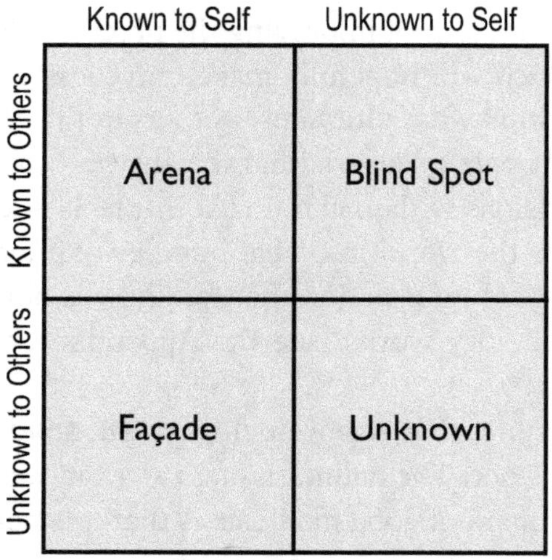

Figure 40. Johari window.

We are at our best when we are in the arena where we are authentic – that is, we know our true selves and it is known to others. But we all have blind spots and façades that we project to the world, too. What's more, there are things that we don't understand about ourselves that others do not understand either; these are the unknown unknowns that informed Rumsfeld's thinking.

But in a more generalized form, we can place all knowledge into one of three categories: things we know, things we know that we don't know, and things that we don't know that we don't know. We make good decisions usually when we deal with things we know and things we know that we don't know. It's the unknown unknowns that cause most of the poor decisions of the world, and that remains the largest category of knowledge.

Psychologically, someone who is comfortable with uncertainty (or comfortable with unknowns) will behave in a way that recognizes just how much knowledge is unknown – both known unknowns and unknown unknowns. But few people are comfortable with uncertainty. It provokes anxiety and cognitive dissonance. Humans develop defense mechanisms to deal with these issues; taken to the extreme, this presents as narcissistic personality disorder. But most physicians just ignore uncertainty and do whatever doesn't directly challenge their egos (that is, they are satisfied by the confirmation bias, etc.). They live in echo chambers where their "knowledge" is unchallenged. They keep doing what they have always done because it seems more certain than something new or different.

Below are two pie charts representing the perception of the distribution of these three categories of knowledge for two different personality types, *A* and *B*.

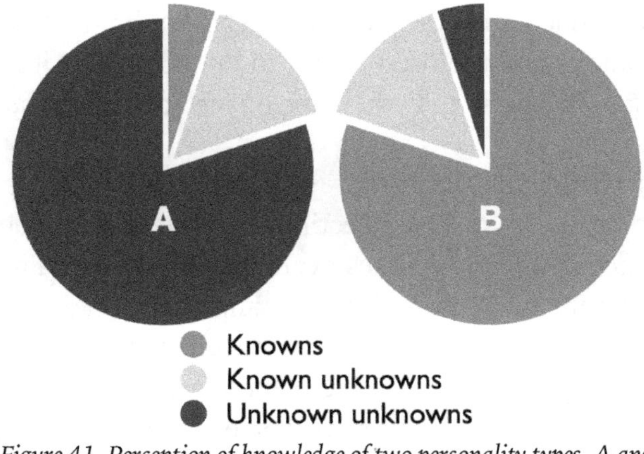

Figure 41. Perception of knowledge of two personality types, A and B.

Note that these charts represent the person's perception of what they know, not what they actually know. The true scientist will perceive her knowledge like person A, being sure of very few things and realizing that most things are unknown, including lots of unknown unknowns. But the pseudo-scientist (and unfortunately the typical arrogant physician) will be more like person B who believes that he knows just about everything there is to know. Person B is more likely to trust his own, anecdotal experiences. Person A is more likely to seek out the scientific literature to learn what she may not know.

Persons A and B will make different decisions given the same information. They have different cognitive dispositions to respond. They will conduct science differently. They will treat patients differently. Imagine a new paper that makes a conclusion about some medical intervention. If person A wrote the paper, its summary will cast doubt on the findings even if the statistics are excellent; she will state that she wasn't able to control for all variables (known unknowns) and that there are likely a number of variables that she didn't know about (unknown unknowns). Person B, on the other hand, will write of the excellence of the P-value and call for immediate and universal adoption of the new intervention, hearkening to science as if it were a supreme being. Person B just doesn't know what he doesn't know.

Person B is less likely to consider alternative diagnoses once he has made a diagnosis. He is less likely to change his practice pattern when new guidelines are published. He is more likely to fall for narrative fallacies built upon shaky facts and less likely to change his mind in light of new evidence.

Medicine is filled with people like Person B who don't know what they don't know, and they don't want to know because it is an attack on their egos to admit that they are unsure of something or don't know something.

In such a world, anecdote and rejection of science reigns supreme. Once a person is comfortable rejecting evidence or not understanding what evidence is, then bias takes over. The lust for certainty leads to absolutist, binary, black-and-white thinking. It also leads to reductionism, where we seek to explain complex phenomena as being controlled by only one or two variables. These are usually variables that are easier to understand or more available or testable, but very often are only a small part of the larger, complex system and rarely are sufficient. An example might be blaming sugar for diabetes or vaccines for autism or meat for heart disease. This phenomenon has been described on an individual level as the Dunning-Kruger effect.

Science and medicine are irreducibly complex and filled with uncertainty. The more aware you are of that – and the limitations of our knowledge – the better you will be at thinking and practicing medicine.

The Dunning-Kruger effect is named for psychologists David Dunning and Justin Kruger, whose 1999 paper described four studies which indicated that incompetent people not only made bad decisions, but their incompetence made them incapable of realizing that they were incompetent.[171] However, as a person's knowledge grows, she gains the metacognitive ability to realize that she is incompetent and eventually begins to underestimate her competency – the exact opposite effect.

Look at Figure 42. Person 1 has learned a little bit about a field, say medicine. Knowing just superficial knowledge, he underestimates how much there is to know; in other words, he has no idea what he doesn't know. He knows there is some stuff he doesn't know (known unknowns) but he has no sense about the rest (unknown unknowns). The light gray box is how much he thinks there is to know in total and the dark gray is how much he currently knows. He perceives himself as knowing most of what there is to know, maybe 80% in this case.

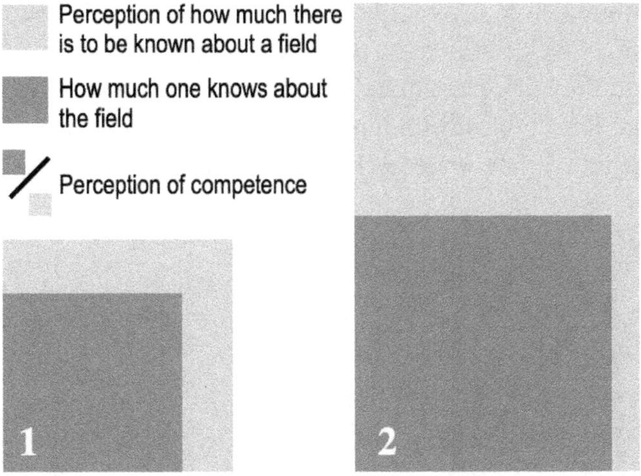

Figure 42. Conceptualization of cause of Dunning-Kruger effect.

Person 2, on the other hand actually knows more stuff; the dark gray box is bigger. She is more competent than Person 1 who has a smaller dark gray box. But she also realizes, by learning more, just how big the field of knowledge is and how much she still doesn't know. Therefore, her perception of her competence is lower, not only because she realizes how much she doesn't know but also because she compares herself to Person 1 who has so much confidence that she assumes he is better than she is. He isn't. Paradoxically, he thinks that he's above average and she thinks she's below average. If something was easy for her, she assumes it was easy for everyone not because she is smart but just because it was easy (the *false-consensus effect*). Some have called this the **Imposter Syndrome**, where a competent but diffident person may view her success as undeserved or by chance and fear being exposed as incompetent because she perceives herself as incompetent.

Dunning and Kruger showed that not only does incompetence cause poor performance, but it also makes one unable to recognize that the performance is poor. Without a realization that the performance is poor, there is no internal drive to improve. A lack of appropriate critical feedback in our professional environment and indeed in medical school and residency (e.g., grade inflation) perpetuates the Dunning-Kruger effect.

A popular graphical representation of the Dunning-Kruger effect is shown below in Figure 43. The Dunning-Kruger effect can be found everywhere and represents the problem with superficial learning about any subject. Many people, even those specifically trained in certain domains, lack the insight to understand how complex their field truly is and they underestimate both known unknowns and unknown unknowns. They typically underestimate how difficult problems in the field are (that is, they underestimate the complexity).

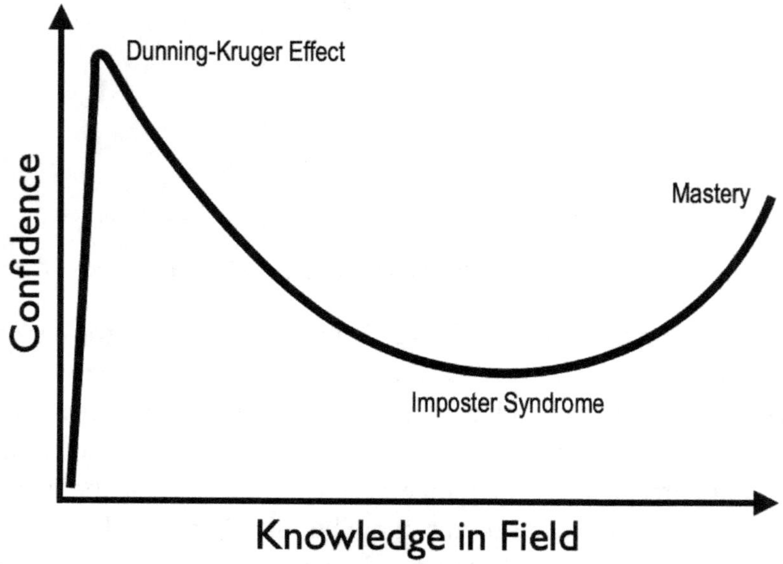

Figure 43. Dunning-Kruger effect and Imposter Syndrome.

These tendencies make them more likely to look for and see simple answers to complex problems, but those solutions are usually inadequate. The more one learns about a field, the more one appreciates the complexity of its problems and how much is truly unknown. This is definitely the case for medicine.

Many of the attacks on EBM come from those who don't truly understand it or who have oversimplified how EBM works, as in the case of the parachute study that we discussed earlier. The tendency to oversimplify how EBM works and even biology itself in the design and interpretation of scientific studies produces erroneous conclusions and apparent contradictions.

Evidence-based medicine involves irreducibly complex systems and higher order problems. Attempts to reduce this complexity or simplify the problem come at the expense of ignoring known and unknown aspects of the problem and inevitably lead to mistakes.

It is unfortunate that EBM has been misunderstood and misused. It is, indeed, time to throw the pirates overboard. EBM, having been hijacked by those with perverse incentives, has sadly contributed to overdiagnosis and over-utilization of medicine and interventions, becoming the *imprimatur* of marketers and making physicians, who don't understand the published literature, increasingly compliant with the will of pharmaceutical companies. Nearly 46% of published trials are financed by pharmaceutical companies. About 90% of published studies probably should not have been published. Less than 0.5% of published papers are cited more than 200 times, and even that doesn't mean they were high quality papers.[172]

Negative findings are rarely published. Replication is practically nonexistent. Peer review is comical. Journals are out of control. Search engines feed the confirmation bias. Contradicted studies are not annotated. Disproven studies are not withdrawn from our databases. Conflicts of interest are rampant and often unstated. Bias in studies and study design is nearly universal. P-hacking is everywhere. Those who make and read our literature suffer from the disease of Dunning-Kruger at an alarming rate. People set out to prove what they already believe or are paid to discover, not to use science as a tool to tackle our most sacred cows.

If all this is what is meant by EBM, it is indeed critically flawed and broken. But none of what I have just stated is EBM. Those are the misdeeds of the pirates and they have planted the Jolly Roger everywhere from corporate boardrooms to public universities to think tanks to the desks of policy wonks in the halls of government.

There is a better way.

Exercises

19.1 Identify something from the list of non-science based treatments presented earlier that you believe in. Perform a PubMed search for outcome based studies of that intervention and look at least at the last twenty years' worth of studies and critically assess your belief. Be sure to search for evidence that disagrees with what you already believe.

AFTERWORD
Where Do We Go Now?

If you've read this far, I want to thank you. It means a lot to me. More importantly, it means a lot to your patients. Our journey together has been at times tedious, revelatory, and even revolutionary. Don't worry, I've put the most boring stuff in the Appendix and I understand if you only use that as a tool to fall asleep or to start a fire.

The journey, though, if you choose to take it up, is just at its beginning. This book should be the start, not the finish. We have a lot of work to do, and I'm including you because you've read this far. We must make medicine a truly patient-centered, non-exploitative, evidence-based, science-driven, accountable endeavor. We must teach our students, our residents, and even our colleagues what science truly is.

This book has been about clinical reasoning; it has been about how we should think in medicine. Thinking is our most important skill. We are not robots who just do (though it seems that way sometimes). Our job is not simply to execute an algorithm (though sometimes we should do just that). Our job is to *think* and to think *well*. Some call what we have been talking about in this book *critical thinking*.

Indeed, good *clinical reasoning* is good *critical thinking*. Critical thinking in medicine requires a variety of tools in our cognitive armamentarium. We must be aware of our biases. We must understand what constitutes good evidence and bad evidence. But perhaps the most important characteristic of a good critical thinker is a large dose of healthy skepticism.

A good scientist is always questioning how things work, why they work, etc. A skeptical empiricist goes one step beyond and questions if things work at all; a skeptical empiricist questions the most fundamental scientific dogmas and demands a certain level of evidence before determining whether an idea is science or pseudoscience.

A great physician is a great skeptic. We should routinely question our assumptions. How many dogmas would exist today in medicine if we didn't dare question the status quo? Bloodletting, humorism, Mesmerism, Hijama, trepanning, etc. have all been displaced because a skeptic dared question whether those practices were effective. Many innovators or questioners of dogma were ostracized for their work, like Ignaz Semmelweis, who was fired for daring to suggest that we should wash our hands before doing pelvic exams on women in labor (even though he was able to demonstrate an amazing decline in death from puerperal fever). In retrospect, Semmelweis hit upon one of the most important discoveries in medical history. Yet, at the time, he was ridiculed because he challenged the status quo. Had his peers held him to science-based scrutiny, he would have been praised while he was still alive.

Scientists, and physicians as scientists, must maintain an atmosphere that welcomes challenging the traditional view. When a new finding is published in a journal, the community of physician-scientists must greet it with a skeptical mindset. Today, most physicians view a new published finding with total acceptance. They assume that folks smarter than themselves shepherded the study through publication and that a bad study or a false conclusion couldn't make it all the way to their mailbox or inbox. Yet, our biological science journals are filled with incompetent peer review, poor statistical methodologies, publication bias, false positive and false negative findings, and, in some cases, outright academic fraud. Remember, most published findings are either false or insignificant. Each new publication, then, should be met with a skeptical view and the physician-scientist must be adept at critically appraising literature.

All these things are required to be a great physician, but there is one more thing that is incredibly important and I want to challenge you with it as we leave our time together: be ethical. We talked about ethics in Chapter 13. Here are some explicit rules you should be aware of if you want ethics to guide your practice:

1. **Follow science-based guidelines.** Guidelines are the best that we, as a group, know. They are not always right and they are often changed. Some guidelines contradict others. I take issue with many of them. But your personal opinions and experiences cannot negate them. If you disagree with a guideline, work to change and influence it, don't just ignore it. Yes, sometimes you should deviate from guidelines. I would guess that about 5% of patients require significant deviation from guidelines. When this occurs, the reasons should be explicitly documented. But, in the real world, guidelines and evidence-based medical recommendations are not followed in around 80% of medical encounters. When a gynecologist deviates from Pap smear guidelines or a cardiologist deviates from coronary stent guidelines, they are guilty of the exact same thing as pediatricians who don't vaccinate children. Worse, the reason for the deviation is often self-interest and economic. Fewer Pap smears mean less money for the practice; fewer mammograms mean less money for the hospital, etc. Following guidelines and evidence-based practice recommendations usually involves doing less, and doing less means making less money for you and your employer. Evidence-based practice guidelines help you provide the best mix of beneficence, nonmaleficence, and justice for your patients. They are an ethical must.
2. **Fully inform patients.** Far too often, autonomy is invoked as the reason for deviation from guidelines or best practices. But, it is unusual that fully-informed patients make poor decisions. It is rare that a patient who truly knew that one treatment pathway was better for her would choose the alternate pathway. For example, does it make sense that a fully informed patient would choose a birth control pill over an IUD if her goal was to remain not pregnant for a number of years? Over five years, a hormonal

IUD will be 22 times less likely to kill her, 42 times more likely to keep her not pregnant, less likely for her to have an ectopic pregnancy, make her 30 times less likely to have an unwanted surgical procedure, make her more likely to have periods she likes, etc. It sounds like a no-brainer, and it is. That's why every important health organization recommends IUDs over birth control pills as first-line contraception. Yet, over 85% of women don't get one (and most of those women aren't even given the choice). I could write a paragraph like this about dozens – hundreds – of things in medicine. If you are doing lots of elective Cesareans, I suspect your patients aren't fully informed. If your patients are getting robotic hysterectomies, they are not fully informed. If your patients are choosing yearly Pap smears or mammography at age 40, they are probably not fully informed. Autonomy without complete and accurate information is not autonomy but coercion and exploitation. Or, if you give a patient antibiotics when they aren't indicated or other interventions just because patients ask for them, you may be negating your own autonomy. Your autonomy counts too.

3. **Don't waste resources.** It is estimated that we waste more money on unindicated labs and imaging tests each year in the US than we spend on the entirety of K-12 education, over $800 billion a year.[173] Add to this cost all the unnecessary procedures and surgeries, plus the unnecessary prescribing of unneeded drugs (or using a branded drug when a generic drug will do), and one realizes quickly that as much as half of US healthcare expenses are wasted. This over-utilization of resources actually leads to worse outcomes and this paradox is generally well-known; we spend about double the amount per patient in the US for inferior outcomes compared to other first-world nations like Finland. This is one way that the ethical principle of justice is subverted. But there is another, more insidious way as well. Sometimes we spend money and resources on patients and we do achieve better outcomes for the expense, but those resources might still have been better spent elsewhere. We spend an incredible amount of money on patients in the last year of life, where the cost per quality year of life is excessively high; when we do, that is money we are not spending on other patients

where the cost per quality year of life is much lower. We must focus on high impact interventions and make sure everyone benefits from these. When an elderly patient with a terminal disease consumes millions of dollars of resources to extend his life by a few months (and miserable, poor quality months at that), we have taken away resources that might have been used to add quality years of life to other patients. Healthcare resources are scarce and finite; it is our duty, through public policy and guideline development, to distribute them as evenly as possible; but this duty also occurs on an individual basis, every day. When you prescribe a branded birth control pill that costs $247 per month when a generic pill with the exact same hormones costing $8 per month would have sufficed (Balcoltra versus Lutera, 2020 prices), you stole birth control from 31 other women. When you do an unnecessary hysterectomy that costs the healthcare system $20,000 dollars, you deprived 200 women of a year's supply of birth control. When you check a CBC or BMP on a healthy 30-year-old patient at her yearly preventative visit – or get a lipid panel or TSH or Vitamin D level – you wasted precious money that could have provided needed services to someone else. I'm sure you can come up with your own examples. Who benefits? Not the patient, but the drug company, the doctor and hospital, and the laboratory company. This is the definition of exploitation.

4. **Don't give bad advice.** Physicians often do more harm with their mouths than with their pens or scalpels. Patients look up to physicians and trust that what they are told is important and factual. We ask patients to change their lives based on our advice, and they pay us for that advice. Patients are incredibly vulnerable in our presence and we wield enormous power over them. But much of the advice that physicians give to patients isn't evidence-based. Don't tell patients anything that isn't clearly based on good science. It seems like 99% of the health information patients read on the Internet is false; but one of the reasons that this erroneous information has digital legs is because somewhere, some irresponsible doctor told a patient that false information. It is your responsibility and ethical duty to

know that what you tell patients is supported by current, replicated, quality scientific data. This means things like don't tell people to participate in dieting fads; don't recommend bed-rest to pregnant women; don't recommend alternative medical interventions that are unscientific, etc.

5. **Be educated.** A corollary to the first four points is that you must be educated. Patients are spending their money on you and entrusting their lives to you because they assume that you are up-to-date and that you are treating them based upon the latest scientific evidence. Clearly, physicians are not commonly doing this (as evidenced by how few physicians follow evidence-based practice recommendations). Your one job is to stay up-to-date and educated about the conditions within your scope of practice. It is not optional. You are not allowed to do things because you have always done them that way or because "that's how I was trained." I went to medical school 20 years ago and I often reflect that virtually nothing in medicine is treated the same way now as it was then, at least not in the hands of competent doctors. Even ten years is an eternity. Since I graduated residency in Obstetrics, we have dramatically changed our labor management guidelines, our fetal heart rate monitoring definitions and algorithms, our Cesarean delivery technique, and the methods of neonatal resuscitation; we have introduced delayed cord clamping as standard practice, revised the GBS management guidelines, made breakthroughs in the most effective methods of induction and augmentation of labor, etc. In short, if I practiced Obstetrics the same way today as I did when I graduated residency, I would call myself a bad doctor. Education doesn't come in the form of detailing from drug or product reps, industry-sponsored dinners, or golf get-aways dressed up as CME conferences. You are paid a lot of money to read a couple of journals a month (and change your practice accordingly).
6. **Don't make medicine a business.** I fear that this is one of the biggest impediments to ethical, evidence-based medical practice. Are you worried about how changing your practice pattern will affect your bottom line? As I said before, there are economic consequences of doing fewer Pap smears, mammograms, and hysterectomies. There are economic consequences of doing

fewer colonoscopies, echocardiograms, coronary stents, CT scans, MRIs, bone density scans, obstetric and gynecologic ultrasounds, cystoscopies, urodynamic studies, first-trimester genetic screens, back surgeries, etc. But forget about not doing unnecessary surgeries and tests for a minute; it gets worse. Have you ever attended a workshop or conference to learn how to "monetize" your practice? Do you offer boutique services to make cash? Do you prescribe hCG injections or phentermine for weight loss despite both practices lacking data that say they are effective? Do you sell vitamins or cosmetics or offer a variety of ineffective laser treatments? Are you practicing cosmetic dermatology or doing cosmetic plastics procedures even though you weren't adequately trained to do them and those things are beyond your scope of practice? Maybe you don't. But do you have patients come back to the office for lab follow-ups when a phone call would have worked? Do you have patients routinely come back after starting a new medicine like birth control just to see how they are doing when you could have just told them to call or make an appointment if they have any problems? Those are all ways of monetizing your practice. If I leave the day and feel like people came to see me for no reason other than for me to bill a 99213, I feel like a complete failure. What sick people didn't have a chance to see me that day? Who couldn't afford insurance that year because we, as a profession, waste so many resources that we have doubled their insurance premiums?

7. **Stand up for what's right.** There is a lot wrong with medicine and healthcare. Most incentives that affect how we act are perverse. Recall our discussion of the Quadruple Aim in Chapter 12: everything we do should serve to improve patient outcomes, improve patient experiences, lower costs of healthcare, and improve the physician/provider experience. Believe it or not, those four aims work together seamlessly and beautifully. It requires integrity to make those aims (and, indeed, the patient) the center of everything you do. It often requires that you go against what might seem to be in your or your institution's best economic interest. Stop making excuses. If physicians want to be respected, they must have this integrity. We must police ourselves against those who pervert our ethical standards. You have

an opportunity to stand up for what's right in every patient encounter, every committee meeting, every conference, etc. Do it.

8. **Stand up for yourself.** We used to talk about the Triple Aim, but now we talk about the Quadruple Aim; the fourth aim is that we should seek to improve the physician experience as well as the patient experience. Patients cannot get better without us. We need to be healthy, both physically and psychologically, to provide effective care to patients. In an age where we spend two thirds of our time doing paperwork or other nonclinical activities, where we worry about our financial futures, where we suffer for years with too-rampant litigation, etc., our patients are the ones who suffer. Medicine, despite it all, is still a physician-led enterprise. Don't cede your power to others. Stand up for yourself and your profession.

Here are some practical things I'd like for you to consider doing every day, starting today, in your practice of medicine.

Consider the differential diagnosis. If you already do, make longer differentials. Challenge your students to list more than 3-5 things on their differential diagnoses. Use the Internet to do this or some other resource and do it every day. If a patient has a chief complaint of dizziness, for example, Google "dizziness AND differential diagnosis" or search for "dizziness AAFP" and read the article that is resulted. Try it right now. Make it a habit. Make sure you know the diagnosis scripts for each diagnosis on the list. Make your learners think in these terms.

Know your guidelines. Whatever field you are in, make sure you are very familiar with the latest guidelines relevant to your scope of practice. If you don't know where to look, try a search. For example, Google "cardiac stents AND guidelines." Go ahead, try it out.

What else can this be? Ask this question of yourself or your learner every day, for every case, if you have the time. Then convince yourself it is not that thing or consider how the treatment would be different if it were. This metacognitive tool will enhance your lateral thinking skills.

Don't interpret studies outside of context. I don't expect you to be an expert on interpreting studies, not without a lot of practice and a lot of rereading the latter chapters of this book. But when you see a new study of interest, never read it out of context. Go to PubMed and search for related articles. What is already known about the question? What other studies have been done? Is this finding new or does it agree with other evidence? No study exists in a vacuum. This practice will enhance your ability to assign a rough prestudy probability to the paper that you are currently reading.

Be leery of novel findings. Novel findings by definition have not been replicated. Most will not be. Most novel findings are false. Be leery.

Look up the prices of drugs you prescribe. Most physicians are dumbfounded to discover how expensive many of the medicines they prescribe every day are. Remember the example of Balcoltra ($247 a month) versus Lutera ($8 a month) a couple of pages ago? They contain the exact same hormones and doses. One is marketed, one is not. Look up the prices and consider if the extra cost is worth it. Try it now; go to GoodRx.com and look up some prices for your favorite drugs.

Know what the predictive value of a test is in a low, moderate, and high probability clinical scenario. You don't need to spend a lot of time considering the positive and negative predictive values of every test you order for every patient. But do look at the tests you commonly use and put in some low, moderate, and high probability scenarios to see the value of the test (for example, 10%, 50%, and 90% pretest probabilities). Or, just learn the likelihood ratios of various tests you order and think in those terms. Try this Google search: "likelihood ratio AND spiral CT pulmonary embolism." No, really; try it.

Ask how a test is going to change what you do for the patient. If you can't specifically answer this question about a test (e.g., *would I do something different if the result were negative or positive?*), don't order the test.

Find out how much benefit someone will get from an intervention in understandable terms. Check out thennt.com or just Google something like "nnt AND statins."

Thanks for reading. Now get to work.

Knowing is not enough; we must apply. Willing is not enough; we must do.

– Johann Wolfgang von Goethe

APPENDIX

Bayes' Theorem and Epistemic Confidence

This book has really been about Bayes' theorem and the use of probability theory as a tool of epistemology, applied specifically to the domain of medicine. In another discipline, we might have talked about the **Mind Projection Fallacy** instead of cognitive biases. There are many such philosophical universalities underlying what we have been discussing. We have touched briefly on many of these, but it is important to spend some more time with Bayes' theorem and epistemic confidence, mainly as a defense against those who will attack Bayes' theorem and its corollaries.

The statistician at your university might object to Bayes' theorem because it sometimes gives different results than the statistical methods which are commonly used, called the 'Orthodox' or 'Frequentist' interpretation. Medical schools do not teach Bayes' theorem well. Even when they do teach it as a way to interpret clinical lab tests, they do not usually refer to it as Bayes' theorem. Many of you have likely never heard of it before reading this book. There is a philosophical difference between the Bayesian and Frequentist interpretations that is rooted in different views of epistemology and ontology. This Appendix will introduce that controversy to you.

Let me first defend the basis of Bayes' theorem here by deriving the formula from basic principles. What follows is a proof of the theorem.

For a Bayesian, probability expresses our confidence in our beliefs. If P (probability of a belief) = 1, we would be 100% certain that our belief is correct; however, in practical terms, we never accept P=1 or P=0 for any postulate in *abductive reasoning*. This latter point is a chief difference between Bayesian and Frequentist conceptualizations. In Frequentism, we think of probability as the likelihood that a future event will occur, given an assumption about reality (a statistical model). In Bayesian inference, we think of probability. Belief is expressed in *degrees of confidence* and this is mathematically related by *probability*. Bayesian methods allow us to deal with belief in degrees of confidence; Frequentism does not. This fundamental difference between these two methods of statistical inference is apparent in the chief and most fundamental relationship expressed in Bayesian logic:

$$P(H) + P(-H) = 1$$

This simply means that the probability that a hypothesis (H) is true plus the probability that it is not true is equal to 1. In other words, it is 100% certain that any given hypothesis is either true or not true. It may be accurate that in an absolute, ontological sense, a given hypothesis is either true or not true; but all rational epistemological interpretations would allow that we are limited in our ability to know such absolute truths. Therefore, Bayesian methods don't accept P=0 or P=1. We might be 99.9999% sure that something is true but there still exists 0.0001% of doubt. These ideas seem like they should be accepted by anyone using any system of logic; I won't present arguments to the contrary. But, if you accept this basic postulate of Bayesian logic, then you must accept all of Bayesian inference and the Bayesian logic that follows.

Logic tells us,

$$P(A\&B) = P(B)P(A|B)$$

This states that the probability of two events, A and B, is equal to the probability of B times the probability of A given that B has occurred (note that a vertical bar '|' is read as *given*). We can see this expressed in the following Venn diagram:

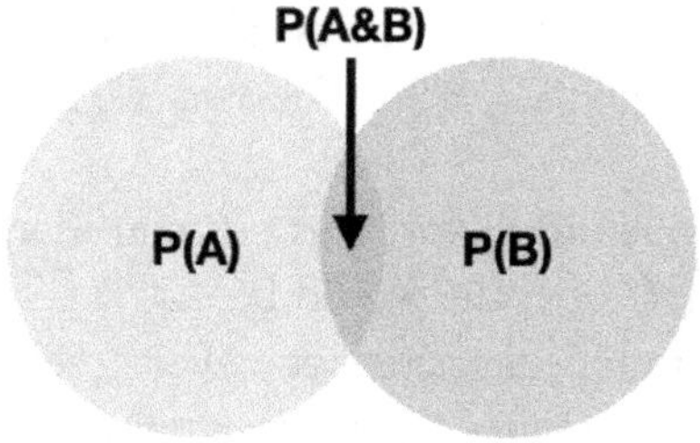

Figure 44. Venn diagram showing the probability of two concurrent events.

Again, this simple logical proposition is irrefutable. What is interesting is that in this one logical proposition alone, we have already established Bayes' theorem and all the principles of Bayesian inference that we have relied upon in this book.

In abductive inference, we use evidence (E) to determine the probability (truthfulness) of a hypothesis (H). Let's substitute these terms into the equation above:

$$P(H\&E) = P(E)P(H|E)$$

This is the same relationship, but now we are dealing with the probability of the truthfulness of a hypothesis (H) in a world where some evidence (E) exists. Similarly,

$$P(E\&H) = P(H)P(E|H)$$

We should also note that,

$$P(E\&H) = P(H\&E)$$

These two statements take up the same space in the Venn diagram in Figure 44; the order doesn't matter. Therefore, by substitution, we get:

$$P(H|E) = \frac{P(E|H)P(H)}{P(E)}$$

This is Bayes' theorem! It is the most powerful (and perhaps most underutilized) equation we possess in science, logic, jurisprudence, and philosophy. Some might call it common sense. It is at once the fundamental key to the scientific method and the key to epistemology.

If you accept Bayes' theorem as stated, thanks for reading. If you reject it on philosophical grounds and seek to substitute a purely Frequentist view instead, keep reading.

Epistemology and ontology

I've mentioned epistemology and ontology and their relationship to the differences in the Bayesian versus Frequentist inferential approaches. Let me exposit with a short introduction to these philosophical ideas as they relate to statistical inference.

Is there an objective nature or reality? Or is it subjective, dependent upon our perception of it? These are the sorts of questions you might recall from your university 'Introduction to Philosophy' course. I will make no argument here other than to say that for the purposes of the practice of medicine and for the considerations of this book, it is assumed that there is an objective reality. Statements of truth belong only in a world where truth is absolute. Still, the problem is that we rarely, if ever, are able to know what reality is in an objectively verifiable way. In other words, the problem is not with truth but with knowledge.

Philosophers talk about these two issues using the terms *ontology* and *epistemology*.

Ontology, in a simple understanding, deals with the question: What is reality? What can be said to be real and what is an illusion? Is there an objective reality that exists regardless of our observation of it?

Epistemology, on the other hand, deals with the question: What constitutes belief and knowledge? How do we justify our belief or knowledge of whatever reality might exist?

In other words, ontology relates to *what is* and epistemology relates to *how we might know it*. The two issues are obviously related but they are not dependent upon one another. Whatever exists does exist regardless of our mechanisms of understanding its existence. That it, there are a certain number of grains of sand on Earth at a given point in time, even though we can't count them; what's more, they would still exist even if humans did not. Note that I am immediately disregarding certain extremes of *solipsism* which argue that the world doesn't exist except in our minds.

Just because we have poor knowledge of truth doesn't mean that such truth doesn't exist – it just means that we may never know what it is for sure. It is easy to convince ourselves of truth subjectively. With too much faith in our subjective realities, and a frustrating inability to prove our subjective realities objectively to others, some have decided that truth itself is subjective (again, a form of solipsism). But this argument is the easy way out and doesn't concern the practical world we live in.

Protagoras, the famous Greek Sophist, stated, "Man is the measure of all things," meaning that reality was subjective, existing only as the individual views it. But Plato rejected such a view and stated that Sophists made their living through deception while Philosophers (like Plato) sought truth. Sophists don't believe in truth (either in an objective reality or a personally useful concept of objective truth of such a reality).

Ironically, many who would call themselves philosophers today are nothing more than Sophists. Subjective truth doesn't require any work to know it (it's what's in your head right now, after all) and sophistry is concerned mostly with making people feel better about themselves (by appealing to the ego). But here we are concerned with how we can attempt to know the objective realities of the universe. This concept seems basic but it is of the utmost importance. It is the foundation of science and we cannot deal with any scientific issues without having a firm foundation.

What is truth? There are many types of truths we might consider, and many famous philosophers (everyone from Plato to Kant to Bertrand Russell) have made careers for themselves (and creating a lot of confusion in the process) simply by developing taxonomies and nomenclature for types of truth. Without relabeling types of truth again, let's think about some of these truths in order to help understand why knowledge of truth is so elusive. Let's consider a few examples (I have no intention of being exhaustive or recreating another taxonomy).

There are historic truths. For example, there is an objective truth (ontological reality) about who killed Nicole Brown Simpson. It happened, someone stabbed her, she died, and no amount of personal bias or subjective views of ontology can change the actual events that happened (don't over-emphasize how certain the statements I just made are – I'm speaking of ontology, not epistemology). At the same time, I don't know for certain who killed her. I believe to a reasonable certainty that O.J. Simpson did it, but I can't prove it (or anything else like this) to an absolute certainty. *My knowledge of this truth is qualified by a degree of certainty.* Even if OJ confessed to the murder, that wouldn't matter; his confession could be a false confession. If you think about it, there is just simply no way of knowing for certain who did it. But it happened: someone did it, and our inability to be certain about it doesn't change that objective reality.

In practical terms, though, I can show with some degree of certainty (that is *probability*) who killed her, and then we must decide whether the probability of the truthfulness of the assertion that OJ did it is sufficient for our purposes. This is the importance of epistemology; it needs to help us understand how we determine what is a useful, *justified* knowledge. This also means we need to define what *use* our knowledge should serve in order to understand what *degree of probability* is acceptable for justification of our knowledge.

Courts say that felony, criminal cases, like the OJ Simpson double-homicide trial, must be proven beyond a reasonable doubt; so, what constitutes a *reasonable* doubt? That is, what degree of probability? How likely does an alternative hypothesis have to be in order to be a reasonable hypothesis? Is a 1 in 100 chance reasonable? How about a 1 in a million chance?

Civil cases demand only that a case be shown to be more likely than not – a much smaller burden. OJ was found to be the killer by a jury in a civil lawsuit. One might argue that this 51% probability standard is too low in many situations. But the point is that both standards attempt to define what degree of certainty is required of a jury; neither require that a jury be *certain*. *More likely than not* is indeed a weak standard. It is more likely than not that I will survive a 45-mph car crash, yet no one will be surprised if the crash kills me. In any event, we must decide in each situation how certain our knowledge of truth needs to be in order to act.

Did the British burn the White House in 1812? I don't know. I can read the historical accounts of many eyewitnesses and examine the evidence and conclude with a high degree of certainty that they did. But I will never know for certain if Dolly Madison didn't do it herself because she really wanted to remodel the place. Yet for all practical purposes, I believe that the British did it. If I am wrong, it doesn't really affect my life in any meaningful way. Thus, the practical truth (and the probable truth) is that OJ killed Nicole and the British burned down the White House. These "facts" are part of my knowledge, and I believe both are reasonably certain (whatever that means). I believe I can *justify* these beliefs with available evidence.

Similarly, there are future truths. I can only state historical truths to a degree of probability. Similarly, I can state future truths to a degree of probability, though for obvious reasons the degree of probability for future truths is often much lower than for historical truths due to different types of evidence. That being said, we might be more certain about some future truths than many historical truths. For example, I believe that the sun will come up tomorrow. Every piece of information I know about this indicates that this is true and that is my knowledge of this truth. I would bet my life on it or any amount of money. But I am not *certain*. There are imaginable (and no doubt some unimaginable) scenarios that could occur which would prevent the sun from rising tomorrow. So, I am incredibly sure but not absolutely certain that the sun will come up tomorrow.

I'm also pretty certain that I will be alive tomorrow, but far less certain about that compared to the sun rising. For practical purposes, I will live my life today assuming that both events will happen tomorrow, even though I know neither for certain and my degree of confidence about the two hypotheses is very different. Again, the point is that my degree of uncertainty has no effect on the future reality. My perception is not the reality. **It is clear that there is a distinction between truth and our knowledge of it.**

Historical and future truths are no different, but what evidence exists and how confidently we assess that evidence varies because of the circumstances. I have more confidence that the sun will come up tomorrow than I do that OJ killed Nicole, but in both cases I have a personal belief that is sufficiently evidenced for me to go about my day. I feel confident in my knowledge of the truth, but allow for the possibility of error.

So, too, current truths are transient and similarly assessed. What is current will soon be past anyway. All truths are similarly known. There really are no aspects of reality (ontology) that can be stated as knowledge (epistemology) with absolute certainty, but that does not mean that the reality or truth itself is not certain. What's more, just because we do not know any realities with absolute certainty does not mean that reality is not knowable. Our human limitations of knowledge do not alter reality. Reality exists with or without us. Similarly, our individual experiences, biases, and perceptions do not alter reality. Reality is *knowable* but with varying degrees of probability or certainty. Our certainty about reality or truth approaches a definable asymptote; the closer it gets, the more certain we are, approaching absolute confidence (but not certainty) in some situations.

In the schematic in Figure 45, a probability (P) = 1 is absolute certainty. The probability of our beliefs is represented by the curve. The curve approaches absolute certainty but remains asymptotic; nevertheless, we can for practical purposes consider the limit of our probability ($P = 0.99\bar{9}$) as certain.

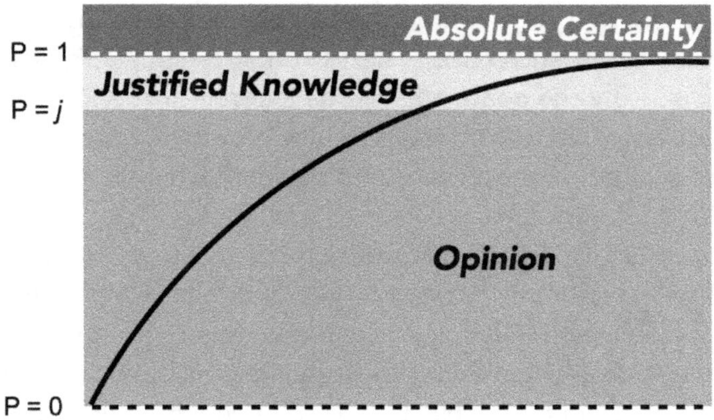

Figure 45. Justified knowledge in probability theory.

At a certain level of probability *j*, we can consider our belief justified and that belief in turn constitutes *justified knowledge*. Everything with a probability less than *j* is opinion. The value of *j* varies, as described above, by its expected utility, though it is difficult to imagine anything being considered justified knowledge with a $P < 0.5$. Most of what we consider knowledge is opinion, and many opinions are useful and should be utilized; still, we must be careful to realize that they are merely opinions. The confusion of opinion with knowledge plagues most fields of human inquiry and modern science is certainly not immune.

There are many things that we state with absolute certainty which are merely constructs and are not predictive of (external) reality. For example, we state with certainty that 2+2=4. But we are able to define this with absolute certainty only by construct. Kant called these constructions "*a priori* forms" in that they are not derived from our experiences; that is, they are not dependent on our ability to accurately understand reality. Similarly, we think of truths like my social security number or hometown as absolute and utterly knowable, but these are not *certain* in the same way as 2+2=4; these "facts" are subject to alternative hypotheses like psychotic delusions or typographic mistakes. They are derived from our experiences and therefore are not *a priori* forms. But we can say that our knowledge of these "facts" approach $P = 1$ well enough to consider them certain for practical purposes. Kant believed that our knowledge of reality was based on both our perceptions (which can be faulty) and *a priori* forms. It is important to make this distinction but also to realize that our *a priori* forms can serve to bias and perhaps alter our perception of reality.

Kant, like many others, failed to allow for different probabilities of certainty. He did, however, believe that an external reality is necessary for the creation of our subjective realities (thus rejecting solipsistic views). And this is really the point: there is an external reality and there is *our personal reality*; we need to understand how best to bridge the gap between these two realities and we need to understand what we practically want to do with our knowledge.

Thus, we need to define what utility we expect from knowledge of our reality. I can never know the exact value of π, but I can know it to sufficient precision to solve all practical and even theoretical problems related to it. I do not know if the chair I'm sitting in is real (maybe I'm psychotic and I'm just imagining it), but I am certain enough that I don't plan on wasting any more time thinking that it might not be real.

Similarly, I can never know the exact date when I first ate peanut butter, but there was a date on which I first ate peanut butter and I can estimate the probabilities of certain days and determine the most probable date. Let's say the most probable date was March 2, 1979 (and estimate this date with a probability of 0.0015%). This type of "truth" is no different than my belief that OJ killed Nicole (which I estimate to have a probability of 99% or greater), but the confidence is dramatically lower. This low-certainty knowledge of the date on which I first ate peanut butter is not useful to me because the probability that this knowledge represents reality is too low. It is therefore opinion at best, and should not be represented as a fact or as justified knowledge. Our knowledge of objective reality must in all cases be qualified with probability.

Our brains are wired to work this way. We gut estimate our level of certainty about anything and everything we believe. But our estimates are skewed by cognitive bias and are usually horribly wrong. It should follow that the more we work to overcome cognitive bias and the more data we have available to analyze (and the better we analyze the data), then the better decisions we should be able to make and the more probable our knowledge of reality will become.

There is an objective reality and usually we have several hypotheses for what that reality is. Thus, confronted with multiple hypotheses, we use the hypothetico-deductive model of hypothesis testing to arrange each hypothesis in order of probability (e.g., it's more likely that the British burned the White House than did Miss Dolly). Occasionally, one hypothesis reaches a level of probability that is acceptable for our use so that we consider it as "true" and therefore it constitutes justified knowledge (I accept it as true that the British burned the White House). This occurs whenever $P > j$.

Think about this: *there is a difference in a fact being uncertain and a fact being unknown.* Imagine that I roll a dice; the result of the roll is not actually *random*. Given enough information (like the velocities and motions of the dice, wind resistance, characteristics of the surface it will land on, etc.) and the correct understanding of all the variables that affect the roll of the dice, the eventual result is actually *deterministic!* The same thing is true for chaos theory and complex systems. The outputs of chaotic systems are actually deterministic, we just don't know enough to know them. The *randomness* describes our knowledge, not reality; it is a *mind projection fallacy* to assign qualities that describe our knowledge onto objective reality.

Thus, when we throw the dice, we are *uncertain* how it will land; but since we can assign an *a priori* probability of each possible outcome, we have *knowledge* of how it will land – just uncertain knowledge (each of our six possibilities are equally likely unless we can find more data to help us refine our probabilities of the outcomes). How the dice will land is actually *knowable* (not *random*), we just don't have enough data to *know it* with a $P > 1/6$ (which is usually not enough to be considered *justified knowledge*, that is, P is not greater than j). Thus, we *know* how the dice will land (one of the six ways), we are just *uncertain* how it will land.

Thus, we need to focus our interest on how we justify our knowledge of reality. We must, therefore, ask the question, *What is knowledge?*

What is knowledge, and how do we justify it? In *The Theaetetus*, Plato records a dialogue between Socrates and Theaetetus regarding the nature of knowledge. Socrates, just before his death, suggests three ideas:

1. **Knowledge is nothing but perception.** This idea, attributed to Protagoras, is easily discounted by the two men ("things are to you as they appear to you, and to me as they appear to me").
2. **Knowledge is true judgment.** This idea is refuted because many people have true knowledge for the wrong reasons or have true knowledge without justification. This gave way to the last idea.
3. **Knowledge is true judgment with an account.** With *an account* means knowledge justified by *logos*. This idea was also rejected and Plato closes the dialogue without a conclusion. Let's deal with this last point in just a moment. But before talking about *logos,* let's consider why any of this is important. There are two main reasons.

First, objective reality and objective truth has come under assault. If truth is only in the eye of the beholder, then intellectual, scientific, and philosophical progress is impossible. If there is no objective truth but only personal truth, then no idea is better than another, and individual experience (you are the author of your own subjective reality) is most important. But some ideas are better than others, and truth does exist. This means that some people are wrong and some people are right about their ideas or views of reality. In medicine, this is the battle between anecdotalism and evidence based medicine.

Second, if objective reality does exist, then we must strive towards understanding it. This was the Platonic view. This means that we must accept the idea that there is a knowable truth and seek to use philosophical and scientific methods that help us best describe that truth. If, philosophically, you are stuck debating whether truth is even real, then it is unlikely that you are making practical progress towards discovering any truth and you truly are a Sophist. Or perhaps you are lost in a fantasy of solipsism and believe that only your mind truly exists and can be known (and only by you). But for those who live in the real world and would like to make philosophical and scientific progress, let's accept objective reality and strive to learn best how to know it (even if it is ultimately unknowable).

Today, few people are able to categorically distinguish between justified belief and opinion. Ten minutes on Facebook is sufficient evidence to justify my proposition. A *justified belief* is what I have called truth – something evidenced sufficiently to afford an adequate degree of probability to serve practical purposes. An opinion is everything else. It is a justified belief that OJ killed Nicole; it is an opinion that Nicole committed suicide. Only in a world of objective reality are these two ideas distinguishable. I have no justified belief about when I first ate peanut butter and, barring new evidence becoming available, I never will.

Much of what is passed off today as "fact" is actually just opinion. I shirk every time I hear someone say, "Science proves…" because that is not the purpose of science. These are the opening words of conmen. Journalists today are unable to distinguish fact from opinion. A brief foray into political "fact checking" shows nothing but fallacy; fact checking is really just opinion checking and any opinion that the reporter disagrees with is a "lie." This type of error results either from deliberate attempts at rhetorical deception or from a reporter who actually believes that his subjective reality is *the reality*. So, epistemology has more relevance than ever and in every aspect of modern life.

Plato records that Socrates failed to support the idea that knowledge was true belief supported by *logos*. He gave three definitions of *logos* and found none acceptable (he considered *logos* to be either a statement or speech, an understanding of the elements of the belief, or an understanding of the salient and unique characteristics of the belief). I suggest that knowledge is a justified belief (or a true belief supported by *logos*); the problem really is what do we use as *logos*? How should we justify our beliefs?

Heraclitus recognized that *logos* linked rational thought to the rational reality of the world; but he did not define what *logos* is in a way that made it accessible to us. In fact, he recognized that *logos* was elusive:

> For this reason it is necessary to follow what is common. But although the *logos* is common, most people live as if they had their own private understanding.[174]

Plato used the word *logos* in a similar way. But Aristotle expanded the use of the word and the concept and tried to make it useful. For Aristotle, *logos* was an argument from reason – that is, *logic*. This was juxtaposed to *pathos* (an argument by emotional appeal) and *ethos* (an argument from moral authority). If *logos* is logic, then what are the rules of logic?

Aristotle, of course, is the father of the traditional system of logic that dominated philosophy for over 2000 years, until being supplanted by more modern *predicate logic*. Aristotelian logic establishes the "truth" of propositions and allows for inferences to be made about other propositions or conclusions. But how can we know if a proposition is true in the first place? Propositions may be true by definition, much like 2+2=4; but, in most cases, we are forced to accept the truth of a proposition without considering the probability of its truthfulness or its falsity. Aristotle's logic did not do well with uncertainty; it prefers absolutes and clear-cut issues. However, most real-world problems, particularly in medicine or other complex systems, are rife with uncertainty.

For "hard sciences," such as mathematics, physics, engineering, and astronomy, Aristotle's logic and the prevailing Greek ideas of "knowledge" led to an unprecedented expansion of technical understanding. Greek geometry, for example, was dominated by axioms, taken as truths, that underpinned all other propositions. Math in general has followed this course.

In any field where a known input results in a known output (a simple system), Aristotle is king. Our modern, scientific process is predicated on these ideas, and as math and physics and astronomy were dominant fields in science, neither the objective reality of the universe nor our ability to use logic to know and predict that reality were questioned.

But other scientific fields have not been quite as simple. In true Platonic fashion, we can understand *why* 2+2=4, and we can demonstrate the consistency and lack of contradiction in this truth. But we don't know why some people who smoke get lung cancer while others do not. We do not know exactly what the weather is going to be like tomorrow in Tampa. That's not to say that there might not be a knowable answer, but we don't fully understand it since we don't understand all the variables involved nor do we understand how they are intertwined in the complex systems of human biology and weather.

Therefore, we cannot define knowledge related to these types of realities in an Aristotelian way. Yet, we must deal with lung cancer and the weather as practical realities, so we still need to describe our knowledge related to these issues. Because Aristotelian logic fails so many aspects of science and philosophy, Socrates would have been the first to point out that the system is fundamentally flawed and that we need a system of epistemology that is generalizable to all fields, not just rhetoric, math, and grammar. Aristotle didn't deal well with uncertainty, but we need to if we are going to deal with complex systems.

How can we deal with uncertainty? Of course, we attempt to deal with uncertainty mathematically with statistics and the theory of probability. But the orthodox view of statistics, Frequentism, grew out of the field of Mendelian genetics. Ronald Fisher, who largely developed the field of Frequentist statistics that has dominated the last century of scientific research, was a biologist who studied Mendelian genetics. This means that he studied discrete genes with predictable and understandable phenotypic effects. He studied simple, discrete systems. His statistics did not need to describe any uncertainty that might exist in the model, only the variation seen as a result of chance when the model (discreet inheritance patterns) was executed.

Let's consider three sorts of scientific problems.

First, there are problems where a known input always produces a known output. For example, an electrical circuit, once all the variables are understood, always produces the same output. It is understandably deterministic and the variables involved are understood so that the whole mechanism is understood; even if the details of the mechanism are not fully elucidated, observation can still yield a predictable output for a given input. We don't need probability theory to understand the outcome. Aristotle and scientists love this type of certainty. It demonstrates cause and effect, and it is immediately observable, predictable, reproducible, and reliable.

Second, there are problems where a known input produces a distribution of possible outcomes in a certain frequency. Such is the case with Mendelian genetics. Seeds of a certain sort, for example, will always produce about 25% red, 25% white, and 50% pink flowers. The exact numbers are not known because there is some apparent "random chance" that has entered into the equation. It is not, apparently, absolutely deterministic, like gravity, but the variables are all known and controlled for so that a model can be used to predict likely outcomes. Frequentist statistical methods are excellent at understanding these types of problems and at understanding experiments related to this type of science. These were the types of problems that Fisher worked with and he could, given the *frequency of outcomes*, determine the probability of observing the frequency of those outcomes given a particular hypothesis. That is, Fischer calculated the probability of the observed data given the hypothesis.

The third type of problem is much more difficult but also more analogous to the complex problems that we usually deal with in modern science in general and medicine specifically. How do we deal with problems where all the variables are not fully understood and the mechanisms controlling the outcomes are not fully elucidated? In other words, what do we do when there are many unknowns and the relationships between those variables is not understood? We must estimate likely outcomes using probability theory.

What's more, we need to update our estimates as new data emerges or new observations are made. This is where we need to use Bayesian statistical methods and Bayesian inference. Cox's theorem is a useful application of this type of statistical inference. What's more, we need to know what the probabilities of our hypotheses are so that we can determine the utility of our hypotheses for constituting a justified belief. It is here where Bayesian inference and Cox's theorem are irreplaceable and where Frequentist methods prove insufficient. Only Bayesian inference allows for continuous updating based on new information and only Bayesian inference allows us to answer the real question: What is the probability of our hypothesis given the data (rather than the probability of the data given the hypothesis)?

Such is the situation for most real-world problems, whether in medicine, biology, meteorology, psychology, sociology, or even machine learning in computer science or quantum computing. Bayesian inference excels at predictive modeling while Frequentist methods excel at descriptive modeling. In real life, whether in health care, stock markets, or weather prediction, the problems we actually want to solve are usually of a predictive nature. For example, What's the probability that a patient will develop a heart attack? What's the probability that one drug versus another will prevent that heart attack? What's the chance that it will rain tomorrow? What's the chance that my hypothesis is true? What's the chance that OJ killed Nicole? These are all problems that need to be answered with Bayesian inference.

Where Aristotle excelled at the first type of problem (a known input producing a known output) and Fisher excelled at the second type (a known input producing a predictable number of different outputs), only Bayes can provide a meaningful *logos* for the third type of problem.

If we are to generalize a system of *logos*, it must work for all three types of problems. Here too, only Bayes works for all three. Frequentism has failed miserably for the third type of problem and this incorrect use of Frequentist statistics is responsible for most of the bad science we see today. Problems of the first type are usually over-simplified, and on closer examination what we thought was a complete understanding seemed complete only because we couldn't detect error on the scale where we normally consider the problem. For example, Newtonian mechanics seems deterministic and fully explains the reality that we normally see; but it is an over-simplification and fails at extremes of scale, where relativity becomes important. It vaguely approximates 99% of reality, but better and more complex theories are now available.

So how does this help us rationalize our understanding of justified beliefs? It affects the way in which we rationalize our beliefs that constitute our knowledge. What is our *logos*? It must be probability theory, or specifically, Bayesian inference. We develop hypotheses that explain our observations and then we use all available information to decide which hypothesis best describes objective reality. If we are sufficiently confident in the leading hypothesis (we often are not), then that constitutes knowledge; it is a justified belief. An example of this that we discussed before was the *diagnostic threshold;* this was the point at which, based on all available evidence, we are confident enough to make a diagnosis (that is, $P > j$). But our knowledge is not absolute (we aren't sure the patient has the diagnosis) even though reality is (the patient does or does not have the diagnosis); rather, it is true to a certain degree of probability, and we revise this probability as new data emerges.

Bayes in turn affects our understanding of logic. Logical propositions are no longer true or false, they are true or false to a certain probability. Of course, it is possible for a probability to equal 0 or equal 1 based on all available data and observations, but even those extreme probabilities might need revision given sufficient new information.

Thinking probabilistically about truth and knowledge helps us be more comfortable with uncertainty (since everything is uncertain to some degree). Donald Knuth, an influential computer scientist, said, "Science is what we understand well enough to explain to a computer. Art is everything else we do." In other words, he is comfortable with classical logic. Computers only work well when true and false propositions are unequivocally true or false. We can program a computer to execute billions of mathematical calculations per second, and the answers are certain and deterministic (I'm obviously not talking about quantum computing). But how do we program a computer to deal with ambiguity and uncertainty? For Knuth, this uncertainty creates discomfort. Humans, however, are not computers, and it is fallacy to try to think like a computer in most circumstances (though in some circumstances, it's ideal). Indeed, we think using a mixture of two distinctly different processes.

We discussed Dual Process Theory (DPT) in detail in Chapter 3. In DPT, System 1 thinking describes decision making that is quick, associative (a lumper), unconscious, implicit (very susceptible to bias), and nonlogical. System 2 thinking is explicit, conscious reasoning that is based on rules and is logical (a splitter). System 1 thinking describes how we make most decisions, through heuristics and gestalt. System 2 thinking requires more work and is slower, requiring more analysis and data.

We all like to think that we are logical, but almost all our decisions are made using System 1 thinking. Where we eat lunch, who we vote for, who we love, how we invest money, and most other major decisions in life are made with System 1 cognition and these decisions are usually illogical and affected by our biases and emotions. Don Knuth doesn't know how to program a computer to make the stupid decisions that humans normally make (but that doesn't make it an art).

This idea of two different ways of thinking is certainly not a modern one. Recent interest in this concept has been sparked from the research and popular psychology work of Daniel Kahneman and Amos Tversky, but Williams James in 1890 laid the groundwork for DPT. He described "associative" thinking, which drew on past experiences, and "true reasoning" which could be used for new experiences. Obviously, associative thinking is subject to all the mistakes and cognitive biases which hang upon our past experiences (or, in modern terms, cognitive dispositions to respond). Forcing "true reasoning" onto familiar situations might mitigate this problem. James recognized the structure that Darwin called "lumpers and splitters." But 250 years before Williams James, Blaise Pascal recognized these same cognitive processes.

Pascal famously said, "The heart has its reasons, which reason knows not." Pascal realized that men are often unable to logically defend their beliefs (and are unwilling to give up those beliefs even though they have no defense). Pascal, in the beginning of his *Pensées,* discusses what he calls the mathematical mind (the logical, rational, System 2 mind) and the intuitive mind (the heuristic-driven, biased, System 1 mind). He writes,

> The difference between the mathematical and the intuitive mind. — In the one, the principles are palpable, but removed from ordinary use; so that for want of habit it is difficult to turn one's mind in that direction: but if one turns it thither ever so little, one sees the principles fully, and one must have a quite inaccurate mind who reasons wrongly from principles so plain that it is almost impossible they should escape notice.
>
> But in the intuitive mind the principles are found in common use and are before the eyes of everybody. One has only to look, and no effort is necessary; it is only a question of good eyesight, but it must be good, for the principles are so subtle and so numerous that it is almost impossible but that some escape notice. Now the omission of one principle leads to error; thus one must have very clear sight to see all the principles and, in the next place, an accurate mind not to draw false deductions from known principles.
>
> All mathematicians would then be intuitive if they had clear sight, for they do not reason incorrectly from principles known to them; and intuitive minds would be mathematical if they could turn their eyes to the principles of mathematics to which they are unused.

Pascal delineated Dual Process Theory long before modern psychology took an interest in it. When Pascal speaks of *seeing clearly,* he is describing in his own words what we now call cognitive bias or logical fallacies. He is right – if one is not subject to any bias or logical error, and if the facts one sees are unquestionably true, then we would all come to the same conclusions about the same knowledge. But, clearly, we do not see facts without bias and our logic is perverted by our *cognitive dispositions to respond*. Further, the facts we see are not always true (but are true to some degree of probability).

Scott Adams, the creator of *Dilbert,* has summed all this up in the following way,

> Smart, well-informed people disagree on nearly all major issues. So being smart and well-informed doesn't help you grasp reality as much as you would hope. If it did, all of the smart, well-informed people would agree. They don't.
>
> ... Facts don't influence decisions. Humans decide first, then rationalize their irrational choices with cherry-picked data. You see this all the time with the people who disagree with your brilliance. Just remember that they see the same irrationality in you that you see in them.

It's not that facts don't matter, it's more the case that most humans don't really consider them. This is the danger of solipsism (why consider facts in the first place if they can't be known or are subjective?). We must recognize that our tendency is to think and decide irrationally but that we also have access to what Pascal called the mathematical mind. Our intuitive minds work better (System 1 thinking) if our mathematical minds are exercised more (System 2 thinking). Our brains cannot work without implicit bias, but the more information that we learn in a logical way, most closely modeling reality, the more accurate our biases will be, and the better our intuitive minds will work. We must practice good thinking. Pascal later says,

> Those who are accustomed to judge by feeling do not understand the process of reasoning, for they would understand at first sight and are not used to seeking for principles. And others, on the contrary, who are accustomed to reason from principles, do not at all understand matters of feeling, seeking principles and being unable to see at a glance.

If the only exercise of our faculties is in the form of System 1 thinking, then we will reject System 2 reasoning altogether; but if we are too caught up in System 2 thinking, then we will be ineffective advocates for our beliefs, using *logos* too much and *pathos* too little. More practically, we would not be able to make all the decisions necessary to live our daily lives if we only employed System 2 cognition. System 2 thinking is too slow and too burdensome to carry us through our daily life and work. *The best decision is not always the most accurate one.*

Thus, we must emphasize learning new knowledge in a structured, System 2 environment, to get the most out of our faulty System 1 decisions later. *The mind is like a parachute: it works best when properly packed.*

Pascal stated, "People are generally better persuaded by the reasons which they have themselves discovered than by those which have come into the mind of others." Aristotle realized that *pathos* was a much more effective tool for persuasion than *logos* and the great rhetoricians from Cicero to Gerry Spence have perfected techniques for persuasion using things other than *logos*. But this is simply because most people don't really utilize System 2 thinking, or, if they do, they do so poorly.

System 1 thinking is what we are stuck with unless there is a deliberate attempt to learn to think in a System 2 way. What separates man from animals is not thought, but System 2 thought. But System 2 thought is not essential to survival and often has no evolutionary advantage. System 2 thought takes work, and that work can only be done in a resource rich environment. It requires education; it requires time; it requires expensive effort – but 99%+ of all people who have ever lived haven't had the resources to dedicate to it.

Today, we still do not emphasize what could be called *critical thinking* in education. We emphasize regurgitation of facts without context, algorithms without understanding, and coloring in the lines without question. We teach people to quote others out of context and we teach children to blindly follow orders without questioning. We emphasize laws and rules that are binary and don't accommodate nuance. We emphasize form over function. We are lumpers rather than splitters. We reduce complicated problems down to simple factors. We find false causation wherever we see association. We focus on effect rather than cause. We judge action rather than intent. We focus on the ends rather than the means. In short, we are primitive, because we think primitively.

We think like children rather than like adults. We think selfishly because we only see facts that self-justify our worldview. We see the world as we want it to be, not as it is.

All of this to say that we need to have an awareness of how we think in order to understand what constitutes knowledge: Knowledge is justified belief, and that justification must come from logical inference. For complex problems, only Bayesian inference will suffice. Knowledge is the result of System 2 reasoning. System 1 thoughts are often correct, but they are not justified and therefore are not knowledge, *even if correct* – they are opinions.

If you have objected to Bayesian methods on philosophical grounds, hopefully you will find that I have answered your objections. If I haven't adequately done so, realize that modern science is embracing Bayesian methods very rapidly in almost every field, including biological sciences. So, it's time to embrace the 300+ year-old formula suggested by Reverend Thomas Bayes, refined by Pierre-Simon Laplace, and perfected by Sir Harold Jeffreys and E.T. Jaynes (who also solidified probability theory as logic and gave us the Mind Projection Fallacy).

Notes

1. Elstein AS. Clinical reasoning in medicine. In: Higgs J, Jones MA, editors. Clinical Reasoning in the health professions. Butterworth-Heinemann; 1995.
2. Falchuk, E. (2012, April 27). Misdiagnosis: Millions of patients are being treated for the wrong conditions. Retrieved October 06, 2020, from https://www.seattletimes.com/opinion/misdiagnosis-millions-of-patients-are-being-treated-for-the-wrong-conditions/
3. Verghese A, Charlton B, Kassirer JP, Ramsey M, Ioannidis JP. Inadequacies of Physical Examination as a Cause of Medical Errors and Adverse Events: A Collection of Vignettes. Am J Med. 2015;128(12):1322-4.e3.
4. Oaksford M, Chater N. Précis of bayesian rationality: the probabilistic approach to human reasoning. Behav Brain Sci. 2009;32(1):69-84.
5. Ormerod RJ. Rational inference: deductive, inductive and probabilistic thinking. Journal of the Operational Research Society. 2010;61(8):1207-1223
6. Clow WM, Joannides T, Saleem AK, Melville-jones GR, Martin I. An unusual cause of postmenopausal bleeding and incontinence of urine: primary lymphoma of the vagina. Br J Obstet Gynaecol. 1995;102(2):164-5.
7. Vlies R. Hypernephroma: an unusual cause of postmenopausal bleeding. N Z Med J. 1981;93(685):375-6.
8. Biswas J, Billingham K, Biswas S, Hirschowitz L, Bisson D. Ligneous cervicitis: an unusual cause of post-coital bleeding in a postmenopausal woman. J Obstet Gynaecol. 2009;29(2):163-5.
9. Oladipo A, Mathew J. Villous papyraceous: an unusual cause of tamoxifen-induced postmenopausal bleeding. Climacteric. 2006;9(3):234-7.
10. Sharma JB, Wadhwa L, Malhotra M, Arora R, Garg A, Singh S. Huge localized vaginal neurofibromatosis: an unusual cause of postmenopausal bleeding. J Obstet Gynaecol Res. 2004;30(2):96-9.
11. Braun HL, Wheelock JB, Amaker BH, Seeds JW. Sonographic evaluation of a uterine angiolipoleiomyoma. J Clin Ultrasound. 2002;30(4):241-4.
12. Gilmore S. Evaluating statistics in clinical trials: making the unintelligible intelligible. Australasian J of Derm. 2008;49:177-186.
13. National Academies of Sciences, Engineering, and Medicine. 2015. Improving Diagnosis in Health Care. Washington, DC: The National Academies Press.

14. Sheets L, Johnson J, Todd T, Perkins T, Gu C, Rau M. Unsupported labeling of race as a risk factor for certain diseases in a widely used medical textbook. Acad Med. 2011;86(10):1300-3.
15. Aberegg SK, Haponik EF, Terry PB. Omission bias and decision making in pulmonary and critical care medicine. Chest. 2005;128(3):1497-1505.
16. Braverman J, Blumenthal-Barby, JS. Assessment of the sunk-cost effect in clinical decision-making. Social Science and Med. 2012;75(2012):186-192.
17. Croskerry P. Diagnostic Failure: A Cognitive and Affective Approach. In: Henriksen K, Battles JB, Marks ES, et al., editors. Advances in Patient Safety: From Research to Implementation (Volume 2: Concepts and Methodology). Rockville (MD): Agency for Healthcare Research and Quality (US); 2005 Feb. Available from: https://www.ncbi.nlm.nih.gov/books/NBK20487/
18. Kohn LT, Corrigan JM, and Donaldson MS, editors. To Err Is Human, Building a Safer Health System. National Academies Press; 2000.
19. Groopman J. How doctors think. Houghton Mifflin, 2008.
20. van den Berge K, Mamede S. Cognitive diagnostic error in internal medicine. European Journal of Internal Medicine. 2013;24(6):525-529
21. Graber M, Gordon R, Franklin N. Reducing diagnostic errors in medicine: what's the goal? Acad Med. 2002;77(10):981-92.
22. Croskerry P, Cosby KS. Patient Safety in Emergency Medicine. Lippincott Williams & Wilkins; 2009.
23. Croskerry P. Diagnostic Failure: A Cognitive and Affective Approach. In: Henriksen K, Battles JB, Marks ES, et al., editors. Advances in Patient Safety: From Research to Implementation (Volume 2: Concepts and Methodology). Rockville (MD): Agency for Healthcare Research and Quality (US); 2005 Feb. Available from: https://www.ncbi.nlm.nih.gov/books/NBK20487/
24. Smith R. Medical journals and pharmaceutical companies: uneasy bedfellows. BMJ. 2003;326(7400):1202-1205.
25. Bates DW, Kuperman GJ, Wang S, et al. Ten commandments for effective clinical decision support: making the practice of evidence-based medicine a reality. J Am Med Inform Assoc. 2003;10(6):523-30.
26. Leonidas, L. Ten Commandments to reduce cognitive errors. Society to Improve Diagnosis in Medicine. Retrieved October 5, 2020 from https://www.improvediagnosis.org/wp-content/uploads/2018/10/ten_commandments_checklist.pdf
27. Graber ML. Educational strategies to reduce diagnostic error: can you teach this stuff?. Adv Health Sci Educ Theory Pract. 2009;14 Suppl 1:63-9.
28. Johnson SB, Park HS, Gross CP, Yu JB. Complementary medicine, refusal of conventional cancer therapy, and survival among patients with curable cancers. JAMA Oncol. 2018;4(10):1375.
29. Matute H, Blanco F, Yarritu I, Díaz-lago M, Vadillo MA, Barberia I. Illusions of causality: how they bias our everyday thinking and how they could

be reduced. Front Psychol. 2015;6:888.
30. Hill AB. The environment and disease: association or causation? *Proc R Soc Med.* 1965;58:295-300.
31. Prasad V, Vandross A, Toomey C, et al. A decade of reversal: an analysis of 146 contradicted medical practices. Mayo Clin Proc. 2013;88(8):790-8.
32. Montgomery K. How Doctors Think, Clinical Judgment and the Practice of Medicine. Oxford University Press; 2005.
33. Gobble RM, Hoang HL, Kachniarz B, Orgill DP. Ketorolac does not increase perioperative bleeding: a meta-analysis of randomized controlled trials. Plast Reconstr Surg. 2014;133(3):741-55.
34. Marik PE, Khangoora V, Rivera R, Hooper MH, Catravas J. Hydrocortisone, Vitamin C, and Thiamine for the Treatment of Severe Sepsis and Septic Shock: A Retrospective Before-After Study. Chest. 2017;151(6):1229-1238.
35. Fowler AA, Truwit JD, Hite RD, et al. Effect of Vitamin C Infusion on Organ Failure and Biomarkers of Inflammation and Vascular Injury in Patients With Sepsis and Severe Acute Respiratory Failure: The CITRIS-ALI Randomized Clinical Trial. JAMA. 2019;322(13):1261-1270.
36. Fujii T, Luethi N, Young PJ, et al. Effect of vitamin c, hydrocortisone, and thiamine vs hydrocortisone alone on time alive and free of vasopressor support among patients with septic shock: the vitamins randomized clinical trial. JAMA. 2020;323(5):423.
37. Holcomb WL, Chaiworapongsa T, Luke DA, Burgdorf KD. An odd measure of risk: use and misuse of the odds ratio. Obstet Gynecol. 2001;98(4):685-8.
38. Berg-lekås ML, Högberg U, Winkvist A. Familial occurrence of dystocia. Am J Obstet Gynecol. 1998;179(1):117-21.
39. Davies HT, Crombie IK, Tavakoli M. When can odds ratios mislead?. BMJ. 1998;316(7136):989-91.
40. Grimes DA, Schulz KF. Making sense of odds and odds ratios. Obstet Gynecol. 2008;111(2 Pt 1):423-6.
41. Maass H. (2013, July 08). The odds are 11 million to 1 that you'll die in a plane crash. Retrieved October 06, 2020, from http://theweek.com/articles/462449/odds-are-11-million-1-that-youll-die-plane-crash
42. Sibbald M, Cavalcanti R. The biasing effect of clinical history on physical examination diagnostic accuracy. Med Educ. 2011;45:827-834.
43. Thomas EJ, Studdert DM, Burstin HR, David M, Burstin HR, Orav EJ, et al. Incidence and types of adverse events and negligent care in Utah and Colorado. Med Care 2000;38:261–71
44. Parsons AS, Wijesekera TP, Rencic JJ. The management script: a practical tool for teaching management reasoning. Academic Medicine. 2020;95(8):1179-1185.
45. Shikino K, Ikusaka M, Ohira Y, et al. Influence of predicting the diagnosis from history on the accuracy of physical examination. Adv Med Educ Pract.

2015;6:143-8.
46. Kruszka PS, Kruszka SJ. Evaluation of acute pelvic pain in women. Am Fam Physician. 2010;82(2):141-7.
47. Ambesh P, Obiagwu C, Shetty V. Homan's sign for deep vein thrombosis: A grain of salt?. Indian Heart J. 2017;69(3):418-419.
48. Verghese A, Charlton B, Kassirer JP, Ramsey M, Ioannidis JP. Inadequacies of Physical Examination as a Cause of Medical Errors and Adverse Events: A Collection of Vignettes. Am J Med. 2015;128(12):1322-4.e3.
49. Gøtzsche PC. Mammography screening is harmful and should be abandoned. J R Soc Med. 2015;108(9):341-5.
50. Thomasino T, Levi C, Draper M, Neubert AG. Diagnosing rupture of membranes using combination monoclonal/polyclonal immunologic protein detection. J Reprod Med. 2013;58(5-6):187-94.
51. De haan HH, Offermans PM, Smits F, Schouten HJ, Peeters LL. Value of the fern test to confirm or reject the diagnosis of ruptured membranes is modest in nonlaboring women presenting with nonspecific vaginal fluid loss. Am J Perinatol. 1994;11(1):46-50.
52. Moyer VA, Kennedy KA. Understanding and using diagnostic tests. Clinics in Perinatology. 2003;30(2):189-204.
53. Mennuti MT, Cherry AM, Morrissette JJD, Dugoff L. Is it time to sound an alarm about false-positive cell-free DNA testing for fetal aneuploidy? American Journal of Obstetrics and Gynecology. 2013;209(5):415-419.
54. Tomas ME, Getman D, Donskey CJ, Hecker MT. Overdiagnosis of Urinary Tract Infection and Underdiagnosis of Sexually Transmitted Infection in Adult Women Presenting to an Emergency Department. J Clin Microbiol. 2015;53(8):2686-92.
55. Gordon LB, Waxman MJ, Ragsdale L, Mermel LA. Overtreatment of presumed urinary tract infection in older women presenting to the emergency department. J Am Geriatr Soc. 2013;61(5):788-92.
56. Hartley S, Valley S, Kuhn L, et al. Inappropriate testing for urinary tract infection in hospitalized patients: an opportunity for improvement. Infect Control Hosp Epidemiol. 2013;34(11):1204-7.
57. Scholes D, Hooton TM, Roberts PL, Stapleton AE, Gupta K, Stamm WE. Risk factors for recurrent urinary tract infection in young women. J Infect Dis. 2000;182(4):1177-82.
58. Bent S, Nallamothu BK, Simel DL, Fihn SD, Saint S. Does this woman have an acute uncomplicated urinary tract infection?. JAMA. 2002;287(20): 2701-10.
59. Simati B, Kriegsman B, Safranek S. FPIN's clinical inquiries. Dipstick urinalysis for the diagnosis of acute UTI. Am Fam Physician. 2013;87(10): Online.
60. Clance PR, Imes SA. The imposter phenomenon in high achieving women: Dynamics and therapeutic intervention. Psychotherapy: Theory, Research

& Practice. 1978;15(3):241-247.
61. Habibzadeh F, Yadollahie M. Number needed to misdiagnose: a measure of diagnostic test effectiveness. Epidemiology. 2013;24(1):170.
62. InformedHealth.org [Internet]. Cologne, Germany: Institute for Quality and Efficiency in Health Care (IQWiG); 2006-. What is overdiagnosis? 2017 Apr 20 Available from: https://www.ncbi.nlm.nih.gov/books/NBK430655/
63. Shaughnessy A, Siwek J. Introducing POEMs. Am Fam Physician. 2003 Mar 15;67(6):1196-1199
64. Schandelmaier S, Briel M, Saccilotto R, et al. Niacin for primary and secondary prevention of cardiovascular events. Cochrane Heart Group, ed. Cochrane Database of Systematic Reviews. Published June 14, 2017.
65. Psaty BM. Surrogate end points, health outcomes, and the drug-approval process for the treatment of risk factors for cardiovascular disease. JAMA. 1999;282(8):786.
66. Gyamfi-Bannerman C, Thom EA, Blackwell SC, et al. Antenatal betamethasone for women at risk for late preterm delivery. N Engl J Med. 2016;374(14):1311-1320.
67. Schrag SJ, Zell ER, Lynfield R, et al. A population-based comparison of strategies to prevent early-onset group b streptococcal disease in neonates. N Engl J Med. 2002;347(4):233-239.
68. Folkman J. Transplacental carcinogenesis by stilbestrol. N Engl J Med. 1971;285(7):404-405.
69. Dieckmann WJ, Davis ME, Rynkiewicz LM, Pottinger RE. Does the administration of diethylstilbestrol during pregnancy have therapeutic value?. Am J Obstet Gynecol. 1953;66(5):1062-81.
70. Eisler P, Hansen B. Doctors perform thousands of unnecessary surgeries. USA Today. Retrieved October 5, 2020 from https://www.usatoday.com/story/news/nation/2013/06/18/unnecessary-surgery-usa-today-investigation/2435009/
71. Gillon R. Medical ethics: four principles plus attention to scope. *BMJ*. 1994;309(6948):184-184.
72. Brown T. 100 Best-selling, most prescribed branded drugs through March. Retrieved October 5, 2020 from https://www.medscape.com/viewarticle/844317
73. Smith, BL. Inappropriate Prescribing. Monitor on Psychology. Monitor on Psychology. 2013;43(6). Retrieved October 5, 2020 from https://www.apa.org/monitor/2012/06/prescribing
74. Schuler J, Dückelmann C, Beindl W, Prinz E, Michalski T, Pichler M. Polypharmacy and inappropriate prescribing in elderly internal-medicine patients in Austria. Wien Klin Wochenschr. 2008;120(23-24):733-41.
75. Fleming-Dutra KE, Hersh AL, Shapiro DJ, et al. Prevalence of inappropriate antibiotic prescriptions among us ambulatory care visits, 2010-2011. JAMA. 2016;315(17):1864.

76. Wall LL, Brown D. The high cost of free lunch. Obstetrics & Gynecology. 2007;110(1):169-173.
77. Swanson A. Big pharmaceutical companies are spending far more on marketing than research. Washington Post. 2015. Retrieved on October 5, 2020 from https://www.washingtonpost.com/news/wonk/wp/2015/02/11/big-pharmaceutical-companies-are-spending-far-more-on-marketing-than-research/
78. Spurling GK, Mansfield PR, Montgomery BD, et al. Information from pharmaceutical companies and the quality, quantity, and cost of physicians' prescribing: a systematic review. Henry D, ed. PLoS Med. 2010;7(10): e1000352.
79. Orwell G. Shooting an Elephant. Penguin UK; 2003.
80. Lieberman T. Low-value medical tests are costly, can lead to harm. Tarbell. 2020. Retrieved on October 5, 2020 from https://tarbell.org/2020/01/low-value-medical-tests-are-costly-can-lead-to-harm
81. Leibovici L. Effects of remote, retroactive intercessory prayer on outcomes in patients with bloodstream infection: randomised controlled trial. BMJ. 2001;323(7327):1450-1451.
82. Unreliable research: trouble at the lab. Retrieved November 29, 2020, from https://www.economist.com/briefing/2013/10/18/trouble-at-the-lab
83. Schroter S, Black N, Evans S, Godlee F, Osorio L, Smith R. What errors do peer reviewers detect, and does training improve their ability to detect them?. J R Soc Med. 2008;101(10):507-14.
84. Bhopal R. Seven mistakes and potential solutions in epidemiology, including a call for a World Council of Epidemiology and Causality. Emerg Themes Epidemiol. 2009;6(1):6.
85. Ioannidis JPA. Why most published research findings are false. PLoS Med. 2005;2(8):e124.
86. Begley CG, Ellis LM. Raise standards for preclinical cancer research. Nature. 2012;483(7391):531-533.
87. Jager LR, Leek JT. An estimate of the science-wise false discovery rate and application to the top medical literature. Biostatistics. 2014;15(1):1-12.
88. Ioannidis JPA. Contradicted and initially stronger effects in highly cited clinical research. JAMA. 2005;294(2):218.
89. Reproducibility Project. Available from https://osf.io/ezcuj/wiki/home/
90. Ioannidis JPA, Allison DB, Ball CA, et al. Repeatability of published microarray gene expression analyses. Nat Genet. 2009;41(2):149-155.
91. Prasad V, Vandross A, Toomey C, et al. A decade of reversal: an analysis of 146 contradicted medical practices. Mayo Clinic Proceedings. 2013;88(8): 790-798.
92. Herrera-Perez D, Haslam A, Crain T, et al. A comprehensive review of randomized clinical trials in three medical journals reveals 396 medical reversals. Elife. 2019;8.
93. Stergiakouli E, Thapar A, Davey Smith G. Association of acetaminophen

use during pregnancy with behavioral problems in childhood: evidence against confounding. JAMA Pediatr. 2016;170(10):964.

94. Vercellini P, Buggio L, Somigliana E, Barbara G, Viganò P, Fedele L. RETRACTED: Attractiveness of women with rectovaginal endometriosis: a case-control study. Fertility and Sterility. 2013;99(1):212-218.
95. Mayo Clinic Statin Choice Decision Aid. Available from https://statindecisionaid.mayoclinic.org/index.php/statin/
96. Head ML, Holman L, Lanfear R, Kahn AT, Jennions MD. The extent and consequences of p-hacking in science. PLoS Biol. 2015;13(3):e1002106.
97. Wasserstein RL, Lazar NA. The ASA Statement on p-Values: Context, Process, and Purpose. The American Statistician, 2016;70:2:129-133.
98. Wasserstein RL, Schirm AL, Lazar NA. Moving to a World Beyond " $p <$ 0.05." *The American Statistician.* 2019;73(supl):1-19.
99. Hurlbert S, Levine R, and Utts J. Coup de Grâce for a Tough Old Bull: 'Statistically Significant' Expires. The American Statistician, 73:sup1, 352-357.
100. Goodman SN. Why is getting rid of p -values so hard? Musings on science and statistics. *The American Statistician.* 2019;73(supl):26-30.
101. Simmons JP, Nelson LD, Simonsohn U. False-positive psychology: undisclosed flexibility in data collection and analysis allows presenting anything as significant. Psychol Sci. 2011;22(11):1359-1366.
102. Plait P. Pareidolia poser. Discover Magazine. Retrieved on October 5, 2020 from https://www.discovermagazine.com/the-sciences/pareidolia-poser
103. Murdoch DJ, Tsai YL, Adcock J. P-values are random variables. *The American Statistician.* 2008;62(3):242-245.
104. Taleb NN. A short note on p-value hacking. Retrieved on October 5, 2020 from https://arxiv.org/pdf/1603.07532.pdf
105. Walsh K, McCormack CA, Webster R, et al. Maternal prenatal stress phenotypes associate with fetal neurodevelopment and birth outcomes. Proc Natl Acad Sci USA. 2019;116(48):23996-24005.
106. Rouse DJ, Hirtz DG, Thom E, et al. A randomized, controlled trial of magnesium sulfate for the prevention of cerebral palsy. N Engl J Med. 2008;359(9):895-905.
107. Lee A. P-values explained by data scientist. Retrieved on October 5, 2020 from https://towardsdatascience.com/p-values-explained-by-data-scientist-f40a746cfc8
108. Chollat C, Enser M, Houivet E, et al. School-age outcomes following a randomized controlled trial of magnesium sulfate for neuroprotection of preterm infants. *The Journal of Pediatrics.* 2014;165(2):398-400.e3.
109. Doyle LW, Anderson PJ, Haslam R, Lee KJ, Crowther C. School-age outcomes of very preterm infants after antenatal treatment with magnesium sulfate vs placebo. *JAMA.* 2014;312(11):1105.
110. Amrhein V, Greenland S, McShane B. Retire statistical significance. Nature. 2019;657:305-307.
111. Ioannidis JPA. Why most published research findings are false. PLoS Med.

2005;2(8):e124.
112. Antonakis J. On doing better science: From thrill of discovery to policy implications. The Leadership Quarterly. 2017;28(1):5-21.
113. Kicinski M. Publication bias in recent meta-analyses. Marinazzo D, ed. PLoS ONE. 2013;8(11):e81823.
114. Choi BCK, Pak AWP. A catalog of biases in questionnaires. Prev Chronic Dis. 2005;2(1):A13.
115. Nelson KB, Grether JK. Can magnesium sulfate reduce the risk of cerebral palsy in very low birthweight infants?. Pediatrics. 1995;95(2):263-9.
116. Becker-brüser W. [Research in the pharmaceutical industry cannot be objective]. Z Evid Fortbild Qual Gesundhwes. 2010;104(3):183-9.
117. Lundh A, Lexchin J, Mintzes B, Schroll JB, Bero L. Industry sponsorship and research outcome. Cochrane Database Syst Rev. 2017;2:MR000033.
118. Vandewiele G, Dehaene I, Kovács G, et al. Overly optimistic prediction results on imbalanced data: flaws and benefits of applying over-sampling. arXiv:200106296 [cs, eess, stat]. Published online January 15, 2020.
119. Nieuwenhuis S, Forstmann BU, Wagenmakers EJ. Erroneous analyses of interactions in neuroscience: a problem of significance. Nat Neurosci. 2011;14(9):1105-7.
120. Beecher HK. The powerful placebo. J Am Med Assoc. 1955;159(17):1602-6.
121. Hróbjartsson A, Gøtzsche PC. Is the placebo powerless? An analysis of clinical trials comparing placebo with no treatment. N Engl J Med. 2001;344(21):1594-602.
122. Marill MC. Why tracking your symptoms can make you feel worse. Wired. Retrieved on October 5, 2020 from https://www.wired.com/ story/why-tracking-your-symptoms-can-make-you-feel-worse/
123. Benedetti F, Lanotte M, Lopiano L, Colloca L. When words are painful: unraveling the mechanisms of the nocebo effect. Neuroscience. 2007;147(2):260-71.
124. Mitsikostas DD, Mantonakis L, Chalarakis N. Nocebo in clinical trials for depression: a meta-analysis. Psychiatry Res. 2014;215(1):82-6.
125. Schroter S, Black N, Evans S, Godlee F, Osorio L, Smith R. What errors do peer reviewers detect, and does training improve their ability to detect them?. J R Soc Med. 2008;101(10):507-14.
126. Baxt WG, Waeckerle JF, Berlin JA, Callaham ML. Who reviews the reviewers? Feasibility of using a fictitious manuscript to evaluate peer reviewer performance. Ann Emerg Med 1998;32:310–7
127. Jefferson T, Alderson P, Wager E, Davidoff F. Effects of editorial peer review: a systematic review. JAMA. 2002;287(21):2784-6.
128. Schroter S, Black N, Evans S, Godlee F, Osorio L, Smith R. What errors do peer reviewers detect, and does training improve their ability to detect them?. J R Soc Med. 2008;101(10):507-14.
129. Altman DG. Poor-quality medical research: what can journals do?. JAMA.

2002;287(21):2765-7.
130. McCook A. Is peer review broken? The Scientist. 2006. Retrieved on October 5, 2020 from https://www.the-scientist.com/uncategorized/is-peer-review-broken-47872
131. Eveleth R. How many people read (and cite) their papers. Smithsonian Magazine. 2014. Retrieved on October 5, 2020 from https://www.smithsonianmag.com/smart-news/half-academic-studies-are-never-read-more-three-people-180950222/
132. Retraction Watch. Retractions rise to nearly 700 in fiscal year 2015 (and psst, this is our 3,000 post). Retrieved on October 5, 2020 from http://retractionwatch.com/2016/03/24/retractions-rise-to-nearly-700-in-fiscal-year-2015-and-psst-this-is-our-3000th-post/
133. Kahneman D, Tversky A. Prospect theory: an analysis of decision under risk. Econometrica. 1979;47(2):263.
134. Kimball JP. Syntax and Semantics. 1975.
135. Wimalasundera N, Stevenson VL. Cerebral palsy. Pract Neurol. 2016;16(3):184-94.
136. Galloway M, Taiyeb T. The interpretation of phrases used to describe uncertainty in pathology reports. Patholog Res Int. 2011;2011:656079.
137. Esserman LJ, Varma M. Should we rename low risk cancers?. BMJ. 2019;364:k4699.
138. NIPT Counseling Strategies: Considering the Test, Reporting the Results. Retrieved October 06, 2020 from https://www.reachmd.com/programs/cme/nipt-counseling-strategies-considering-test-reporting-results/11688/
139. Sever PS, Dahlöf B, Poulter NR, et al. Prevention of coronary and stroke events with atorvastatin in hypertensive patients who have average or lower-than-average cholesterol concentrations, in the Anglo-Scandinavian Cardiac Outcomes Trial—Lipid Lowering Arm (Ascot-lla): a multicentre randomised controlled trial. *The Lancet*. 2003;361(9364):1149-1158.
140. Diamond DM, Ravnskov U. How statistical deception created the appearance that statins are safe and effective in primary and secondary prevention of cardiovascular disease. *Expert Review of Clinical Pharmacology*. 2015;8(2):201-210.
141. Burson, K. A., Larrick, R. P., & Lynch, J. G., Jr. (2009). Six of one, half dozen of the other: expanding and contracting numerical dimensions produces preference reversals. Psychological Science, 20(9), 1074-1078.
142. McCall CA, Grimes DA, Lyerly AD. "Therapeutic" bed rest in pregnancy: unethical and unsupported by data. Obstetrics & Gynecology. 2013;121(6):1305-1308.
143. Galas DJ, Sakhanenko NA, Skupin A, Ignac T. Describing the complexity of systems: multivariable "set complexity" and the information basis of systems biology. Journal of Computational Biology. 2014;21(2):118-140.
144. Loffing J. Paradoxical antidiuretic effect of thiazides in diabetes insipidus:

another piece in the puzzle. Journal of the American Society of Nephrology. 2004;15(11):2948-2950.

145. Abdo WF, Heunks LM. Oxygen-induced hypercapnia in COPD: myths and facts. Crit Care. 2012;16(5):323.

146. The ARISE Investigators and the ANZICS Clinical Trials Group. Goal-directed resuscitation for patients with early septic shock. N Engl J Med. 2014;371(16):1496-1506.

147. Dickinson K, Roberts I. Medical anti-shock trousers (pneumatic anti-shock garments) for circulatory support in patients with trauma. Cochrane Review, The Cochrane Library, 2002, p. 4.

148 Partsch H. Ambulation and compression after deep vein thrombosis: dispelling myths. Seminars in Vascular Surgery. 2005;18(3):148-152.

149. Vain NE, Szyld EG, Prudent LM, Wiswell TE, Aguilar AM, Vivas NI. Oropharyngeal and nasopharyngeal suctioning of meconium-stained neonates before delivery of their shoulders: multicentre, randomised controlled trial. The Lancet. 2004;364(9434):597-602.

150. Anderson GL, Chlebowski RT, Aragaki AK, et al. Conjugated equine oestrogen and breast cancer incidence and mortality in postmenopausal women with hysterectomy: extended follow-up of the Women's Health Initiative randomised placebo-controlled trial. The Lancet Oncology. 2012;13(5):476-486.

151. Abernethy AP, McDonald CF, Frith PA, et al. Effect of palliative oxygen versus room air in relief of breathlessness in patients with refractory dyspnoea: a double-blind, randomised controlled trial. The Lancet. 2010;376(9743):784-793.

152. Nasu K, Yoshimatsu J, Anai T, Miyakawa I. Low-dose dopamine in treating acute renal failure caused by preeclampsia. Gynecol Obstet Invest. 1996;42(2):140-141.

153. Nagaraja V. Management of scleroderma renal crisis. Curr Opin Rheumatol. 2019;31(3):223-230.

154. Hauswald M, et al. Out-of-Hospital Spinal Immobilization: Its Effect on Neurologic Injury. Academic Emergency Medicine, 1998; 5(3): 214–219.

155. Halm M. Trendelenburg Position: "Put to Bed" or Angled Toward Use in Your Unit? American Journal of Critical Care, 2012; 21(6): 449–452.

156. Goldberger AL. Is the normal heartbeat chaotic or homeostatic? News Physiol Sci. 1991;6:87-91.

157. Hoekstra R, Morey RD, Rouder JN, Wagenmakers EJ. Robust misinterpretation of confidence intervals. Psychon Bull Rev. 2014;21(5):1157-64.

158. Oakes, M. (1986). Statistical inference: A commentary for the social and behavioural sciences. Chicester: John Wiley & Sons.

159. Sackett DL, Rosenberg WM, Gray JA, Haynes RB, Richardson WS. Evidence based medicine: what it is and what it isn't. BMJ. 1996;312(7023):71-2.

160. Ioannidis JP. Evidence-based medicine has been hijacked: a report to David

Sackett. J Clin Epidemiol. 2016;73:82-6.

161. Ioannidis JPA. Hijacked evidence-based medicine: stay the course and throw the pirates overboard. J Clin Epidemiol. 2017;84:11-13.
162. Wilson K. Evidence-based medicine. The good the bad and the ugly. A clinician's perspective: A clinician's perspective on evidence-based medicine. J of Eval in Clin Prac. 2010;16(2):398-400.
163. Morrison DA. Each generation improves on the last (In cardiology): Bayesian thinking, the randomized clinical trial, and directed evolution. Cathet Cardiovasc Intervent. 2011;77(7):1018-1019.
164. Herrell HE. The role of spirituality in Hippocratic medicine. South Med J. 2011;104(4):289-291.
165. Stolberg HO, Norman G, Trop I. Randomized controlled trials. AJR Am J Roentgenol. 2004;183(6):1539-44.
166. Greenhalgh T, Howick J, Maskrey N, for the Evidence Based Medicine Renaissance Group. Evidence based medicine: a movement in crisis? *BMJ*. 2014;348(jun13 4):g3725-g3725.
167. Smith GC, Pell JP. Parachute use to prevent death and major trauma related to gravitational challenge: systematic review of randomised controlled trials. BMJ. 2003;327(7429):1459-61.
168. Timmermans S, Mauck A. The promises and pitfalls of evidence-based medicine. Health Affairs. 2005;24(1):18-28.
169. Iannone P, Montano N, Minardi M, Doyle J, Cavagnaro P, Cartabellotta A. Wrong guidelines: why and how often they occur. Evid Based Med. 2017;22(1):1-3.
170. Healy D, Mangin D, Antonuccio D. Data based medicine and clinical judgement. International Journal of Risk & Safety in Medicine. 2013;25(2):111-121.
171. Kruger J, & Dunning D. (1999). Unskilled and unaware of it: How difficulties in recognizing one's own incompetence lead to inflated self-assessments. Journal of Personality and Social Psychology, 77(6), 1121–1134.
172. You S. Perspective and future of evidence-based medicine. Stroke Vasc Neurol. 2016;1(4):161-164.
173. Gawande A., & Groopman J. (n.d.). America's Epidemic of Unnecessary Care. Retrieved October 06, 2020, from https://www.newyorker.com/magazine/2015/05/11/overkill-atul-gawande
174. Heraclitus. Fragments. Diels-Kranz 22B2.

Index

D

E

F

G

H

I

J

L

M

N

O

P

Q

R

S

T

U

V